I0492320

THE REALITY OF RELATIVITY

EINSTEIN'S HOAX ON HUMANITY

Ken Gonder

Independently Published (kgonder0@gmail.com)

ISBN: 9798687387477 (RR-EHH 7.4 240606 300 6X9 KPB O1A 1a, last revision 2025)

Original Library of Congress Control Number: 2020901711

The cover photo is an altered Hubble Deep Field: ST Scl OPO January 15, 1996. R. Williams and the HDF Team (ST Scl) and NASA.

In the Beginning...

For reasons mostly other than science, relativity continues to form the basis of our entire cosmology. We consider it the pinnacle of scientific insight, the height of human intellect. Some even revere it as if it were religion. But with all of its inherent contradictions and elemental discrepancies a few are now beginning to awaken to its inutility. They've begun to legitimately question its fundamental viability, even wondering whether it has any practical relevance at all. That in reality, it may be nothing more than just a delusive theoretical narrative that we've credulously mistaken for reasoned physics.

Despite that its invalidating flaws are blatantly obvious, easy to demonstrate, and just as easy to understand, we're totally incapable of facing this dire possibility. It'd be disastrous and unbearably humiliating. So we turn a blind eye to what could be characterized as a looming crisis for cosmology while continuing to paint ourselves into an ever-shrinking corner with manic confirmation bias, unconsciously misconstruing invalidating evidence and inventing palatable ad hoc rationalizations that have to become increasingly more fantastic and evermore complicated to stay within the constraints of our compulsory big bang-relativity dogma.

If you're a follower, you'll certainly be challenged but probably more offended by this exposé. It will undermine all you believe and hold dear. But if you've been able to retain enough objectivity and courage to question your own big bang-relativity conditioning and allow for the possibility that our universe may be completely misinterpreted, are willing to accept basic demonstrable facts that defy orthodoxy, but most important have retained your innate longing for truth then this objective commonsense analysis may be for you.

You won't find wild theories about time travel, multiple universes, higher or hidden dimensions, or other such science fiction fantasies. Nor will you find any seemingly profound esoteric revelations or privileged insights or genius of any kind. And you won't find any kind of religious agenda or a covert attempt to elevate stature by subverting relativity.

But what you will find is a comprehensive no-nonsense inquiry that's based on publicly available evidence and the most fundamental physics out of which emerge consistent realistic conclusions that can hardly be rationally refuted. These findings that are simple and become obvious when brought to the forefront are clearly explained in a nontechnical way without complex math through elemental logic and basic geometry using simple graphic illustrations and a plainspoken language that can be easily understood by anyone.

1

So proceed expectantly knowing that what you'll find is altogether unique but decidedly rational founded on sound physical principles, not fads, gimmicks, or hidden agendas. And remember, it's only our preconceived notions that are obscuring reality - nothing else.

Part One

The Relativity Ruse_____

Relativity is not the masterstroke of genius we've all been led to believe. The truth is, none of it actually works. Even at its most fundamental level it fails. Still, almost everyone believes it to be science's greatest achievement. The simple explanation is that we've been duped and have been indoctrinating ourselves to its dogma ever since.

Having fallen for relativity's ruse (only the wise and intelligent can comprehend it) this has created an elitist, tightly bound groupthink whose unspoken goal is to preserve and justify its prescribed doctrine above all else. Albeit insidious, some might consider that this is relativity's real achievement, where its true genius lies. It's the all-important setup for its sleight of hand that allows it to perpetrate its unprecedented, self-perpetuating fraud.

Anyone who disagrees with its canon is preemptively judged to be dim-witted, labeled a crackpot and heretic, and banished from the fold. So no one dares to question any of its outrageous claims. Instead, we compliantly embrace our inability to understand, blindly accept its ideology as gospel, then rush to affirm its scripture as orthodoxy while smugly proclaiming its eminence.

We all go about rotely parroting its precepts like experts. But fearful it'll be discovered that we really don't grasp any of it. Totally unaware that in reality, there's nothing there to grasp. It's complete nonsense.

The more honest among us may admit to some lack of understanding, but they'll still assure you of relativity's sensibility. The math proves it, they'll insist. Every enlightened person knows that. Besides, it's what we all believe. We can't all be wrong. It must be - it has to be correct, they secretly hope. We have no idea that we've become nothing more than lemmings. And you know how that turns out.

If you were audacious enough to actually question its validity, even less zealous advocates never debate the question. They too quickly assert dismissively that it's been proven beyond any doubt through repeated experiments. Every intelligent person knows that too.

This is also not correct. Those experiments and their predetermined outcomes are the product of the extreme confirmation bias we've conditioned ourselves to. But there are perfectly natural and rational explanations for the results. They just don't support, validate, or confirm relativity in any way. In some cases, exactly the opposite is true.

Any normal, sensible (nongenius) individual with enough courage to confront their indoctrination and risk questioning its legitimacy can easily see through its deceptive gimmickry and perceive its invalidating failures. But that won't be your most difficult problem.

Your greatest challenge will be accepting that all of those highly-educated experts who we hold to be our most intelligent, whose judgment we trust most, who may even be admired as mentors and role models, have been blindly crusading for an inconceivable, make-believe fantasy for more than a century now while confidently proclaiming its genius with virtually no understanding or even a hint of its dubious nature. You too will want to believe that they can't all be wrong. But they are.

If you're honest with yourself and can consider the human psyche objectively, you'll have to agree that generally most people prefer fantasy over fact and reason. Reality is the last thing we want. Most of the time we're also easily swayed by what's mistakenly perceived as self-interests, which makes us all that more susceptible to hucksters eager to exploit our delusions. Some of us are also highly susceptible to suggestion as well.

For those with a more scholarly slant, there's the additional burden of years of intense academic conditioning. And the ones who go on to work in the field are also subject to subverting financial pressures and more peer groupthink. Their judgment's been even more compromised and there's no way for them to know. They just keep grinding away with their heads down researching the subtle implications of its presumed effects that don't and can't actually exist in reality.

Many of us are also easily spellbound by the lure of science fiction and metaphysics. This is a big issue. So like Einstein who's envied by all, the ultimate role model, we often unconsciously postulate fantasies, not analytically propose rational scientific hypotheses.

But also like Einstein, and what may be just as problematic, many of us are also entranced by mathematical puzzlement. This has us compulsively pursuing numerical solutions to every problem even before a sound conceptual basis is established. Forget logic. "Follow the math" has become the obligatory edict for investigation. This is responsible for much of our confusion.

Light's constancy, relativity's founding premise, is a perfect example. It only works theoretically, mathematically, only in the one abstract dimension of linear motion. It's conceptually impossible in our real world of three actual dimensions. We're so smitten with relativity's sophisticated appearance and impaired by groupthink, we can't see the obvious: light simply has to compound with the motion of its and other sources, which ultimately invalidates all of it. This is only the beginning of its conceptual failures.

6

Einstein again inadvertently invalidates his own theory, in its entirety, with his realization of light's variability in gravity fields. But just like with its compounding, a varying velocity for light is fundamentally at odds with it being fixed, which also nullifies it, in its entirety, from the outset.

He also asserts that any motion is relative motion and that any object's velocity is not absolute but is determined by the subjective choices of each observer, which he also has deciding its mass. But different choices made by different observers would have different velocities, times, lengths, and masses impossibly coexisting together at the same time. Varying the mass and velocity of massive bodies would of course also wreak havoc with their gravitational relationships.

The legendary mass-energy equation, $E=mc^2$, attributed to Einstein, is also inherently flawed. It has the same intrinsic problem as so much else of relativity. It's a purely theoretical, mathematical result that's not based on physical reality. It has accelerating objects, whose rate is not absolute but whimsically decided by each observer, magically acquiring more matter out of nothingness only because of light's presumed but absolutely impossible fixed velocity. This has them somehow not only becoming infinitely large at the speed of light, but also at the same time contracting to the infinitely small in the direction of their motion.

He also asserts that those same accelerating objects create gravity despite that it's a gravity that doesn't coalesce. It only acts opposite the motion, in only one dimension. He also argues that the centrifugal force of an object's rotation is gravity as well. This supposed gravity wouldn't coalesce either. It would actually disperse objects outward while increasing in strength, the opposite of real gravity. And it would act in only two dimensions. For a rotating body experiencing linear acceleration, both of these conjectured forms of gravity would be in conflict with one another while also conflicting with the body's natural, mass-created gravity that coalesces three-dimensionally.

Gravity is another of relativity's obvious but unrecognized failings. Einstein first melds a nonexistent space to a nonexistent time to concoct an inconceivable, four-dimensional "space-time." Then he has that nonexistent, inconceivable, four-dimensional abstraction somehow curve two-dimensionally as a nonexistent plane that impossibly dents underneath three-dimensional massive bodies to somehow cause their attraction as they supposedly fall like they're rolling downhill toward one another without actually rolling or being uphill.

Another blatant failing is relativity's absurd finite yet somehow unbounded, non-Euclidean curving universe that with expansion has become the big bang. Despite that physical existence in two dimensions isn't possible at any scale, it also metaphysically expresses two-dimensionally but like the surface of a sphere so it can remain homogeneous and isotropic, just as we observe but three-dimensionally, and not condense exponentially from gravity (or diffuse exponentially from expansion) as it otherwise would in a normal three-dimensional but finite universe.

These and the rest of relativity's nullifying flaws are easily debunked by exposing Einstein's incorrect assumptions, conceptual errors, and delusive logic. But we refuse to or can't face them, despite that they're all right out in the open for everyone to see. We're wholly unprepared for the catastrophic consequences of admitting our mistake and conceding our gullibility.

So it may take generations before relativity's ruse is fully recognized, accepted, and embraced by institutional academia. Accordingly, our cosmology continues to flounder and the Einstein-relativity myth endures, sown into the fabric of cultural folklore, self-perpetuated by a fictitious canon that cunningly employs groupthink to rotely self-indoctrinate while declaring its concocted fantasy to be factually verified reality.

Relativity is our only remaining obstacle. It's obscuring reality. Realizing its fallacy dissolves all of the universe's mysteries and lays bare its true nature. Everything at every scale falls neatly into place, including unification in a surprisingly simple and unexpected way. Exotic math, exclusive knowledge, or genius is not necessary. All it takes is a willingness to remain open to the possibility that we may all be victims of habitual self-indoctrination to the self-perpetuating dogma of a self-deluded work of fiction, a smidgen of common sense, and a genuine desire for reality.

For the most part, this section attempts to follow the elusive organization of Einstein's book, Relativity: The Special and the General Theory. It could be argued that it's the best overall and most up-to-date resource on relativity. It includes a comprehensive compilation of his work that he apparently updated and edited periodically for more than 35 years up to within a few years before his passing in 1955.

In the Preface, he states, "The author has spared himself no pains in his endeavour to present the main ideas in the simplest and most intelligible form, and on the whole, in the sequence and connection in which they actually originated." He continues, "On the other hand, I have purposely treated the empirical physical foundations of the theory in a 'step-motherly' fashion [negligently], so that readers unfamiliar with physics may not feel like the wanderer who was unable to see the forest for the trees."[1]

1. Albert Einstein, *Relativity: The Special and the General Theory*, 15th ed. Trans. Robert W. Lawson (New York: Three Rivers Press, 1961), v-vi.

Let that sink in for a moment. What is he really saying here? Is he actually telling us that he's gone to great lengths to dumb-down his lofty theory for us ignorant simpletons, even going so far as to intentionally omit the basis of it because we're incapable of comprehending it, thereby sparing us the realization of our own ineptitude? That's considerate of him.

What it sounds like he's really doing is fabricating an excuse up front for relativity's abject failure and incomprehensibility: Complete and valid explanations exist that clarify all of its invalidating defects. But my gifted intellect's superior sagacity has me withholding them for your benefit. It's better that I provide you with a lesser, counterfeit version instead, something you're able to grasp that will cause you to believe you actually comprehend relativity so your self-esteem won't suffer from discovering your true lack of intelligence.

That's how I decipher it. Decide for yourself. This type of elliptic rhetoric is typical of Einstein. When you allow yourself to really see what's actually behind and driving his discourse, it truly is stunning.

The truth is, not unlike this dialogue, his "explanations" are deliberately garbled with deceptive, condescending narration that's intended to belittle, confound, obstruct, and misdirect to obscure incoherent, nonsensical rationalizations of a contrived and illusive ideology that's completely untethered from reality.

Speed of Light

Einstein initially underpins his reasoning with the assumption that the speed of light is fixed everywhere for everyone regardless of their relative motion. Meaning that its velocity does not vary or compound with the motion of its source or other reference frames as we'd naturally infer. It's always 186,000mi/s.

This is relativity's founding premise that he continuously refers to as "law": "the simple law of the constancy of the velocity of light c (in vacuum)... this simple law... general law of nature, the law of the transmission of light in vacuo... the simple law of the propagation of light in vacuo... The law of the constancy of the velocity of light in vacuo."[2] And that's only three pages. You get the feeling he wants us to accept that light's constancy is and has been for some time an established law of nature that's beyond criticism.

He never offers an explanation or evidence for why light's velocity is fixed. He just states that it is and everyone knows it, admonishing us that even, "Every child at school knows, or believes he knows, that [its] propagation takes place in straight lines with a velocity c = 300,000 km./sec... In short, let us assume that [this] simple law... is justifiably believed by the child at school."[3]

He continues: "in the theory of relativity the velocity c plays the part of a limiting velocity, which can neither be reached or exceeded by any real body... the velocity of the transmission of light in vacuo has to be considered equal a constant c... The law of the transmission of light, the acceptance of which is justified by our actual knowledge... According to the theory of relativity, action at a distance with the velocity light always takes the place of instantaneous action at a distance... [special relativity's] results [that are dependent on light's constancy] hold only so long as we are able to disregard the influences of gravitational fields... [being] able to make use of space-time co-ordinates as four-dimensional Cartesian co-ordinates was possible on the basis of the law of the constancy of the velocity of light... the principle of the constancy of the velocity of light [is] valid only with respect to an inertial system [not a gravity field]."[4]

The point is, Einstein clearly believes light's velocity is fixed. And he wants us to believe it too. Special relativity is, "the idea... [that] every motion must be considered only as relative motion... [where the] general laws of nature (e.g. the laws of mechanics or the law of the propagation of light in vacuo) have exactly the same form in [all] cases."[5] General relativity states that, "All bodies of reference... are equivalent

---

2. Einstein, *Relativity,* 21-23.
3. Einstein, *Relativity,* 21.
4. Einstein, *Relativity,* 41, 47, 53, 85, 104, 171.
5. Einstein, *Relativity,* 67-68.

for the description of natural phenomena (formation of the general laws of nature [mechanics and light's constancy]), whatever may be their state of motion."[6]

(A Cartesian coordinate system is a two-dimensional rectilinear grid where any point can be specified with two numerical values. It was developed by René Descartes, a French philosopher & mathematician, 1596-1659. "Inertial" generally means relating to or arising from inertia. Inertia is the property of matter by which it remains at rest or in uniform motion in a straight line unless acted upon by some external force. A gravitational field can be broadly defined as the region surrounding any physical body, including that of subatomic particles, that exerts an "attractive" influence proportional to their mass. Mass is the property of a body that is commonly taken as a measure of the amount of material or matter it contains and causes it to have weight in a gravitational field. Matter is physical substance. Physical substance is a relative term, though. At what point does a subatomic particle's electromagnetic field stop and its physical substance begin? It doesn't. It has no surface. It's only field. And it keeps condensing to some maximum density at its center that's continuously changing depending on its environment.[7])

He argues that relativity's application to the mechanical addition of velocities (the compounding of motion) is necessary to prevent moving objects from attaining/exceeding the speed of light. Light's fixed velocity (mathematically) forces an object's (or reference frame's) time to slow and its contraction in the direction of motion (and the increasing mass of accelerating objects).[8] This is (theoretically) accomplished between reference frames through use of the Lorentz transformation (a system of equations developed by Hendrik Lorentz, a Dutch physicist, 1853-1928,.that Einstein adopted for relativity that translates the space and time coordinates from one reference frame to another).[9]

The problem is, relativity's adjustment to the mechanical compounding of motion doesn't work. It's conceptually flawed. It only works in the one abstract dimension of linear motion. In our real three-dimensional environment, it's inherently conflicted. At one point, Einstein seems to acknowledge this: "we then obtain the equation... which corresponds to the theorem of addition for velocities in one direction [forward, one-dimensionally] according to the theory of relativity."[10]

6. Einstein, *Relativity,* 69.
7. Those processes, suppositions, and beliefs referenced without citation were extrapolated and paraphrased from a diverse spectrum of popular sources that can be found in the Bibliography. An exhaustive qualification of their definition that in the end will still vary with the subjective sentiment of each reader and over time is not necessary to the argument.
8. Einstein, *Relativity,* 40-44.
9. Einstein, *Relativity,* 34-39.
10. Einstein, *Relativity,* 44.

This suggests that he may have been aware that it only worked in one dimension. If he was, why didn't he state it up front as a limiting qualifier? The reason is, it would have made all of relativity nothing more than a theoretical exercise that doesn't actually work in our real world. If he didn't fully comprehend the implications of its one-dimensionality, what does that tell you? But if he actually did and deceitfully maintained its feasibility anyway, that's even worse. Either he's delusive or deceptive. It has to be one or the other. There's no way around it.

As evidence of relativity's validity as applied to the compounding of motion, he cites Armand Fizeau's experiment (French physicist, 1819-1896). We have to assume it's his well-known 1851 experiment. Einstein doesn't say. The way he characterizes it, Fizeau measured an increase in light's velocity when it's shone through flowing water in the direction of its motion as compared to when it's still. He interprets that rate of increase as better matching his formula than that of classic Galilean mechanics.[11] ("Galileian-Newton mechanics" refers to physical relationships between force, matter, and motion where the basic laws of physics remain the same everywhere. Galileo Galilei was an Italian astronomer, physicist & engineer, 1564-1642. Isaac Newton was an English philosopher and mathematician, 1642-1727)

But since in Fizeau's experiment there's no difference in motion between the light's source and the observer, a compounding of velocity between them is not possible. They're of the same reference frame. So Einstein's employment of relativity to the addition of velocities theorem, even if it were correct, is not applicable.

All that appears to be happening is that the water first slows the light's velocity. Its speed in water is about 140,000mi/s. When it's flowing, it's then freed up to increase in the direction of flow. The issue seems to be that the increase does not appear to match the speed of the flowing water as expected. There could be any number of technical reasons that don't involve light's compounding.

The fact is, light has to mechanically compound with all relative motion. Its constancy is simply not possible. Nor is Einstein's application of relativity to the theorem of addition of velocities. This can be easily established with just simple, commonsense logic.

Imagine you're in one of Einstein's thought experiments riding a train with a flashlight that you're pointing directly forward. He'd have us believe that to maintain its fixed velocity, the speed of its light would be 186,000mi/s less the train's speed, that the train's rate of time would be running slightly slower, and that it and you would be physically contracting but only in the direction of its motion all to satisfy his assumption of light's fixed velocity. Most of us believe this to be true. It's our academic conditioning.

11. Einstein, *Relativity*, 45-46.

But what would happen if you then pointed another flashlight perpendicular (or at any angle) to its motion? With no contraction or motion in that direction, and with time's "slower" rate, that light's velocity would not only differ from the forward pointing light but it'd exceed 186,000mi/s, the universe's supposed maximum speed limit.

This ordinary circumstance that's impossible to deny, that should be obvious to everyone but isn't, reveals the unresolvable conflict inherent in light's presumed constancy. Conceptually, in two or the three real dimensions of our nontheoretical world, it cannot be fixed.

Light's metaphysical contraction in the forward direction, along with time and length's corresponding metaphysical contraction, that'd be necessary to maintain light's constancy for objects (or reference frames) in motion will always be in conflict with their natural noncontracted condition to the sides or their necessary metaphysical expansion to the rear. Light's constancy only works theoretically in only one dimension.

This is confirmed by the Lorentz transformation as well. It's conceptually limited to the one dimension of linear motion. But it's also limited to only one occurrence at a time between two reference frames. Two pair of reference frames arranged one-dimensionally that share a common reference frame is the same unworkable condition as in two dimensions. The common reference frame ends up with two conflicting rates of time.[12]

Light's velocity is mechanically required to compound with the motion of its source and that of other reference frames. This completely undermines any argument for its constancy. Without light's constancy, relativity loses its founding premise. Without that premise, it's completely unworkable.

Despite its obviousity, Einstein, with all of his presumed insight and reputed intellect, appears to have failed to simply perceive light's, and time's, inherent three-dimensionality. He reasons instead only in the one abstract dimension of linear motion. I know, it's hard to believe. How could that happen? No one could be that cognitively impaired. But it's either that or he's misleading us intentionally. He does concede though that if it were found that light's velocity was not constant in all cases then relativity would out of necessity completely unravel.[13] (See Diagram **3.1** Light's Constancy 7a, **3.2** Light's Compounding 4a & **51** SR&LT Dim Lim 5a)

---

12. David Bower, "A New Paradox Involving the Lorentz Transformation," viXra open-access archive, Jan 16, 2024, https://vixra.org/abs/2401.0070. Ken Gonder, "Resolving Relativity's Unresolvable Paradoxes," viXra open-access archive, Jan 16, 2024, https://vixra.org/abs/2402.0042.
13. I first referenced this quote years ago but have since lost its source. It may be paraphrased, but I doubt it. In any case, it would not be conceptually inconsistent with other pronouncements. He expresses similar sentiment on page 85 and 151 of *Relativity*.

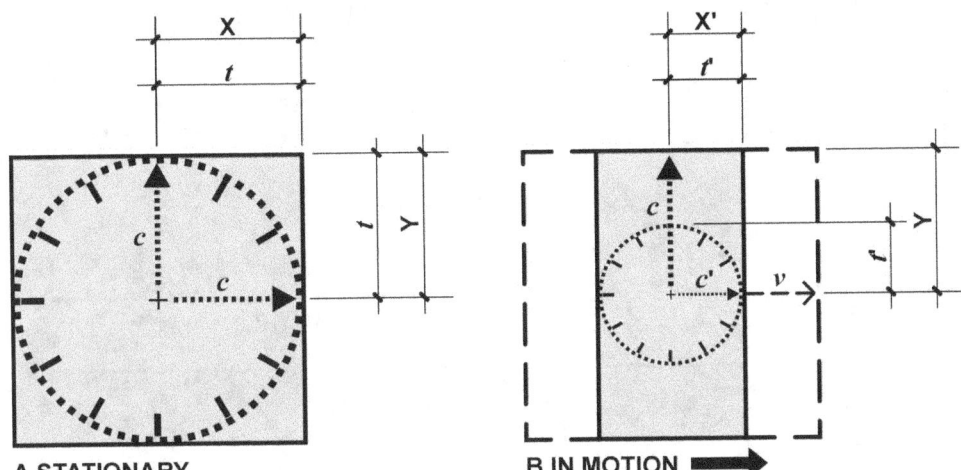

A STATIONARY **B IN MOTION** ➡️

LIGHT'S CONSTANCY

A simple way to illustrate the impossibility of light's fixed velocity (relativity's founding premise) is by establishing a two-dimensional square reference frame that could be of any size, as depicted in diagram **A**. When theoretically stationary, its **X** and **Y** dimensions from its center would correspond to light's constant velocity, indicated by the arrows at c, and time's constant rate, symbolized by the clock-like circle that fills the entire reference frame equally that equates to t.

Einstein would contend that when the reference frame is put in motion, let's say moving from left to right at velocity v, as depicted in diagram **B**, for light's velocity to maintain its constancy in the direction of motion, it would have to slow in that direction by the amount of the reference frame's velocity to c'. This would require the reference frame to contract correspondingly in the direction of motion to the distance **X'** while its rate of time also contracted ("slowed/dilated") equivalently to t', as implied by the smaller clock-like circle.

But since there's no motion in the perpendicular direction, the reference frame's **Y** dimension and light's velocity, c, are not required to contract to maintain its constancy. And since time's contracted rate, t', has to apply equally over the entire reference frame, this creates an unresolvable conflict in every other direction, as indicated by the smaller clock-like circle. Its contracted time, t', corresponds to the contracted **X'** dimension and light's contracted velocity, c', in the direction of motion. But in the perpendicular direction, its contracted rate conflicts with the noncontracted dimension at **Y** and light's noncontracted velocity at c, which would cause it to exceed 186,000mi/s by v.

This unresolvable conflict clearly shows how light's velocity can only remain fixed, theoretically, in the one abstract dimension of linear motion. Even if time had a real existence and was a viable constituent of the universe, light's constancy would still be conceptually impossible in two or the three real dimensions of our physical environment. Which means, its velocity has to compound with the motion of its source and that of other reference frames. Left without the possibility of light's velocity ever being fixed, relativity loses its underlying premise and becomes altogether untenable.

(3.1 Light's Constancy 7a)

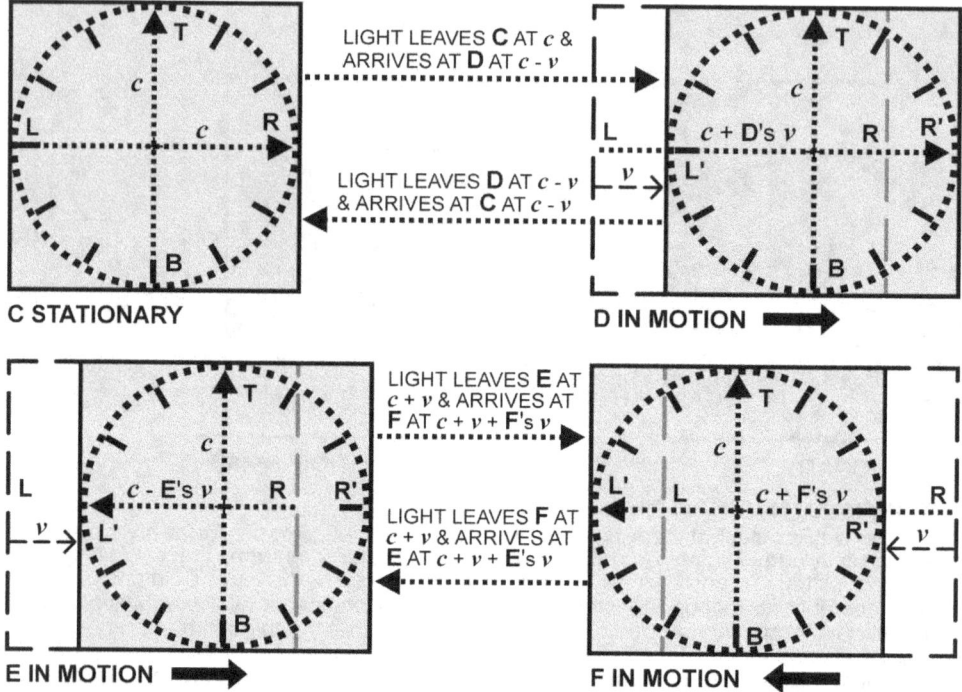

C STATIONARY

D IN MOTION ➡

E IN MOTION ➡

F IN MOTION ⬅

LIGHT'S COMPOUNDING

Light compounds with the motion of its source and that of other reference frames just as we'd naturally infer. The four conditions represent generic reference frames theoretically free of gravity fields to avoid light's variability. The clock-like circle in each symbolizes time's theoretical rate that remains constant throughout the entire reference frame. The dotted arrowed lines denoted with c indicate light's velocity. The dashed grey lines indicate the reference frame's original location prior to motion.

Reference frame **C** is portrayed as theoretically stationary. For its observers, light moves from left to right and from the bottom to the top at c. An outside observer also theoretically stationary would record the same.

For reference frame **D**, it's depicted as moving from left to right with velocity v. For those observers, light moves normally as if it were stationary from **L** to **R** and **B** to **T** at c. This is what all of the Michelson-Morley and Sagnac type experiments show. Light always leaves its source at the same rate in all directions at the same time regardless of motion.

For an outside observer that's theoretically stationary, light begins from its initial position at **L** and arrives at **R'**. Light travels a longer distance in the same amount of time, leaving **L** at c + D's v. This is a compounding of velocities. Light projected between **C** and **D** also indicates its compounding as noted for the different directions.

For **E**, this time light is shown as projected from right to left, opposite the direction of its motion. Its observers again record the light's progress but this time from **R** to **L** as if stationary. But a stationary, outside observer records it traversing a shorter overall distance from **R** to **L'** in the same amount of time. This compounded velocity would be slower than c by E's v because the light leaves **R** opposite the direction of E's motion at c - E's v.

F is the same as **D**, just in the opposite direction. The light projected between **F** and **E** indicates the compounding conditions for the other circumstances of relative motion.

(3.2 Light's Compounding 4a)

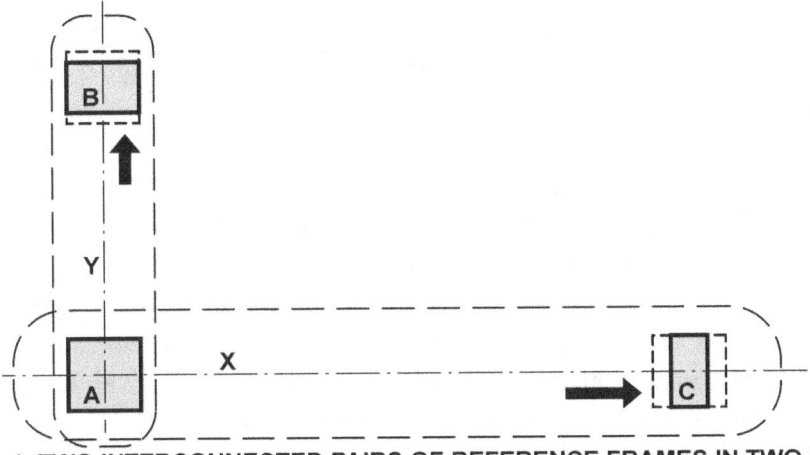

1. TWO INTERCONNECTED PAIRS OF REFERENCE FRAMES IN TWO DIMENSIONS

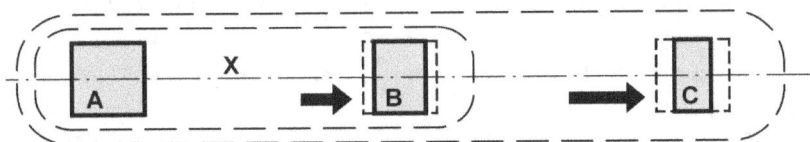

2. TWO INTERCONNECTED PAIRS OF REFERENCE FRAMES IN ONE DIMENSION

THE ONE-DIMENSIONAL LIMITATION OF SPECIAL RELATIVITY & LORENTZ TRANSFORMATIONS

Diagram **1** shows how special relativity & Lorentz transformations only work in the one dimension of linear motion and only in one direction at a time. In two, or our three real dimensions, they're conceptually impossible.

The direction and rate of motion of reference frames (or objects) **B** & **C** is suggested by the arrows. **A** is assumed to be stationary. When they're arranged two-dimensionally and paired through Lorentz transformations, **A**&**B** and **A**&**C**, as implied by the dashed ovals that share the common reference frame, **A**, special relativity & Lorentz transformations are inherently unworkable. **A** will always have two simultaneously conflicting rates of time.

This is essentially the same invalidating contradiction that's inherent in a single reference frame. It too only works theoretically, one-dimensionally. In two, or our three real dimensions, its necessary metaphysical contraction in the forward direction of linear motion will always be in conflict with its natural noncontracted condition to its sides and its necessary metaphysical expansion to the rear, as implied by reference frames **B**&**C** that are contracted one-dimensionally. The dashed square rectangles represent their original noncontracted size prior to motion. This clearly indicates how it's impossible for light's velocity to ever be fixed in our real nontheoretical three-dimensional environment.

Diagram **2** shows how two pairs of reference frames moving in one direction, one-dimensionally, have the same two-dimensional contradiction. When reference frames **A**&**B** and **A**&**C** are similarly paired through Lorentz transformations that share the common reference frame **A**, it will also always have two simultaneously conflicting rates of time.

This demonstrates how special relativity and Lorentz transformations are not only limited to one direction, one-dimensionally, but also to only one pairing occurrence at a time, which again confirms the impossibility of light's constancy. In reality, light's velocity can only compound with the motion of its source and that of other reference frames.

(51 SR&LT 1 Dim Lim 5a)

Light's compounding is also clearly indicated by the renowned Michelson-Morley experiment (1887) and all the others like it. (Albert Michelson, 1852-1931, & Edward Morley, 1838-1923, were American physicists.) It failed to establish the existence of an aether (a theorized universal medium that light was thought to propagate through). What it did show was that the speed of light remained constant when comparing its velocity in the direction of the Earth's orbital (and rotational) motion to that in the perpendicular direction.[14]

This demonstrates that light always leaves its source at the same rate in every direction at the same time. Einstein employed the Lorentz transformation to accommodate the negative results to calculate relativity's presumed contraction in the direction of motion to maintain light's presumed constancy.

The construct of their experiment basically consisted of sets of mirrors perpendicularly arranged an equal distance from a central beam splitter in a cross fashion on a table that can be rotated so that a recombined beam of light would show an interference pattern if its velocity changed when aligned in the direction of the Earth's orbital motion.[15]

To conclude from this experiment that light's velocity is fixed, as many do, is just not possible. The opposite is actually true. It plainly shows that whatever the source's relative motion, light always leaves it at the same velocity in every direction at the same time. This indicates a compounding of velocities, which suggests that every velocity for light has to be faster than the 186,000mi/s that we record here on Earth: 186,000mi/s plus the speed of the Earth's rotation plus the speed of its orbital motion plus the speed of our solar system through our galaxy plus the speed of our galaxy through the universe.

If the speed of light was actually fixed, its velocity in the direction of the Earth's rotation would be 186,000mi/s minus the Earth's rotational velocity minus the speed of its orbital velocity that would be different from its velocity in the perpendicular direction, which should have produced an interference pattern but didn't. The motion of our solar system and galaxy would have to be subtracted as well.

Einstein agreed with Lorentz and others that it's the experiment's contraction in the direction of motion and time's corresponding "slowing" because of light's fixed velocity that's responsible for the negative result. But assuming "time" is something that actually exists and that its rate can actually change, its presumed "slower" rate, which would correspond to the experiment's contraction in the direction of motion, would have to be applied equally over the entire experiment along with the entire Earth. They're of the same reference frame.

14. Einstein, *Relativity*, 58-60, 167-168.
15. "Michelson-Morley Experiment," Wikipedia: The Free Encyclopedia, last modified Dec 26, 2022, https://en.wikipedia.org/wiki/Michelson-Morley_experiment.

And every reference frame can only have one rate of time. "Every reference-body (co-ordinate system) has its own particular time; unless we are told the reference-body to which the statement of time refers, there is no meaning in a statement of the time of an event."[16] So with time's innate three-dimensionality but motion's one-dimensional direction, this causes light's velocity in the perpendicular direction, or any angle other than directly forward, to increase. This creates an unresolvable conflict between the two, along with a velocity that's supposedly impossible that exceeds 186,000mi/s.

This demonstrates that objects cannot be contracting in the direction of their motion. Nor can their time be "slowing." Which clearly indicates light's compounding with the motion of its source. (See Diagram **1.1**, **1.2**, **1.3**, **1.4** M-M Exp 6a & **16.1**, **16.2** MM 7a. There will be more on the Michelson-Morley experiment in an upcoming chapter.)

16. Einstein, *Relativity*, 31.

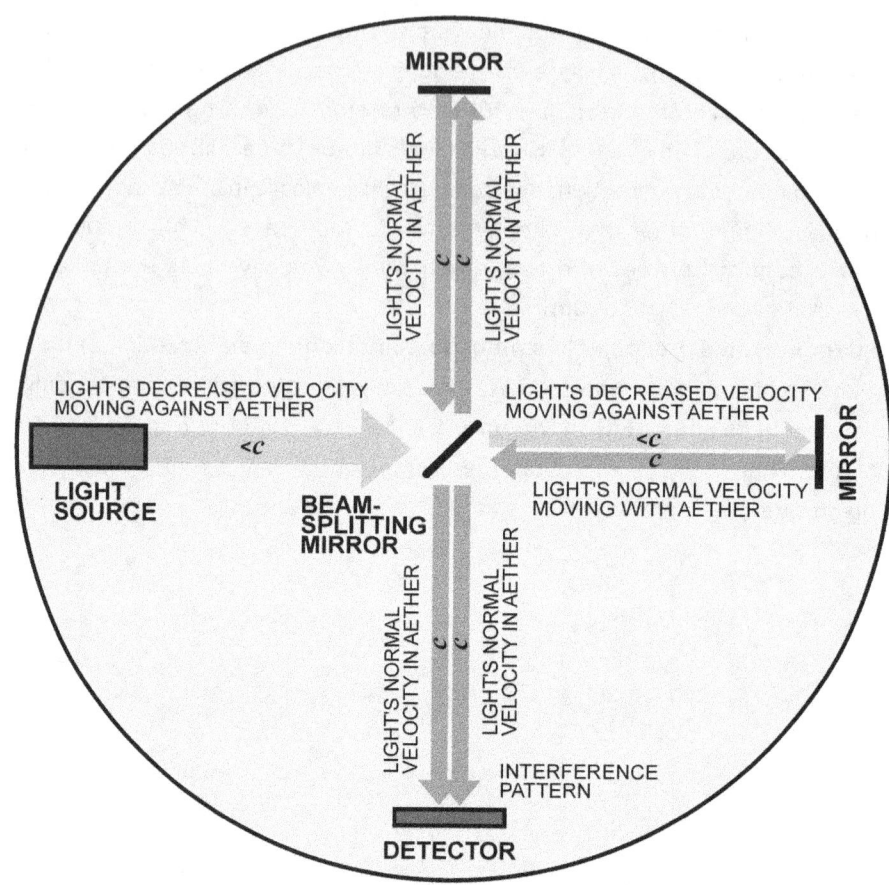

DIRECTION OF EARTH'S ROTATION & ORBITAL MOTION ➡️

MICHELSON-MORLEY - CONCEPTUAL DIAGRAM
EXPECTED RESULT

The experiment essentially consisted of a light source projected onto a series of mirrors arranged perpendicular equal distances from a central beamsplitter mounted on a rotating table oriented with one beam projected in the direction of the Earth's orbital motion and the other perpendicular. When the light was recombined, it was expected to produce an interference pattern due to its decreased velocity from the theorized aether "headwind." This would confirm the aether's existence. But no interference pattern was found.

(1.1 M-M Exp 6a)

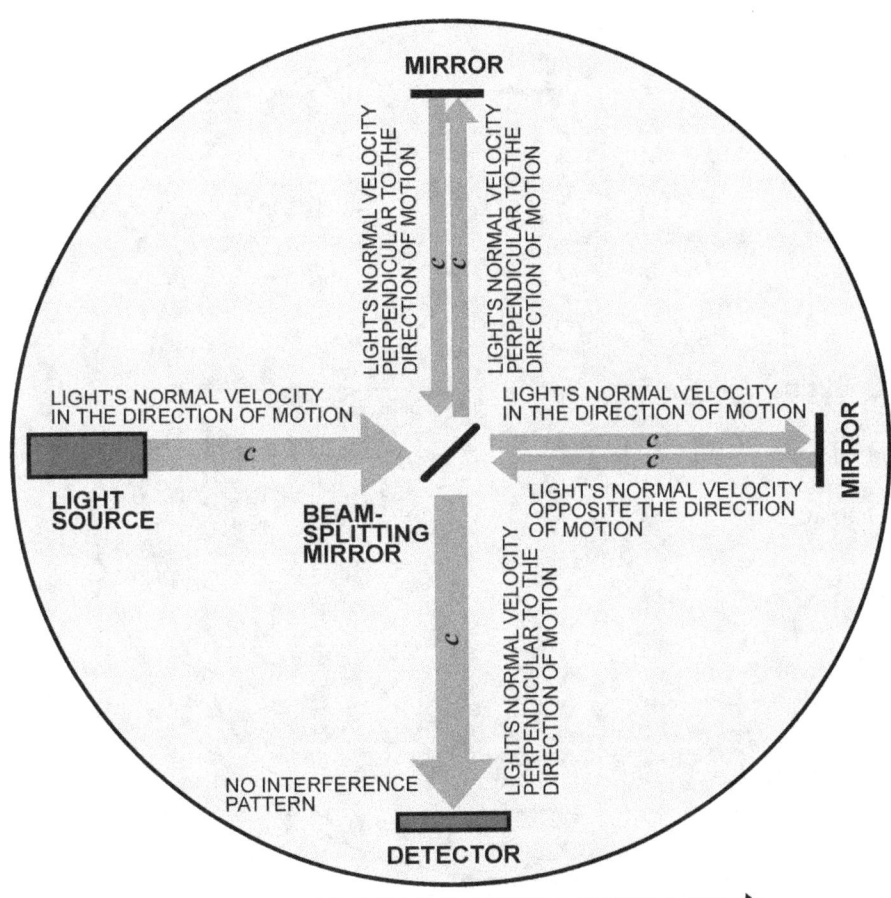

DIRECTION OF EARTH'S ROTATION & ORBITAL MOTION ➡️

MICHELSON-MORLEY - CONCEPTUAL DIAGRAM
ACTUAL RESULT

What the experiment actually showed is that light always leaves its source at 186,000mi/s in every direction at the same time as we'd naturally expect. This indicates its compounding with the motion of its source and implies its compounding with other reference frames. Which means that because everything's in motion, its velocity can never be fixed at 186,000mi/s but will always be some slower or faster rate that can be any velocity up to instantaneous.

If someone was out in space stationary with respect to the solar system, they'd be in a different reference frame recording a compounding of light's varying velocity, which is determined by the field density at their location, plus/minus the Earth's rotational and orbital velocity or some vector angle of it.

(1.2 M-M Exp 6a)

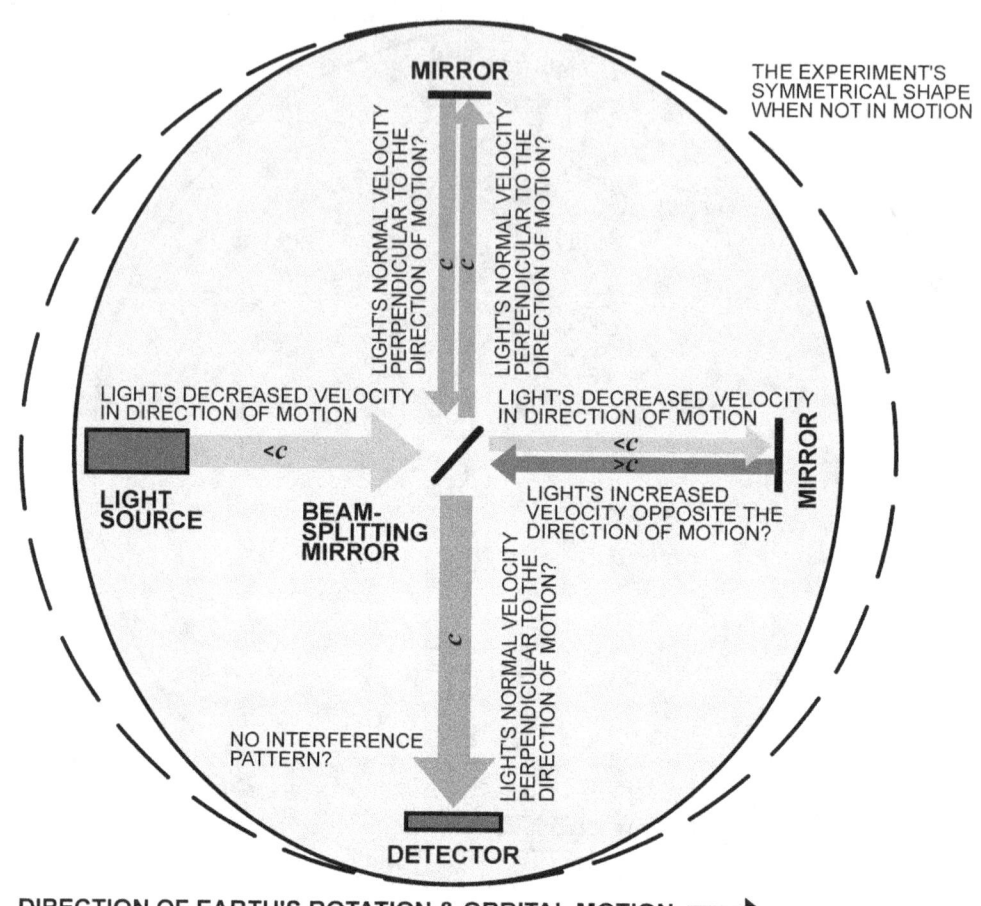

THE EXPERIMENT'S SYMMETRICAL SHAPE WHEN NOT IN MOTION

MIRROR

LIGHT'S NORMAL VELOCITY PERPENDICULAR TO THE DIRECTION OF MOTION?

LIGHT'S NORMAL VELOCITY PERPENDICULAR TO THE DIRECTION OF MOTION?

LIGHT'S DECREASED VELOCITY IN DIRECTION OF MOTION

LIGHT'S DECREASED VELOCITY IN DIRECTION OF MOTION

$<c$

$<c$
$>c$

MIRROR

LIGHT SOURCE

BEAM-SPLITTING MIRROR

LIGHT'S INCREASED VELOCITY OPPOSITE THE DIRECTION OF MOTION?

LIGHT'S NORMAL VELOCITY PERPENDICULAR TO THE DIRECTION OF MOTION?

NO INTERFERENCE PATTERN?

DETECTOR

DIRECTION OF EARTH'S ROTATION & ORBITAL MOTION ➡

MICHELSON-MORLEY - CONCEPTUAL DIAGRAM
LORENTZ'S EXPLANATION ADOPTED BY EINSTEIN

What Lorentz proposed to explain the result, and Einstein later adopted for relativity, was that objects contract in the direction of their motion while time slows to maintain light's fixed velocity so that an interference pattern is not produced. But what happens to the experiment for the reflected light moving opposite the direction of motion? To maintain light's constancy wouldn't it have to expand while time's rate increased? And how about in the perpendicular direction? Without any contraction wouldn't time's "slower" rate cause light's velocity to exceed 186,000mi/s? These irresolvable conflicts confirm that light's velocity cannot remain fixed in our real nontheoretical environment of three actual dimensions but must compound with motion, which invalidates any Lorentz contraction and undermines nearly all of relativity.

(1.3 M-M Exp 6a)

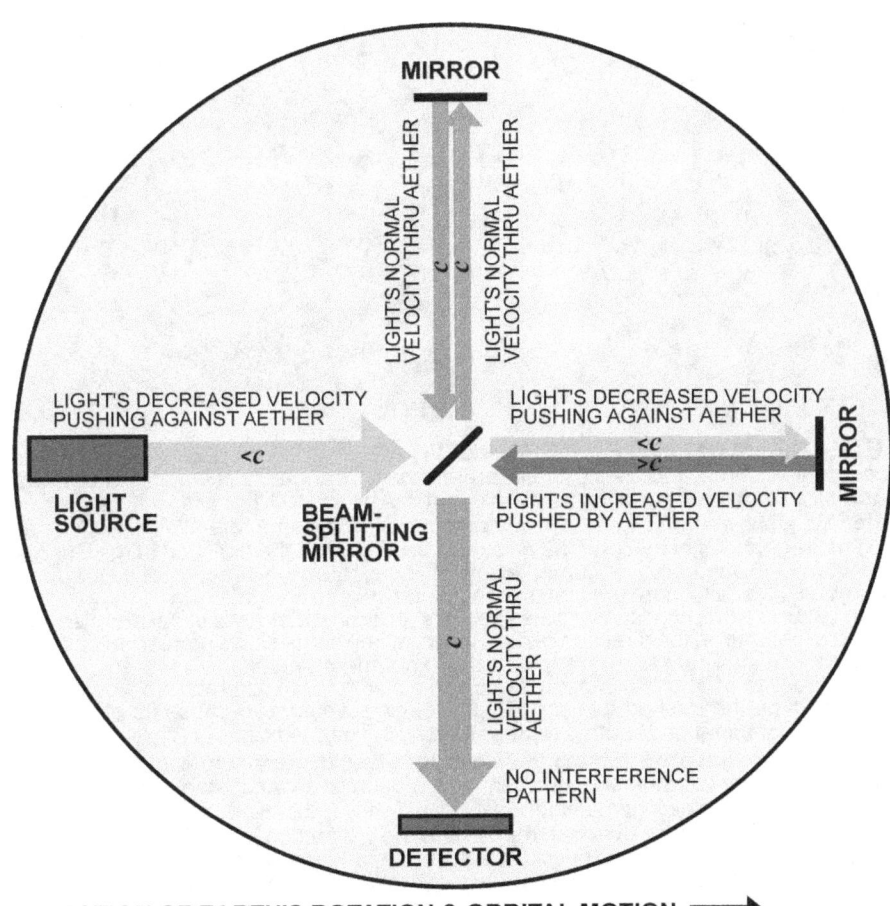

DIRECTION OF EARTH'S ROTATION & ORBITAL MOTION ➡

MICHELSON-MORLEY - CONCEPTUAL DIAGRAM
AETHER EXPLANATION

Instead of contriving the fantastic, self-conflicted notion that objects physically contract but only in the direction of motion while time's rate slows to maintain light's constancy, why wouldn't you simply reason that light's velocity is first slowed by the aether's "headwind" then is increased by the same amount from its "tailwind" after it's been reflected backward? It's not the correct explanation. But at least it's rational.

(1.4 M-M Exp 6a)

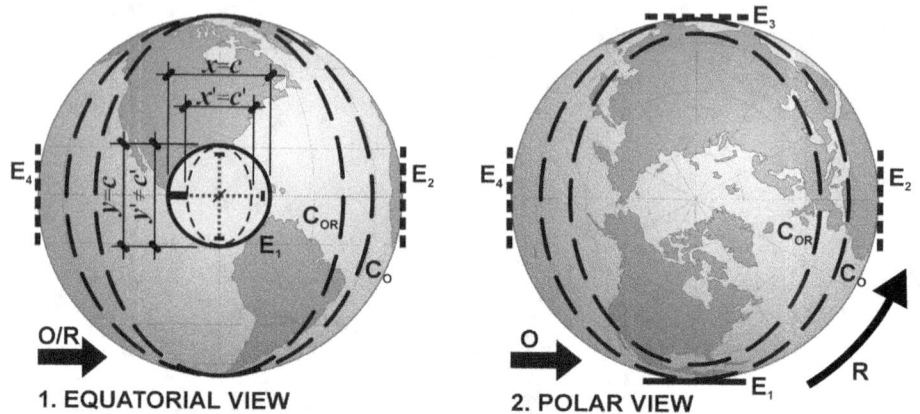

1. EQUATORIAL VIEW **2. POLAR VIEW**

MICHELSON-MORLEY EXPERIMENT - 1

The experiment essentially consists of several mirrors arranged rectilinearly in a cross pattern around a central beamsplitter on a table that can be rotated. For simplicity, ours is at the equator, E_1. It showed that light, c, always radiates at the same rate in all directions at the same time regardless of motion. The arrow at **O/R** indicates the direction of the Earth's orbit & rotation. At **O**, its orbit. The curved arrow at **R** indicates the direction of its rotation. For our purposes, we can ignore the motion of our solar system and galaxy.

Einstein assumes that light's velocity is fixed. It's (special) relativity's basal premise. So objects have to contract in the direction of motion while time's rate slows correspondingly to maintain its constancy. That's how he explains the experiment's result.

So in the direction of its orbital motion, the Earth would have to contract into an ellipsoid shape, indicated by the dashed ovals at C_o. Its rotation would also cause its ellipsoidal contraction. The combination of both is represented by the dashed ovals at C_{OR}.

As the experiment revolves around the Earth, this would cause it to contract in the x direction, as implied by the dashed oval at E_1, which decreases the distance between the mirrors while time's rate slows to maintain light's constancy (assuming "time" is something that actually exists and that it slows with motion). In the y direction, there is no motion. So there's no contraction. The distance between the mirrors remains the same.

Since time's slower rate has to apply equally over the experiment (and over the entire Earth, they're of the same reference frame), it affects light's velocity, c', the same in both the x and y directions. But with contraction only in the x direction, not in the y, this creates different velocities for light. y' will always be faster than x'. This would cause different arrival times, which should have produced a negative interference pattern. But it didn't.

For the Earth's contraction from its orbital motion, the distance between the mirrors in the y direction would also remain unchanged, while the distance in the x direction would be constantly fluctuating. The Earth's rotation would pass the experiment through its orbital tangents, causing maximum contraction at E_1 & E_3 with no contraction at its perpendicular positions, E_2 & E_4, which would compound with its constant contraction from its rotation. The result would be C_{OR}. This should have also produced a negative interference pattern, but one that was expanding and contracting every twelve hours. But this didn't happen either.

Einstein also asserts that special relativity's effects can't occur within gravity fields where light's velocity varies. But the experiment rests in the Earth's gravity field, and all others. They extend indefinitely. So how can it even be considered to explain the results?

The experiment clearly demonstrates that objects do not contract in the direction of motion. Nor does (nonexistent) time slow with motion. So its velocity can't be fixed. It compounds with the motion of its source and that of other reference frames, and that's in addition to its variability. With a founding premise that's untenable, relativity has no viability.

(16.1 MM 7a)

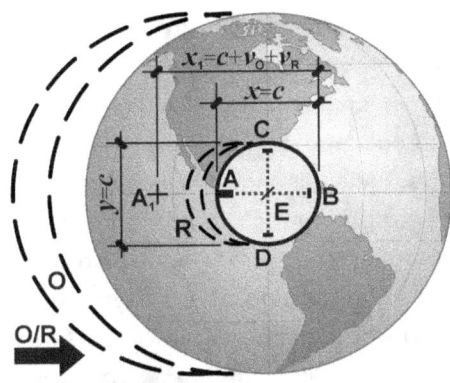

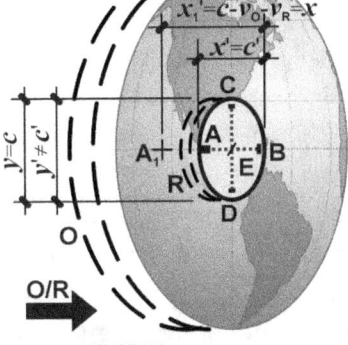

1. REALITY:
LIGHT'S COMPOUNDING
TIME IS NONEXISTENT

2. EINSTEIN:
LENGTH CONTRACTION
TIME DILATION

MICHELSON-MORLEY EXPERIMENT - 2

Light always radiates at the same rate in all directions at once regardless of the motion of its source. So someone at the experiment, E in diagram **1**, measuring light's velocity from **A** to **B**, distance x, and also from **C** to **D**, distance y, would get 186,000mi/s in each direction, $x=c$ and $y=c$, despite their motion. They're of the same reference frame, moving in unison with the Earth's orbit and rotation, implied by the dashed ovals at **O** and **R**.

If we positioned ourselves just outside the Earth's orbit, stationary with respect to the solar system, which is a different reference frame, we'd see that the measurement they made in the x direction actually covered a longer distance from A_1 to **B**, distance x_1. But for them, they still measured from **A** to **B**, covering the shorter distance x due to their motion.

If we measured light's velocity from **A** to **B** as the Earth sped by, our measurement would also begin at A_1 and end at **B**. But light would traverse the longer distance, x_1, in the same time it took for it to go from **A** to **B**, distance x, for the person at **E** because we're not moving with the Earth. So for us, light's velocity would have to include the Earth's orbital velocity, v_0 (66,000mph), and its rotational velocity, v_R (1,000mph). Their compounded velocities over distance $x_1=c+v_0+v_R$. That works out to be about 186,020mi/s.

With his underlying assumption that light's velocity is fixed regardless of motion, Einstein reasons that time slows for moving objects while their length contracts to preserve light's constancy. So for the Earth, light's velocity has to decrease by its orbital and rotational velocity, v_0 and v_R (excluding our solar system and galaxy's motion), while time's rate has to slow and the Earth has to contract correspondingly, along with everything else in its reference frame, in the direction of those motions, as portrayed in diagram **2**. The distance x_1' has to contract to distance x to maintain light's constancy where $c'=c-v_0-v_R$.

But "time" doesn't actually exist. It's not a property of the universe. So there's nothing there to slow with motion. And even if it did exist, its rate wouldn't slow. It'd increase. A unit of time, like a second, that corresponded to distance x that was theoretically contracted to the shorter distance x' would be a faster second. Compressed "time" is a faster running "time."

Einstein also "ignores" light and time's innate three-dimensionality. Time's changing rate and length's contraction only work in the one abstract dimension of linear motion. In the three actual dimensions of our real environment, they're inherently conflicted. There's no motion in the y direction. So there's no contraction. y' and x' can never be the same length. But time's "slower" rate still has to apply in the y direction. It's the same reference frame. So light's velocity can never be the same. y' 's will always exceed x' 's. This invalidates Einstein's basal premise, which undermines relativity in its entirety, and that's without even considering his contradictory but correct assertion of light's variability that does so as well.

(16.2 MM 7a)

Sagnac's experiment also confirms light's compounding with motion but more succinctly. Georges Sagnac (French physicist, 1869-1928) devised an experiment in 1913 that he thought would prove the existence of an aether, while also disproving special relativity. He believed he succeeded.

The construct was not that dissimilar from Michelson-Morley's. In concept, it essentially consisted of a source that sent light through a beam splitter that separated it in opposite directions, routing it to several mirrors located around the perimeter of a rotating platform that formed the corners of a closed loop that returned the light back to its entry point where the recombined beams would create an interference pattern if their velocities were different. The primary difference from Michelson-Morley's was the closed loop.[17]

When the platform was not rotating, no interference pattern was observed. The light took the same amount of time to reach the detector in each direction despite all of the Earth's motions (its rotational and orbital, our solar system's motion through the galaxy, and our galaxy's motion through the universe). This was the same result as Michelson-Morley's.

When it was rotating, the recombined beams did produce an interference pattern. Sagnac concluded that light's velocity is independent of the motion of its source. That's actually not correct. Light always leaves its source at 186,000mi/s in all directions at the same time. That makes it very much dependent on the its source. Its velocity always gets added or subtracted to its source's velocity and that of other reference frames. It compounds. His and Michelson-Morley's experiment clearly demonstrate this.

When the platform is not rotating, the light departs its source at 186,000mi/s and it remains the same in both directions after it's split. The moment it leaves its source, its motion defines it as a different reference frame. But it's moving in unison with the platform (and the platform is moving in unison with the Earth). There's no compounding of velocities. So the light in both directions reaches the detector at the same time and no interference pattern is created.

When the platform is spun, though, light's velocity is compounded with its rotation. This is what's responsible for the interference pattern. The light still leaves its source at 186,000mi/s and it still acts as an independent reference frame. But the platform's rotational or angular velocity, ω, is added/subtracted to the light's velocity.

When the light gets split in opposite directions, it in essence creates two different reference frames from the initially emitted light. The light split in the forward direction travels at $c+\omega$. The light split to the rear travels at $c-\omega$.

17. "Sagnac Effect," Wikipedia, last modified Dec 24, 2022, https://en.wikipedia.org/wiki/Sagnac_effect.

Both beams reach the detector at the same time. But their different velocities cause them to be out of phase. So an interference pattern is created.

Another way to express the same idea is that a Doppler shift occurs between the two beams. (A Doppler shift is a change in frequency due to the motion between a source and an observer.) The forward split light's faster compounded velocity causes it to be slightly blueshifted relative to the rearward light. Or the other way around, the rearward split light's slower compounded velocity causes it to be slightly redshifted as compared to the forward split light. Any way you look at it, their relative shift in wavelength has them out of phase at the detector, which produces an interference pattern.

In another one of his many invalidating contradictions, Einstein apparently came to the same compounding-of-velocities conclusion when investigating the effect. He decided that for accelerating frames of reference "the principle of the constancy of light must be modified."[18] In other words, it doesn't work and needs to be scrapped.

The most common explanation for the Sagnac effect does not incorporate a compounding of velocities. It never addresses the emitter and beamsplitter's constant rotational velocity that would normally be imparted to its light. It's just ignored. This causes different arrival times that produce an interference pattern. But light always leaves its source at 186,000mi/s in all directions at once. Michelson-Morley and Sagnac when not rotating definitively establish this. And every source is in motion. So that motion has to be accounted for. Either it's compounded or light's velocity has to be metaphysically modified similar to what special relativity does.

But special relativity fails completely as an explanation of the effect for both the rotating and nonrotating conditions. It's inherently flawed. Conceptually, it can only address an environment in one dimension, the direction of motion. In every other direction, it's insurmountably conflicted.[19]

For the nonrotating condition, light's velocity in the perpendicular direction (or at any angle other than directly forward) would be greater than the forward direction, exceeding 186,000mi/s. Time dilation's innate three-dimensionality and length's one-dimensional contraction can maintain its fixed velocity only in one dimension, the direction of linear motion.

In the other two dimensions of our real world, it's unworkable. Light's velocity in those dimensions would contradictorily be increasing. If relativistic effects were actually feasible, this would create conflicting velocities that would produce an

18. "Sagnac effect," Wikipedia, last modified Oct 23, 2022, https://en.wikipedia.org/wiki/Sagnac_effect, footnoted [18] A. Einstein, 'Generalized theory of relativity', 94; the anthology 'The Principle of Relativity', A. Einstein and H. Minkowski, University of Calcutta, 1920.
19. Einstein, *Relativity*, 44.

interference pattern for the nonrotating condition just like what was demonstrated in the diagrams for Michelson-Morley.

The term "relativistic" refers to effects resulting from relativity, but mostly special relativity. In theory, light's presumed constancy (mathematically) forces a moving object's (or reference frame's) time to slow, its contraction in the direction of its motion, and the increasing mass of accelerating objects.[20]

For the rotating condition, special relativity would theoretically compound the platform's rotational velocity with light's velocity. But it enforces the assumption of light's constancy by reducing its velocity by the amount of the rotational velocity in the forward direction and increasing it by the same amount in the rearward direction. This maintains light's fixed velocity and produces the same result, different arrival times that create an interference pattern.

But relativistic effects always produce the same contradictory results. The spinning platform's time is required to slow, it's one reference frame, while its perimeter around its circumference is required to contract. But its interior does not. Its radius remains the same.[21] That's not possible.

Moreover, Einstein asserts that, "The special theory of relativity [is only valid where] no gravitational field exists" because of light's variability in them.[22] (More on light's variability in gravity fields shortly.) So if his "principle of equivalence" (acceleration/braking and rotation's reactions are the same as gravity's[23]) were actually true and rotation's centrifugal force actually did produce real gravity then light's constancy, time's dilation (a larger or slower rate of time), length's contraction, and the increasing mass of accelerating objects cannot even be considered as an option to explain the Sagnac effect. Its associated rotation would be producing centrifugal gravity where light's velocity varies, which would preemptively nullify its constancy and special relativity's relativistic effects.

Trying to explain the results through his "principle of equivalence" doesn't work either. It's also entirely unfeasible. Light's slower velocity in the rotating experiment's centrifugal gravity field would presumably account for the disparity that causes the interference pattern. But it can be easily shown that rotation doesn't create gravity. (More on this in the chapter on equivalence.) So Einstein's "principle of equivalence" isn't an option.

Relativity's simultaneity has also been proposed as a possible explanation. But it also doesn't work. It's fundamentally flawed as well. Any factual review quickly

20. Einstein, *Relativity*, 21-42, 52.
21. Einstein, *Relativity*, 90-91.
22. Einstein, *Relativity*, 85, 104, 109, 171.
23. Einstein, *Relativity*, 75-79, 172.

reveals its obvious failure in logic. It's also dealt with in an upcoming chapter.[24]
(See Diagram **49.1**, **49.2**, **49.3**, **49.4** Sagnac Exp 5a)

24. Einstein, *Relativity*, 29-31.

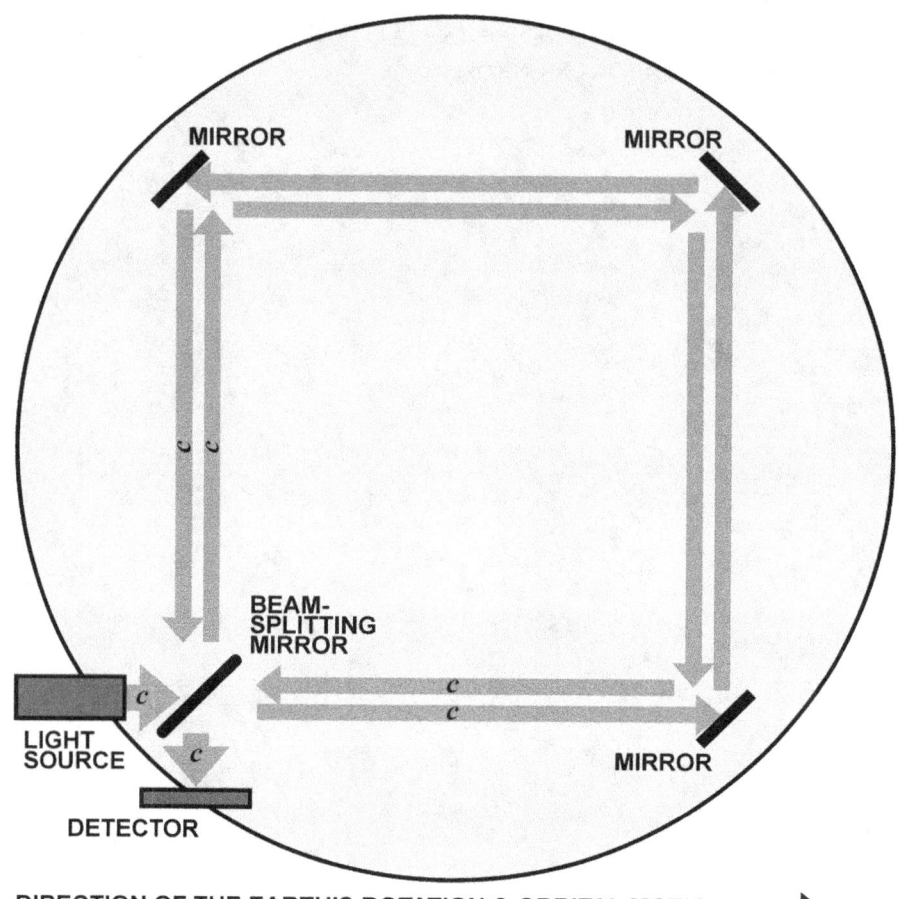

DIRECTION OF THE EARTH'S ROTATION & ORBITAL MOTION ➡️

SAGNAC EFFECT - CONCEPTUAL DIAGRAM
NO ROTATION

Sagnac's experiment essentially consists of a source of light that's projected onto a beamsplitter that sends it in opposite directions around a series of mirrors arranged in a closed loop at the perimeter of a platform form that can be spun that recombines the beams back at a detector, as suggested by the grey linear arrows labeled as c that indicates light's velocity. The inside row of arrows indicates light's clockwise path. The outside counterclockwise.

When the platform is not rotating, no interference pattern is produced. This is essentially the same result as the Michelson-Morley experiment. Both show light always leaving its source at 186,000mi/s in all directions at the same time. Light's independent motion could qualify it as a separate reference frame. Because the platform and the light it's emitting move with all of the Earth's motions, rotational and orbital, our solar system's motion through our galaxy, and our galaxy's motion through the universe, this suggests light compounds with the motion of its source and that of other reference frames.

(49.1 Sagnac Exp 5a)

30

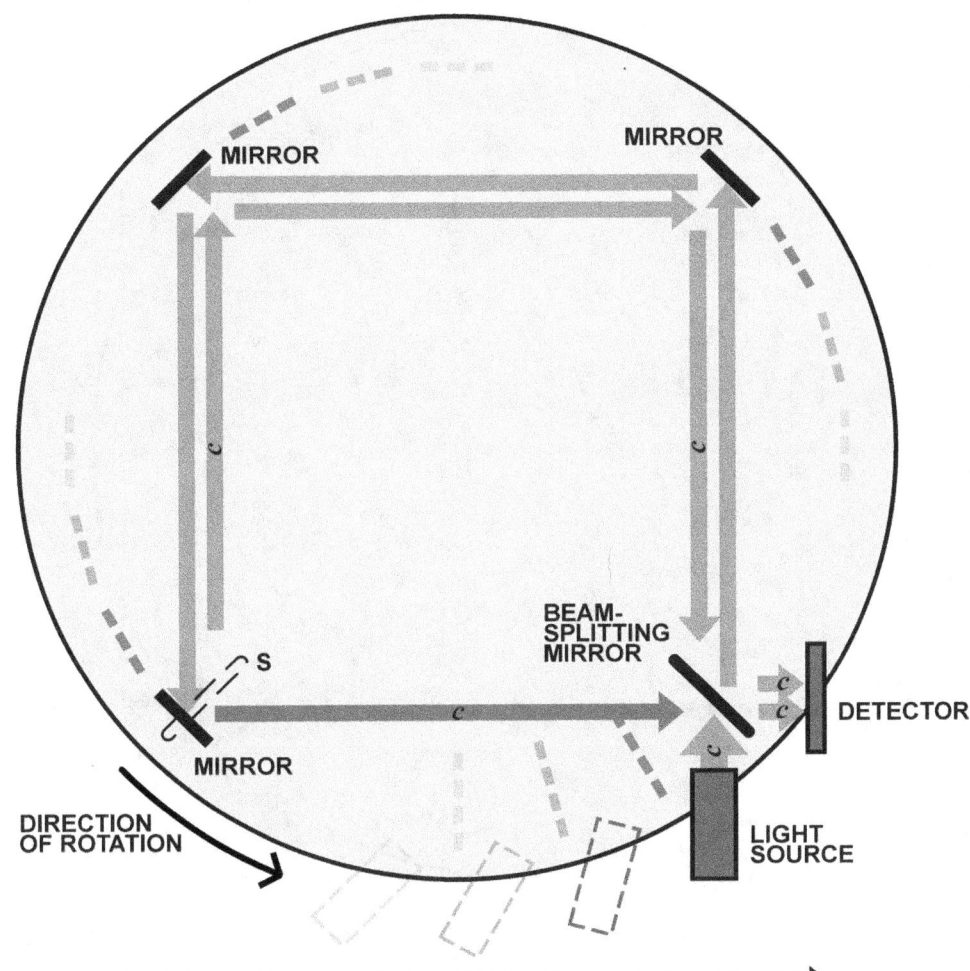

DIRECTION OF THE EARTH'S ROTATION & ORBITAL MOTION ➡

SAGNAC EFFECT - CONCEPTUAL DIAGRAM
WITH ROTATION - ORTHODOXY
The interference pattern produced when the platform is spun has several possible interpretations. For convenience, let's establish the platform's rotation as constant with an angular velocity that completes one-quarter rotation for three-quarters of light's.

The conventional explanation assumes that the light still leaves its source at 186,000mi/s and is split from **S** at the same velocity in both directions. It's thought that because the platform is rotating into it, the rearward split light arrives at the detector first, which for our diagram is 3/4 of one revolution. While the forward split light arrives later, in 11/4 revolutions, the overlap indicated by the darker arrow. The difference in arrival times would produce a phase shift that creates an interference pattern.

Sounds reasonable enough, but it's inherently flawed. It fails to account for the platform's constant rotation. It departs from it at 186,000mi/s as if there were none.

(49.2 Sagnac Exp 5a)

31

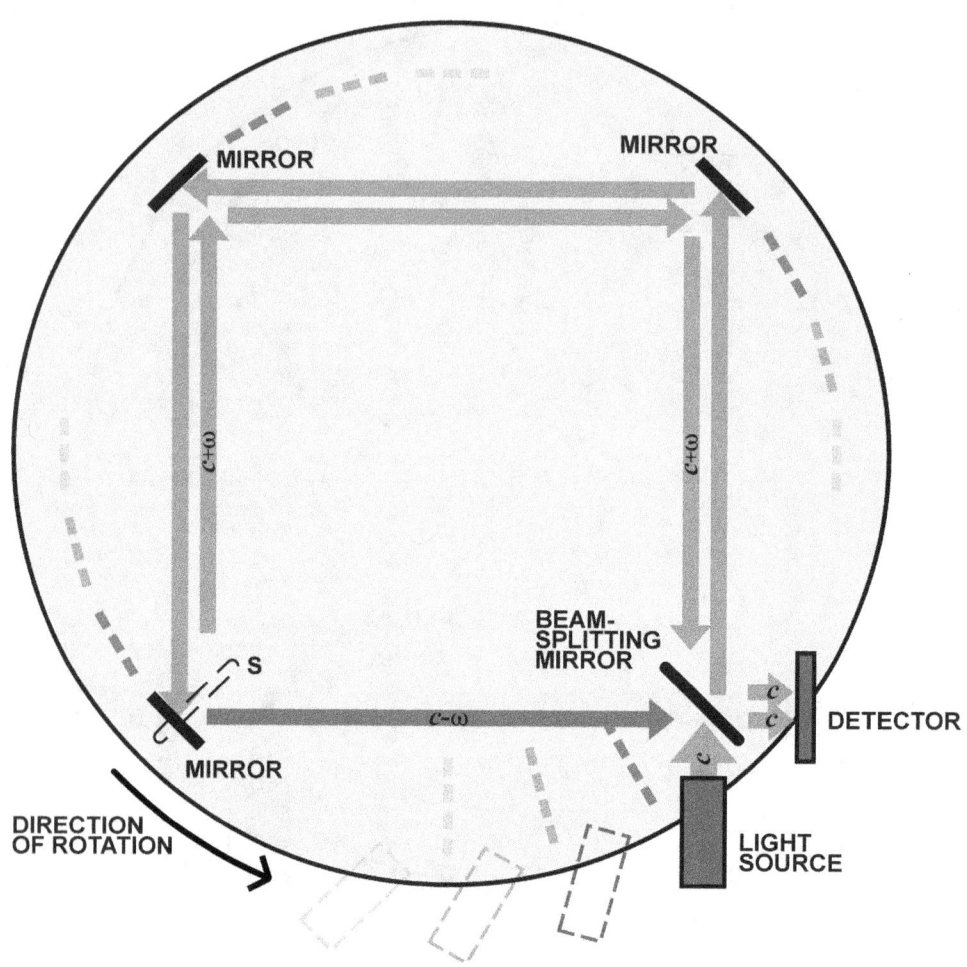

DIRECTION OF THE EARTH'S ROTATION & ORBITAL MOTION ➡

SAGNAC EFFECT - CONCEPTUAL DIAGRAM
WITH ROTATION - SPECIAL RELATIVITY

Special relativity does account for the platform's rotation. But it doesn't work either. The split light would leave at c-ω in the direction of rotation and c+ω opposite the direction of rotation to maintain light's fixed velocity. The light would arrive at the detector at different times because of the platform's rotation, creating an interference pattern.

But special relativity is inherently self-conflicted. It would have the platform's perimeter contracting while its radius remains constant and its time dilates for the entire platform. That's not even remotely feasible.

It would also conflict with the results when the platform is not rotating. It would have to be contracting in the direction of the Earth's motions to enforce light's constancy but not in the perpendicular direction while time's slowing would again have to be applied equally over the entire platform. It's one reference frame. So it fails in every respects.

(49.3 Sagnac Exp 5a)

32

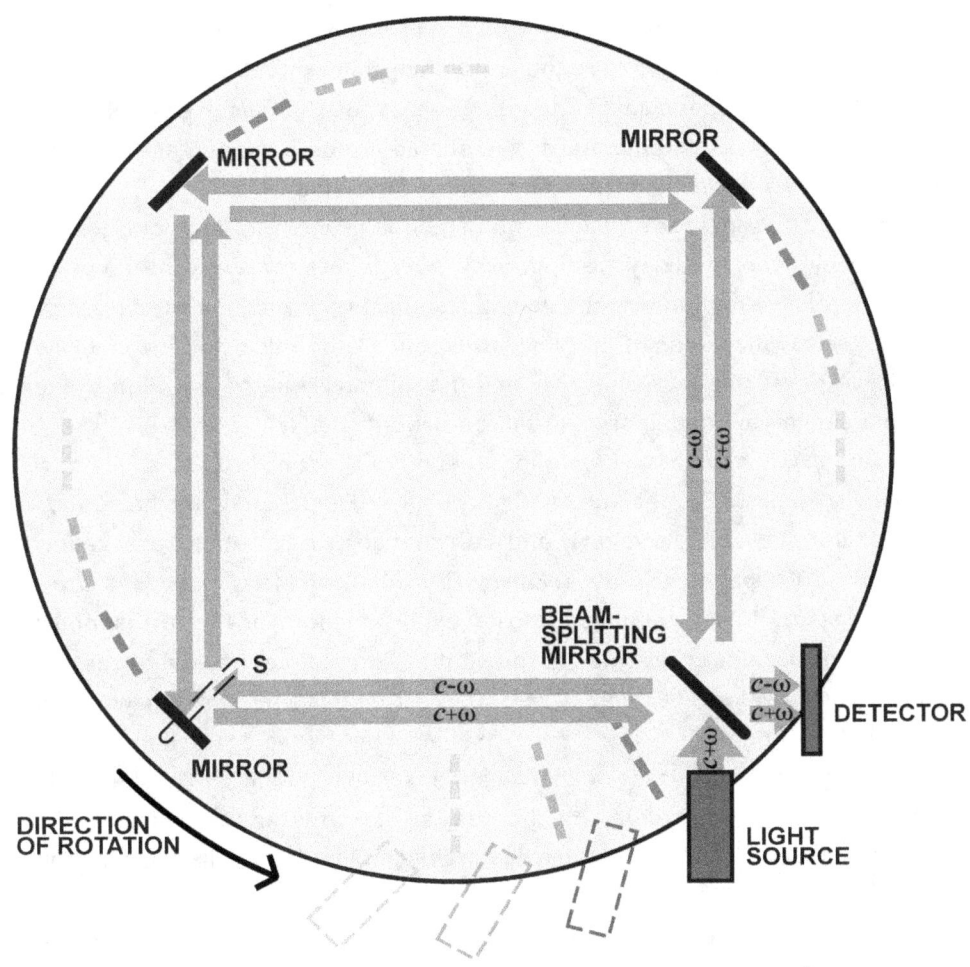

DIRECTION OF THE EARTH'S ROTATION & ORBITAL MOTION ➡

SAGNAC EFFECT - CONCEPTUAL DIAGRAM
WITH ROTATION - REALITY

The only consistent way to explain the effect is if light compounds with the motion of other reference frames. It always leaves its source at 186,000mi/s in all directions at the same time. The non-rotating condition and Michelson-Morley clearly confirm that.

The platform's angular velocity, ω, (or some vector angle of it, which for this diagram would be a 45°, or .707) has to be imparted to the light. Its emitter is moving with it. So it has to be added/subtracted to light's velocity: $c+\omega$ for the forward split light and $c-\omega$ for the rearward split light. The result is that both beams reach the detector at the same time. But it's their different velocities that puts them out of phase and produces the interference pattern, not their different arrival times.

Sagnac's experiment unequivocally establishes light's compounding with the relative motion of other reference frames.

(49.4 Sagnac Exp 5a)

So if relativity can't account for the Sagnac effect, the compounding of light with the relative velocity of any other reference frame must be a fact. And the universe's ultimate speed limit cannot be fixed at 186,000mi/s. It essentially can be any velocity up to instantaneous. We should not be surprised at all with this conclusion. It's a perfectly rational and natural inference.

What is it in a person's psyche that would lead them to first conclude that light's velocity should always remain fixed everywhere for everyone no matter what their relative motion (which Einstein also asserts is subjectively decided by each observer) then reason that this would cause their time to slow and their impossible physical contraction only one-dimensionally in the direction of their motion, despite time and light's overt three-dimensionality?

Apparently, a couple of years after presuming light's constancy as the basis for relativity in 1905, Einstein modified his position. He decided that the speed of light is actually variable. Its velocity and path change as it traverses a gravitational field. In his 1907 paper, *On the Relativity Principle and the Conclusions Drawn from It*, he states, "These equations too have the same form as the corresponding equations of the nonaccelerated or gravitation-free space; however, c is here replaced by the value $c\left[1 + \frac{\gamma \xi}{c^2}\right] = c\left[1 + \frac{\Phi}{c^2}\right]$. From this it follows that those light rays that do not propagate along the £-axis are bent by the gravitational field;"[25]

Many have difficulty believing that he asserts light's variability. They're holding on to the popular narrative that he proved its constancy. And why not? Einstein believed he did. He (delusively) maintains that it's still fixed despite the unresolvable conflict.[26]

In this same paper prior to this quote, he spends five pages (302 to 307) making an argument for gravitational time dilation. That's time's (presumed) slowing in a gravity field. More on that later. The problem is, it's contradictorily based on light's fixed velocity, in gravity fields no less. He's using light's velocity both ways, constant and variable, as the need arises. Either it's being done unknowingly or deceptively.

For his explanation of starlight's displacement observed during the 1919 eclipse that supposedly confirmed general relativity, he correctly concludes that: "A curvature of rays of light can only take place when the velocity of propagation of light varies with position [in gravity fields]." He also explains that, "the general theory of relativity cannot retain this law, [the law of the constancy of the velocity

25. Albert Einstein, ON THE RELATIVITY PRINCIPLE AND THE CONCLUSIONS DRAWN FROM IT [Jahrbuch der Radioaktivität und Elektronik 4, (1907): 411-462] from the Princeton University Press' THE COLLECTED PAPERS OF ALBERT EINSTEIN, Vol 2: The Swiss Years: Writings, 1900-1909 (English translation supplement), Doc. 47, 310.
26. Einstein, *Relativity*, 85-87.

of light]... the velocity of light must always depend on the co-ordinates when a gravitational field is present... The special theory of relativity has reference to Galilean domains, *i.e.* to those in which no gravitational fields exist... the principle of the constancy of the velocity of light [is] valid only with respect to an inertial system [not a gravity field]." All of this mounts to nothing more than light's refraction.[27] (Refraction can be defined as light's displacement due to a change in its velocity due to a change in the density of the medium it's traversing.[28])

The dilemma he's faced with, just like with its compounding, its variability fundamentally invalidates relativity's founding premise. But Einstein never reverses his position on its constancy. Just the opposite, he maintains that both are true despite the nullifying contradiction.

He tries to argue that special relativity is still valid despite light's variability because "its results hold only so long as we are able to disregard the influences of gravitational fields... The special theory of relativity has reference to Galileian domains, *i.e.* to those in which no gravitational field exists."[29] How does that work? (The term "Galileian domains" refers to conditions where the basic laws of physics remain the same everywhere.)

Where are the locations where gravity fields don't exist or the conditions under which the effect of gravity fields can be ignored? Whether it's at the subatomic level or the self-gravity of our entire (presumed) finite universe, gravity fields everywhere. They surround and permeate every object and they extend indefinitely. So there's no place where they aren't. So there's no way they can be disregarded.

Which means light's velocity has no possibility of ever being fixed (if it weren't already conceptually impossible). It has to vary everywhere. And that's in addition to its compounding. Without its underlying premise, how can relativity, or any of its ancillaries like the Lorentz transformation or Einstein's application of relativity to the theorem of addition of velocities, have any validity? They all become nothing more than theoretical contrivances that have no practical relevance.

Einstein could not disagree. He qualifies his assertion of light's variability: "[Relativity's] results hold only so long as we are able to disregard the influences of gravitational fields on the phenomena (*e.g.* of light) [that causes its variability]." If we're unable to disregard gravity fields, as we just reasoned is impossible because they're everywhere so light's velocity has to vary everywhere then,

---

27. Einstein, *Relativity*, 85, 104, 109, 171; In his article, "The speed of light is not constant," on his website, The Physics Detective: https://physicsdetective.com/the-speed-of-light/, John Duffield assembled a total of eight Einstein quotes from eight papers beginning from 1907 through 1920 from THE COLLECTED PAPERS OF ALBERT EINSTEIN from the Princeton University Press that affirm his evolved belief in and contradictory assertion of light's variability.
28. "Refraction," Wikipedia, last modified Feb 5, 2023, https://en.wikipedia.org/wiki/Refraction.
29. Einstein, *Relativity*, 85, 109.

"as a consequence of this, the special theory of relativity and with it the whole theory of relativity would be laid in the dust."[30] Our entire cosmology, including the big bang, is rooted in a theory whose originator would have to concede is altogether untenable. (See Diagram **15.1.1** Light's Bending 5a, **15.1.2** Light's Refraction 5a & **15.2.1**, **15.2.2**, **15.2.3**, **15.2.4** Refraction 3a)

30. Einstein, *Relativity*, 85.

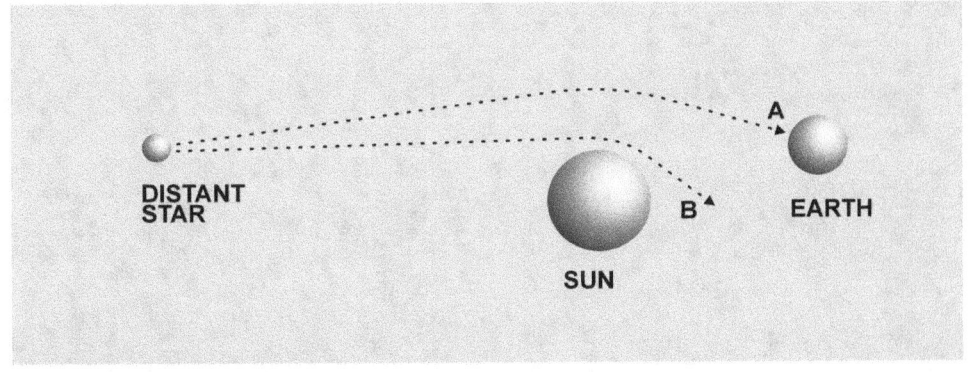

DISTANT STAR

SUN

A

B

EARTH

LIGHT'S BENDING

Our current belief is that a "ray" of light from a star or any distant object passing near a massive body like our sun is being pulled by gravity, that it's being bent from its otherwise straight path in the direction of **B** as it follows space's geodesic that somehow curves two-dimensionally in the vicinity of mass. And when viewed from Earth, its position is distorted in the direction of **A** due to an optical illusion.

Even though Einstein contends that light's distortion is actually due to its slowing through gravity fields, which is nothing more than refraction, which contradicts relativity's founding premise, light's fixed velocity, we reject his explanation. Instead, we hold to our belief that a photon, which remember is only a hypothetical quantum of massless energy, is subject to gravity's influence. We first mistakenly assumed that a photon is a particle. And then we incorrectly reason that because it's in motion it must have momentum. If it has momentum, it must have inertial mass. And then because of relativity's principle of equivalence, if it has inertial mass it must also have gravitational mass. And if it has gravitational mass, it must then be affected by gravity.

We're highly motivated to retain this convoluted logic because if we use light's refracted slowing like Einstein, we're abruptly confronted with the total collapse of relativity, which is wholly dependent on light's constancy. Incredibly, Einstein actually agrees that relativity would completely unravel if it were found that light's velocity was not fixed but variable.

(15.1.1 Refraction 5a)

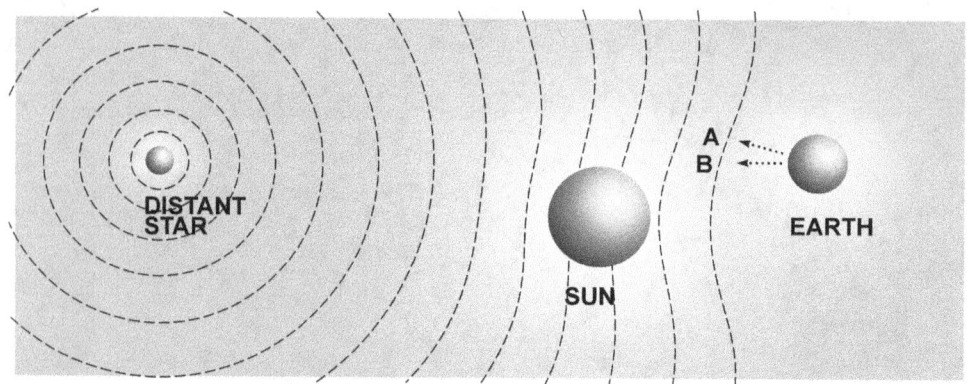

LIGHT'S REFRACTION

Light refracts through gravity fields. The distant star appears displaced in the direction of **A** not because light rays follow the impossible curvature of two-dimensional space or a resultant optical illusion but because the light in that direction reaches us slightly before the light coming directly straight from the star in the direction of **B**. Light's velocity slows through the decreasing density of the Sun's gravitational field, depicted in section as the diffusing background, just as any wave travels slower through a less dense medium, as portrayed by the series of circular and waving dashed lines that indicate the varying velocity of the incoming light emanating from the distant star.

It's also light's refracted slowing that's responsible for the gravitational lensing of distant galaxies or quasars that are split into two or more images that are assumed to be the product of the mass of some unseen foreground galaxy that's closer but fainter. But more often than not, it's just the common center of mass of any number of galaxies or galaxy clusters that is located between us and the object along its line of sight that's responsible for the lensing effect, which is why the refracting mass is so often never identified.

(15.1.2 Refraction 5a)

38

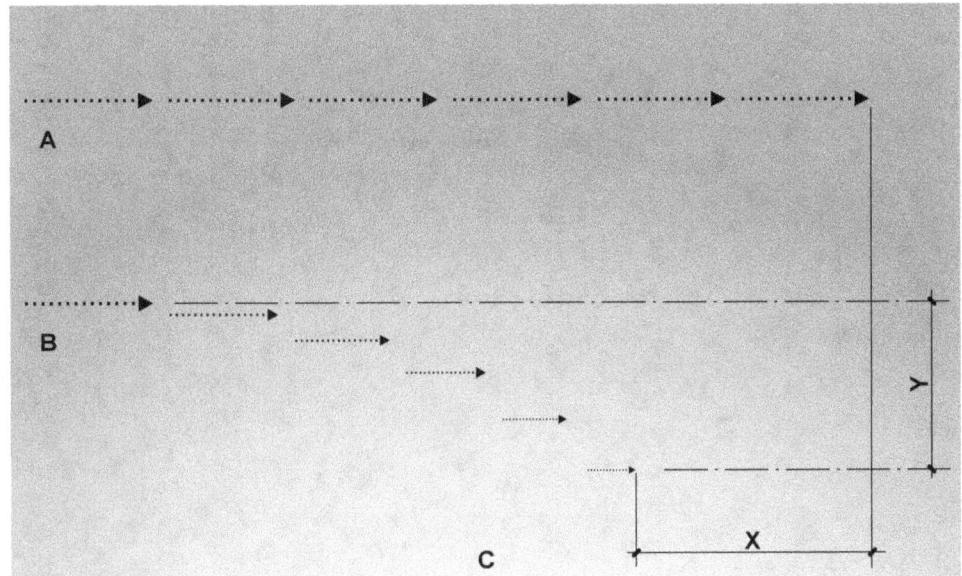

REFRACTION: LIGHT

To conceptually portray light's refracted slowing, let's say we could theoretically quantize it into a "ray" and represent it with a dotted arrow of fixed length. And then let's set two parallel rays out in space somewhere and project them so that the ray in row **A** would pass through a region where gravity essentially remained constant. Its velocity and direction would remain constant as well, as indicated by the sequence of unchanging arrows.

Let's set the other ray in row **B** to simultaneously pass through the gravity field of a nearby massive body like our Sun or a galaxy located in the direction of **C**. That field represented in section view by the diffusing background. As that ray's quantized light traverses the field, it would begin to slow in its exponentially decreasing density, as suggested by the decreasing length of the arrow. The difference expressed by the distance **X**. The light's course would also begin to deviate from its parallel path toward the body as it followed the field's decreasing density. That difference expressed as **Y**. Any other distortion is omitted for clarity. Anyone at the receiving end of our hypothetical rays would see the light in row **A** first because of its faster velocity.

(15.2.1 Refraction 3a)

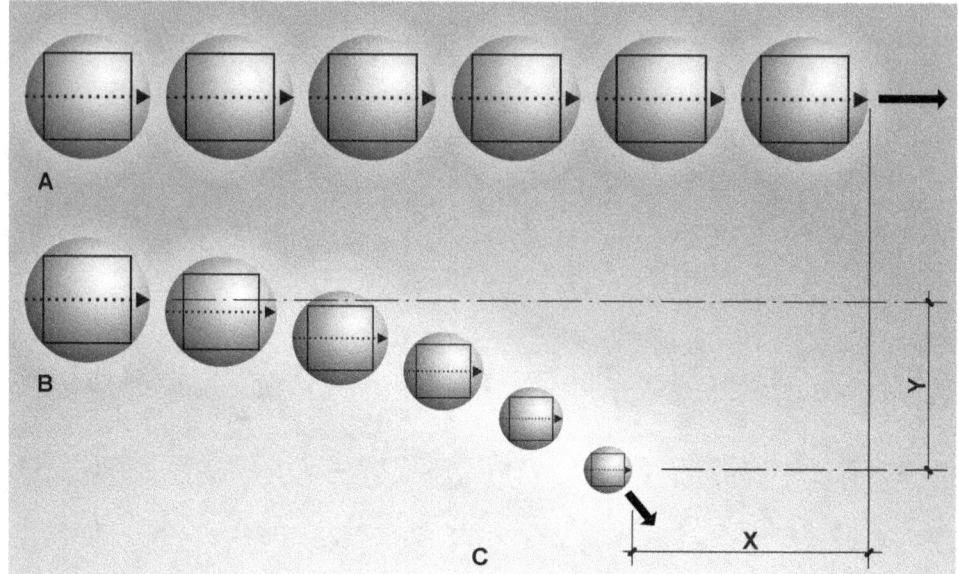

REFRACTION: OBJECTS

Objects themselves can also be interpreted as refracting in the decreasing density of a gravity field. Let's say we could associate a massive body, the shaded sphere, with our theoretical quantized light ray. The body in row **A** would remain unchanged just like its ray as it passes through gravity fields that we're assuming are uniform.

But in row **B**, as it enters **C**'s gravity field, it also begins to slow in the forward direction, indicated by **X**, while accelerating in **C**'s direction, indicated by **Y**. Because the universal field of radiant energy that comprises the electromagnetic spectrum is essentially the support structure for matter, its inherent exponential thinning at any mass, which is a gravity field, causes the body's synchronized contraction, and distortion, just the same as the ray's, as both our body and the body at **C** naturally pursue equilibrium in the decreasing density of their combining gravity fields. Any other distortion is still omitted for simplicity's sake.

If we were to put an object on our moving bodies like a box, symbolically represented by the corresponding rectangle, it too would theoretically contract, and distort, the same as the body and the same as the light ray.

(15.2.2 Refraction 3a)

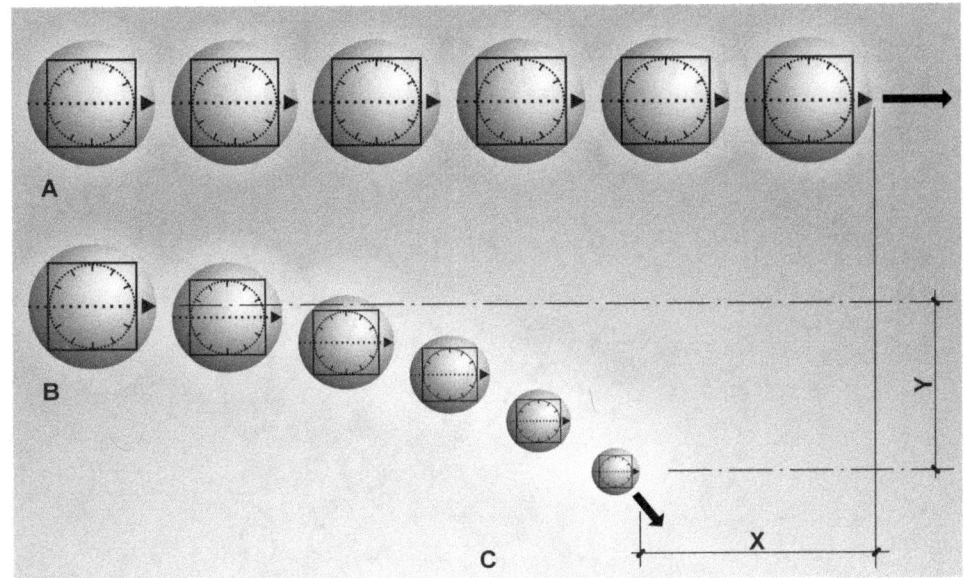

REFRACTION: TIME

Now let's establish a rate of time for our moving bodies by setting it to the natural frequency of their rotation just like we do for ourselves with the Earth's rotation, and let's represent that with a clock-like circle within the box on the bodies so that it too is associated with our body and light's velocity where its changing rate is represented by the changing size of the clock. In row **A**, the clock, the box, the body, and light's velocity all remain consistent in our assumed uniform gravity fields.

In row **B**, they all contract, and distort, in unison while both our body and that of **C**'s naturally draw together because of their reactive search for equilibrium in their combining gravity fields' decreasing density, which is always less between them.

As our body condenses, its mass is redistributed inward, which increases its rate of rotation, which in turn increases time's rate, as signified by the clock's smaller size. A smaller time being a faster running time that theoretically exactly compensates for light's decreasing velocity, as represented by the synchronized contraction of its arrow.

(15.2.3 Refraction 3a)

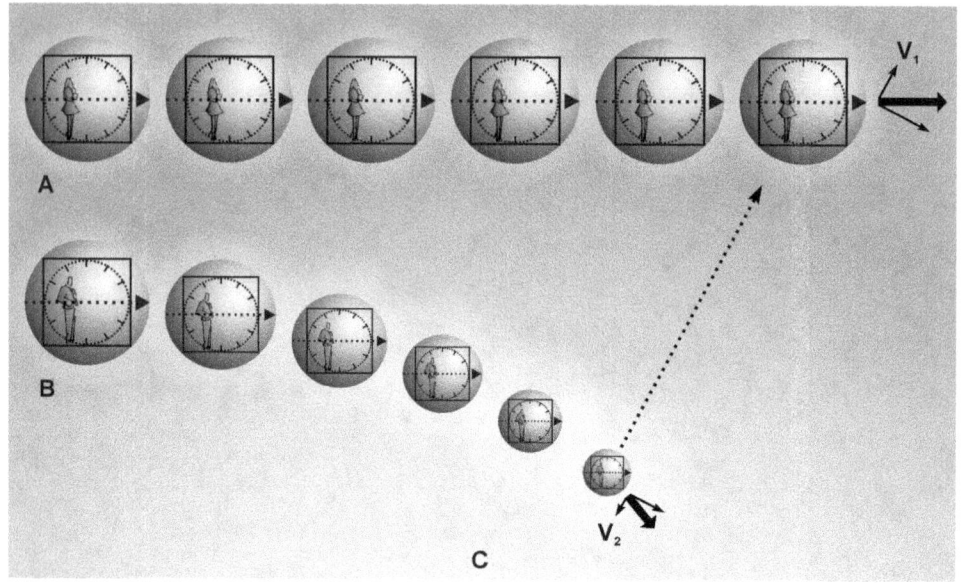

REFRACTION: PERCEPTION

With light, objects, and time all in place, let's put someone inside our box, or we'll call it a crate like Einstein. But we'll make ours transparent. For row **A**, everything still remains the same, just as the person perceives.

But it's also the same for row **B**. The person perceives no change despite that everything would be condensing. That's because the contraction, and any other distortion, omitted for clarity, all occur in a synchronized way. With faster time compensating for light's slower velocity and all measuring devices condensing, and distorting, the same as the objects they're measuring, there's no way, theoretically, to compare a pervious measurement to a current measurement to perceive a difference. Everything's contracting in unison. So the surroundings appear to remain unchanged. And for all practical purposes, row **B**'s Euclidean environment remains intact, no different from **A**'s.

If row **B**'s person were to look out at the object in **A**, it might appear to be maintaining a consistent distance or even enlarging while higher redshift observations confirmed an increasing recessional velocity. That increasing recessional velocity would express as a function of the three-dimensional line-of-sight vectors associated with each object's motion, diagramed two-dimensionally as V_1 & V_2. A perpendicular perspective could confirm its farther distance and ever-growing separation.

We could apply these same circumstances as if our objects were in separate galaxies that were a fixed distance apart. Let's say the object in **B** was infalling towards its galaxy's center in the direction of **C** while **A** was at the outer reaches of its own galaxy so its trajectory was nearly circular. Both objects' condensing would occur in the same manner, they'd have the same recessional velocity based on the same line-of-sight vectors, and they'd record the same redshift.

The problem is for us here on Earth, we misinterpret our observations of recessional velocity as every entire galaxy moving away from every other because of the big bang's presumed expansion. When in reality, it's simply the recessional velocity of each galaxy's continuous infall of material along with our own infall velocity that's a natural consequence of gravity's inherent runaway nature.

(15.2.4 Refraction 3a)

Because of the observational fact of light's displacement in gravitational fields, we'd naturally agree that light must be propagating at varying rates throughout the universe. It slows because of its refraction as it approaches a massive body due to a gravity field's decreasing density. It speeds up farther away from the body as the density of the field increases. So light's velocity can only be interpreted as being both variable and compounding everywhere. It can never be constant but has the capacity for any velocity.

It's difficult to explain why Einstein would first assume light's constancy and then later assert its variability while maintaining that both are true when it's perfectly clear that their coexistence is not feasible. It seems that he's fully cognizant of the implications of light's variability, but he insists on maintaining special relativity anyway. It appears at any cost. He can't let go of it.

So he manufactures an embarrassing workaround to try to salvage it, and the rest of relativity. It feels like desperation and/or deception. Either way, he's unwilling to fully accept the contradiction. Light's variability, and its compounding, clearly invalidate his special relativity reasoning. Time and dimension can't contract for objects in motion unless light's velocity is fixed. But it also invalidates his general principles of relativity that also rely on its constancy. How can time and dimension change for objects in gravity fields if light's velocity is changing?

Beyond that, light's variability has dire implications for his renowned mass-energy relation, $E_{kinetic} = mc^2 / \sqrt{1 - v^2/c^2}$ (simplified as $E = mc^2$ for us of normal intelligence).[31] It's conceived on the incorrect premise as well. It's no longer valid either.

Despite its overt implausibility, it's only our continuing belief in relativity's untenable doctrine that gets at the heart of why we insist that faster-than-light travel is not possible. We're constantly lectured on how we can never exceed the speed of light. Our mass would become infinitely large. But with no limit on light's velocity, the mass of objects (outside of an accelerator) cannot increase with increasing velocity, much less become infinite at the speed of light.

And without the impossible, and contradictory, contraction of objects in the direction of motion and associated "slowing" of time that also become infinite at the speed of light, both a mathematical consequence of light's presumed constancy, there's no way to conclude any limitation on our velocity at all. Our speed could be anything up to instantaneous without violating any law of physics or sound logic like cause and effect.

So right from the outset relativity is completely nullified, not from a critic, but from its originator no less. Its viability has no chance of ever being (re)established.

31. Einstein, *Relativity*, 49-54; "Mass-energy equivalence," Wikipedia, last modified Feb 20, 2023, https://en.wikipedia.org/wiki/Mass-energy_equivalence.

Our entire cosmology, including the big bang, is rooted in a theory whose own author reasons is entirely unworkable. Just incredible. But mesmerized by relativity's metaphysical mysticism and rendered impotent by unrestrained groupthink, we continue rushing headlong like lemmings completely blind to our folly.

For those readers who have difficulty, as I initially did, believing that Einstein actually judged light's velocity to be variable, contrary to the myth, I suggest you peruse his book (*Relativity: The Special and the General Theory*). And I strongly encourage you to investigate this or any of these other issues yourself. But I must warn you. You may be in for a bit of a shock. Directly confirming for yourself how baseless and irrational his suppositions actually are and worse having firsthand knowledge of how gullible we've actually been accepting his nonsense can be an unsettling experience. The implications are vast and profound.

Time

Even though he claims to, Einstein never actually defines time. He only offers us an explanation for how an individual clock behaves near any given reference system, which is no definition.[32] Essentially, he fails to recognize that time is a concept we've created. It's nothing more than a reference that we use to measure motion.[33] Its existence is wholly dependent on our selection of a physical object with harmonic motion that we use to establish its rate of passage.

In reality, there's no time component or time property of the universe. There exists only a continuous present moment that's perpetually changing with motion. What we've labeled "time" is only an expression of the natural frequency of objects that's usually stated as a unit of the cyclic motion of the Earth's rotation or its orbit around the Sun.

Since time can only be expressed as a function of the reference object's natural frequency, it cannot vary in rate independent of that natural frequency. And any variation in that natural frequency can only be the product of a change in the object's mass or the distribution of its mass. This occurs naturally with gravitation when the density of an object's gravity field changes, which alters its size while the amount of its material remains constant, or when it acquires more material through gravitational coalescing. Both of which affect the object's periodicity.

A natural change in natural frequency can also occur when objects like subatomic particles or the cesium atoms of atomic clocks are subjected to a charge. This increases their mass by increasing the amount of material they contain, which in turn slows their periodicity, causing their clocks to run slower. This is often mistaken as a relativistic slowing in the rate of (nonexistent) time when the atomic clock's slower rate is simply due to an induced charge or a change of field density.

To show how the rate of time supposedly varies with velocity, Einstein contends that a person on a moving train passing the midpoint of two simultaneous lightning flashes would see the one in the direction of travel before the other. This seems logical. The train is moving in that direction. But he concludes that it's the passenger's motion that causes a change in the rate of their time which has them seeing the lightning strike in the direction of motion first.[34]

But by simply compensating for the distance traveled by the train or its time duration, which would be a necessary requirement, both lightning strikes would remain simultaneous at their midpoint for both the moving passenger on the train

32. Einstein, *Relativity*, 11-12.
33. "Time," Merriam-Webster Dictionary, last modified Mar 28, 2023, https://www.merriam-webster.com/dictionary/time.
34. Einstein, *Relativity*, 29-31.

and any observer at the same location on the adjacent embankment. So there's not any change in time's rate because of the train's motion.

But if time really could change with motion then the passenger would still see both flashes simultaneously no matter their speed or direction but only at a different time than someone at the same location stationary on the embankment. Any change in time's rate would have to be applied equally in all directions from the moving passenger's. It could not change only in the direction of motion or only in the direction they were looking. Multiple times for a single reference frame is conceptually impossible. We have to be astonished at how elemental this error actually is.

But even beyond that, if the person on the moving train was really seeing the lightning flash in the direction of motion first because of a change in the rate of their time then that rate of time would had to have increased, now running faster. But Einstein insists that all moving clocks run slower because of light's constancy. So, he's fundamentally at odds with his own supposition and argument.

Still, he employs this misguided reasoning to establish a definition of simultaneity that he then uses to argue for the principle of relativity, that each object or coordinate system with motion has its own particular rate of time. And because of light's fixed velocity, that rate of time has to change, slowing with motion, rather than light compounding with motion of its source.

It's hard to believe that this conflicted reasoning underlies almost all of Einstein's assertions. But it's even more difficult to believe that it remains unrecognized and unchallenged. I also find it very revealing of his personality, indicative of his inability to accurately perceive reality, but also characteristic of his skill to fabricate relationships that do not and cannot exist to rationalize his delusive conclusions at any cost. (See Diagram **2** Simultaneity 3a)

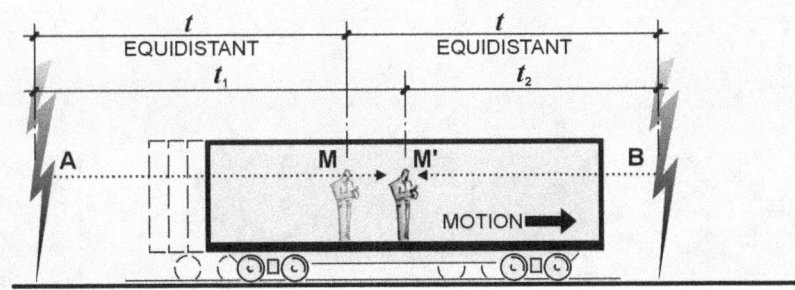

1. RELATIVITY

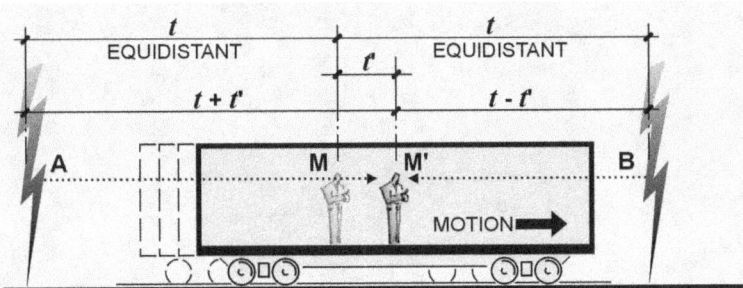

2. REALITY

EINSTEIN'S SIMULTANEITY

Einstein would have us believe that a person on a moving train, at **M** in diagram **1**, midway between two simultaneous lightning strikes, at **A** & **B**, would see strike **B**, the one in the direction of travel first, at location **M'**, because the person's motion causes their time to speed up, represented by the shorter duration t_2, while a motionless person on the adjacent embankment at the same midpoint, **M**, would see both strikes at the same normal time, represented by the duration t.

He acknowledges that the passenger travels the distance **M** to **M'** during light's trip from the strikes to the passenger but still somehow reasons that they'd see strike **B** first because of their faster rate of time. He doesn't adjust time's duration as you'd have to by simply subtracting t' from t, as indicated in diagram **2**. But even without this invalidating error, how can time's rate be quicker only in the forward direction? Wouldn't that mean it'd have to be slower in the backward direction, t_1? How can someone's rate of time change depending on the direction they're looking? And how can a person, or any reference frame, have multiple rates of time? That's not possible either. And if the moving passenger could actually have a quicker rate of time, doesn't that really mean they'd still see both strikes simultaneously but only at an earlier time, compensating of course for duration t_2?

As Einstein conceives relativity, time's rate slows for all moving objects due to light's assumed fixed velocity that also forces their contraction in the direction of motion. So how is it that time's rate slows for objects in motion but when our passenger is moving they experience a faster rate of time? And if time is not an independent constituent of the universe but only exists through our selection of an object with cyclic motion that's used as reference, how can its rate change without a change in the natural frequency of our object?

All of these insurmountable conflicts get right to the heart of the fundamental problems, including light's presumed constancy, that plague relativity and invalidate it in its entirety.

(2 Simultaneity 3a)

Distance and Objects in Motion

Einstein mistakenly reasons from his incorrect premise of light's constancy that because it must always measure 186,000mi/s regardless of any relative motion, objects must contract in the direction of their motion while their time slows.[35] As if the implausibility of an object contracting only in the direction of its motion isn't enough by itself, its length contraction without a physical catalyst but only to mathematically enforce light's constancy, argues well enough alone for its outright dismissal.

Time cannot change independent of the cyclic motion of the object selected by us to establish its rate, like the Earth's rotation and orbit. There is no actual time element, property, or constituent of the universe for moving objects or for any other condition that can actually change time's rate without first altering the size or distribution of the mass of the object that we've selected to use for reference.

Aside from time's independent existence being conceptually impossible, all of the Michelson-Morley and Sagnac type experiments establish that light always leaves its source at 186,000mi/s in all directions at once no matter the source's velocity. This demonstrates that light compounds with the motion of its and other reference frames. And because of light's refraction through the varying density of a gravity (or electromagnetic) field, its velocity also has to vary. So, its speed cannot in any way be constant.

Apparently, Einstein also failed to recognize that it's not conceptually possible to meld time's changing rate for the one-dimensional condition of linear motion together with our actual three-dimensional condition of our real environment. If we were at the center of any moving object, or let's say any three-dimensional reference frame, and we had a light with us that was radiating in all directions then our reference frame would supposedly be contracting in the direction of its motion while our time's rate also contracted correspondingly to enforce light's constancy. But since time's contracted rate would have to apply equally over our entire reference frame, in every direction, and since the reference frame can only move in one direction at one time, this creates an impossible conflict with the light radiating in every other direction that cannot be resolved.

For light's velocity to remain constant, time for our moving reference frame would have to impossibly distort differently in all directions simultaneously. The distance directly forward of the center of the reference frame would contract to its smallest size where time would run at its smallest rate. The distance from the center to its sides would remain their original length where time would run at its normal rate. And the distance from the center directly to the rear would elongate to its maximum where time increased to its largest rate. Obviously, this is unworkable.

35. Einstein, *Relativity*, 32-46.

It clearly illustrates the impossibility of light's constancy and how it's required to compound with motion.

What's more, if our rate of time really did slow with motion then we'd have to perceive everything outside of our reference frame, which we could assume to not be in motion, as either contracting in size to maintain light's constancy or light's exterior velocity would be increasing, exceeding 186,000mi/s by the same velocity as our reference frame. All these inherent conflicts, no matter how we view them, explicitly demonstrate that light's velocity cannot remain constant, but that it has to be either added to or subtracted from all relative motion.

Einstein also holds that any length contraction (and time dilation) can only be perceived from a different reference frame.[36] But this isn't correct either. In our real three-dimensional world, we can observe more than one dimension at a time and make a comparison. He intentionally ignores this fact or is unable to fathom it.

Because time's rate would correspond to our contraction in the forward direction, it would appear as if everything perpendicular was growing, becoming larger with our increasing velocity. An easy way to compare dimensions would be to place a right-angled triangle (a triangle where one angle is 90°) such that one leg is parallel to the direction of our motion and then watch the angle change as our velocity increases.

Despite all of the scientific experimentation, mathematical formulations, and highly technical arguments that seem to legitimately contend otherwise, our innate common sense quickly exposes how ill-conceived the premise of light's constancy actually is. We should be concerned at how extensive our academic indoctrination must be to allow such a tenuous assumption supported by such flawed logic to first become and then remain an accepted tenet of mainstream physics. (See Diagram **3.3** Dimension 3a)

---

36. Einstein, *Relativity*, 30, 33, 60, 69, 88-91... Every reference frame used as an example has to be qualified.

DIMENSION OF OBJECTS IN MOTION

Beginning with diagram **1**, we have a three-dimensional object at rest without motion represented by a rectilinear coordinate system where time is running normally, as indicated by the dotted clock-like circle centered on the origin. Each leg of the object corresponds to an axis of equal length. If we were to measure the speed of light as it traversed any one of the axises, timing it from end to end, we would of course find it to be 186,000mi/s.

Referring to diagram **2**, Einstein asserts that because the speed of light is fixed, our object when in motion, in this case moving from left to right in the direction of the x axis, would have to contract in the direction of motion while time's rate slowed so that light's constant velocity is maintained. Time's rate would actually increase. A smaller unit of time would be a faster running time. So for this example, let's refer to it as being a smaller rate of time. The heavy dashed line indicates the original length of the x axis. Time's smaller rate is indicated by the smaller clock at the origin. But because time's smaller rate would have to apply over the entire reference frame, this creates an impossible conflict. The speed of light would have to increase by the reference frame's velocity, exceeding 186,000mi/s perpendicular to the direction of motion to maintain the original length of the y and z axises.

Einstein also contends that it would not be possible for someone in motion to perceive their own contraction. This is not true. They could easily establish the reduced length in the x direction by simply measuring the increased angle defined by the end location of each axis. For them to not perceive their contraction, the y and z axes would have to compensate for time's smaller rate by also contracting with motion, as shown in diagram **3**. But this would be another impossible condition. The speed of light would now exceed 186,000mi/s at the y and z axes as there is no motion in those directions or slowing in light's velocity while time's rate increases.

Now referring to diagram **4**, let's consider the consequences of light radiating outward in all directions, as it normally does, from the origin of our moving object, as indicated by the arrowed lines. To maintain its constancy, our object would have to deform incrementally, scribing an ellipsoidal shape around the x axis, as represented two-dimensionally in the x-y plane by the lighter dashed line around its perimeter, and time would have to impossibly vary in all directions simultaneously, running at its smallest rate for the light projected directly forward, normal to the sides, and largest to the rear that if charted would also scribe an ellipsoidal shape around the x axis, as represented by the deformed clock-like circle. This plainly illustrates all of the insurmountable conceptual conflicts that render light's constancy physically impossible, which unequivocally establishes its compounding with motion that in turn undermines all of Einstein's absurd notions of contraction and time's variability for objects in motion.

(3.3 Dimension 3a)

An object's size can change with motion, which could affect its natural frequency and potentially its rate of time depending on how its time was established. Everything contracts three-dimensionally in an omnidirectional manner becoming more dense when gravitating toward other objects because of the ever-decreasing density of their ever-combining gravity fields. Since their fields are everywhere, an object's contraction cannot just stop at its surface. It has to continue on through the object, becoming thinner in between its molecules and even in between the subatomic particles of each of its atoms.

This means that as objects coalesce their molecules and subatomic particles naturally draw inward toward one another, causing their omnidirectional condensing as they move toward their common center of mass. This naturally increases their natural frequency, which could be rotation, which increases their rate of time if that's what's used to establish time's rate.

Einstein also asserts that because all motion is relative motion, it's a personal and subjective choice for each observer as to which object is moving and its rate of motion.[37] But we can personally observe all sorts of objects whose relative motion is definitely not subjective. We'd never conclude, as he actually does when discussing the Michelson-Morley experiment, that its contraction in the direction of motion and its rate of time is determined by our choice of reference frame: At the experiment, no motion, length contraction, nor time dilation. Away from the Earth, motion, length contraction, and time dilation.[38]

By his criteria, we could just as well decide that the Sun could revolve around the Earth. Or our galaxy could be revolving around our solar system. Obviously, that's theoretical nonsense. We'd be arbitrarily violating the laws of gravitation while deciding an object's motion(s), its size, shape, rate of time, mass, its natural frequency, and whether or not it has an acceleration-created gravitational field or not, all at our whim. And our choices would disagree with the choices of others.

Einstein again contradicts himself when he remarks that, "only experience can decide as to [the] correctness or incorrectness [of an object's motion]."[39] By that, we can take him to mean that it's reality that actually determines an object's true motion. But this undermines the whole idea of its subjectivity. Subjectivity does not exist if it's always trumped by an observable objective reality. If subjectivity does not exist then relative motion does not exist. If relative motion does not exist, then where does that leave "relativity"?

---

37. Einstein, *Relativity*, 16-18, 30, 67, 68... The subjective relativity of motion is intrinsic to his "principle(s) of relativity." He refers to it continuously.
38. Einstein, *Relativity*, 60.
39. Einstein, *Relativity*, 68.
52

Still working from light's assumed constancy, he reasons that the maximum speed of any object has to be restricted to just under that of light. Nothing can attain light's velocity.[40] Any accelerating object, like a spacecraft, could only approach the speed of light. Its mass would become infinitely large while its measurement in the direction of motion would become infinitely small if it were to ever attain it. This is also incorrect.

First, as just demonstrated, light's constancy and the linear contraction of moving objects along with their slowing rate of time is a conceptual impossibility. It's self-conflicted in three-dimensions. This is confirmed by all of the Michelson-Morley and Sagnac type experiments. Light can only compound with motion.

And with its observed refraction around massive bodies, its velocity is variable as well. Even Einstein (contradictorily) asserts its variability. This means that it could range anywhere from almost zero to nearly instantaneous as it traverses the varying density of gravity fields. With light being both variable and compounding, there can be no restriction at all on its velocity. With a nonfixed velocity, there can be no increasing mass with acceleration or decreasing dimension in the direction of motion. So, there can be no limitation on an object's velocity either.

Second, every gravitating body is capable of naturally exceeding the speed of light. It's occurring everywhere. Light's velocity slows to a very small rate in the thinnest density of gravity fields of very massive bodies while that same field would produce very high velocities for the bodies that it's coalescing. This would naturally exceed light's given velocity for any particular body.

We could use the Michelson-Morley experiment as a simple example. The velocity of the light measured at the experiment is about 186,000mi/s. Its velocity varies at every other body based on the density of their gravity field. But to our light's velocity, we have to add the Earth's rate of rotation, about 1,000mph at its equator, plus the rate of its orbit, about 66,000mph, plus our solar system's rate of rotation around our galaxy, about 500,000mph, plus our galaxy's perceived motion. It's thought to be about 1,300,000mph. That's a lot. The total is about 520mi/s. So the actual speed of light will always be more than what we measure here on Earth, 186,520mi/s.

And third, Einstein misinterprets his mass-energy relation as applicable to objects with kinetic energy acquired from motion. He should have realized instead that the mass of particles only increases when induced with a charge, as when being accelerated by or traversing an electromagnetic field. The mass of moving objects not subjected to electromagnetic energy doesn't increase. So they can't become infinitely large at the speed of light. Without the artificial (mathematical)

40. Einstein, *Relativity*, 40-42.

limitation of light's constancy, there's no conceivable reason why an object's velocity could not continue to increase indefinitely. (Kinetic energy is the capacity to do work due to the motion of mass.)

Apart from all this, if the mass of accelerating objects really did increase, becoming infinitely large at the speed of light, but also were simultaneously contracting in the direction of their motion, becoming infinitely small at the speed of light, both the mathematical product of light's fixed velocity, this would create another unresolvable conflict. To even seriously consider that something could actually become both infinitely large and infinitely small simultaneously is ludicrous. Just becoming infinite is bad enough. Yet both of these contradictory precepts remain fact for those focused exclusively on the mathematical result, not on what the math is actually revealing.

Experiments performed by accelerating particles, like electrons, with electromagnetic fields that supposedly evidence the impossibility of attaining the speed of light, in reality, only demonstrate the field's limit for light's velocity. Because the electromagnetic field and light are one and the same, light's velocity can never exceed that particular field's velocity for light that's naturally established by the density of the Earth's ambient electromagnetic and gravity fields.

So the particles being pushed by the electromagnetic field can't exceed the field's velocity for light either. It's the electromagnetic field's increasing strength that induces a particle with an increasing charge that in turn increases its size and mass. This decreases its natural frequency, which we mistakenly infer to be a slowing in its rate of time.

If we were to theoretically place a couple of particles into a field of uniform density, the effect would be free fall motion (the condition of unrestrained motion in a gravity field) in the direction of the thinnest density of their compounded gravity fields, which is the location of their common center of mass. This would move them toward each other proportional to each of their masses until the repulsive effect of their electromagnetic fields neutralized their approach. And even though they'd be moving through the electromagnetic energy of their combined gravity fields, their free fall state would maintain them in a balanced state.

So they wouldn't be induced with a charge and their mass wouldn't be increasing. But their inertia would be increasing despite that their mass remained constant because of their increasing velocity. Their natural frequency would be increasing as well due to their ongoing contraction in the ever-decreasing density of their ever-compounding gravity fields.

If we were to substitute an object like a heavenly body of any size, mass, or composition in place of the particles, their inertia would increase in the same way. They'd also contract in the same way due to their compounded gravity fields' decreasing density that would likewise condense their mass, causing an increase in their rotation and natural frequency. But that increasing rotation wouldn't necessarily be imparting an increasing charge. Their ambient field density would be decreasing proportionally.

(The term "energy" generally refers to an object or system's ability to do work. But more often in our context, it's used generically when referring to the radiation of the electromagnetic spectrum. I try to stick to the term radiation. The electromagnetic spectrum is the entire range of frequencies of radiant electromagnetic energy. It's essentially a universal field that extends from the shortest wavelengths, gamma rays, through X-rays, ultraviolet, visible, infrared, and microwave, to the longest wavelengths, radio waves. It's important to note that there's no limit to the spectrum at either end. It extends indefinitely. See Diagram **8** EM Spectrum 9a.)

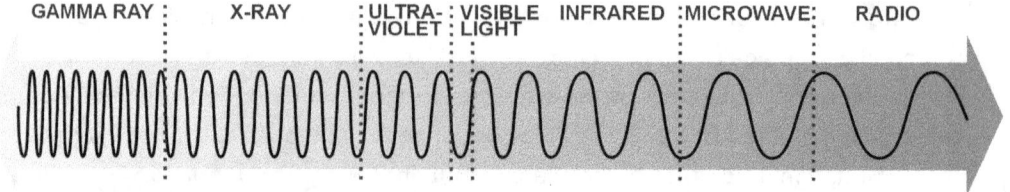

GAMMA RAY | X-RAY | ULTRA-VIOLET | VISIBLE LIGHT | INFRARED | MICROWAVE | RADIO

1. ELECTROMAGNETIC SPECTRUM

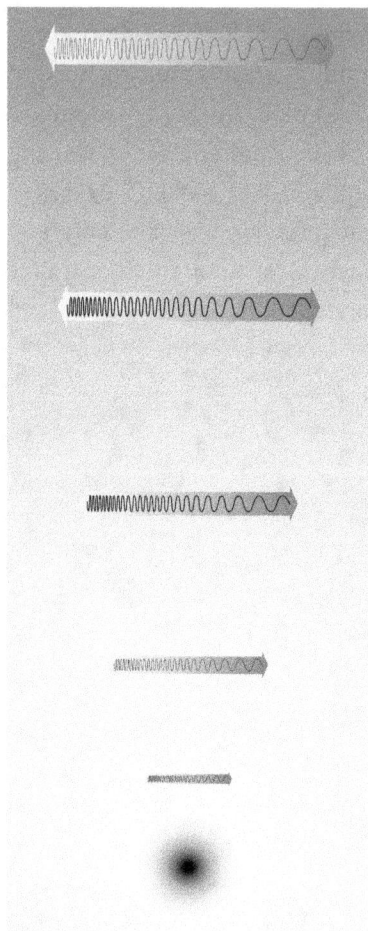

2. WAVELENGTH & AMPLITUDE CONDENSE EXPONENTIALLY IN A GRAVITY FIELD

(8 EM Spectrum 9a)

ELECTROMAGNETIC SPECTRUM

The universe is essentially comprised of only one thing, radiation, which is the radiant energy of the universal field of the electromagnetic spectrum. Wavelength or frequency increases or decreases without limit at either end of the spectrum, as portrayed in diagram **1**.

When this radiation "spontaneously" condenses toward a single point, it forms a subatomic particle. The remaining ambient radiation not drawn into the particle dissipates inward, not outward, but still diffuses exponentially per the inverse square law because of the simple geometry of a sphere, as depicted in section view by the diffusing background in diagram **2**. This inward diffusion defines every particle's or object's gravity field that's literally but not practically felt over an infinite distance. The inward exponential decrease in field density, which is always less in between particles or any object, causes them to naturally move toward one another as they mechanically seek equilibrium, constantly pushed by the higher density toward the lower.

The condensed radiation that's been drawn together to form the particle does not exhibit a surface but increases in density exponentially to a maximum at its center also because of the simple geometry of a sphere. This defines its charged field that lies at the center of its much larger gravity field. Its increasing density naturally repulses, pushing away or holding other particles at bay.

Like everything else, the wavelength and amplitude of the universal electromagnetic spectrum expand or contract depending on its density where the measurement is recorded. Stronger gravity fields that are less dense naturally result in shorter wavelengths and smaller amplitudes at any given location. The problem is, without a baseline you can never compare a measurement from one location to another. Only someone at third location receiving the other two at the same time could perceive a difference.

Clocks in Motion

Einstein reasons that clocks in motion have to run slower and their associated reference frames have to contract in the direction of motion so that light's velocity can remain fixed. To justify this logic, he employs Lorentz's notion of one-dimensional linear contraction for objects in motion.[41] But light's variability nullifies the whole notion.

The unresolvable conflict resulting from mixing light and time's inherent three-dimensionality with the one abstract dimension of linear motion does the same. It confirms light's compounding with the motion of its source and that of other sources. Its velocity cannot be fixed.

But beyond that, remember time itself does not exist. It's not real. It's not a property of the universe. It's something we create. Any discussion of its rate changing due to motion is strictly theoretical.

Experimental results that appear to support his contention, like when particles seem to have a slowing rate of time when passing through electromagnetic fields or when an electromagnetic field accelerates a particle, are in reality only demonstrating a decrease in their natural frequency caused by an increase in their mass and/or size that's a natural consequence of the charge imparted to them by the field.[42]

What he should have argued, is that any object that we've selected to represent time's passage that has an increasing velocity naturally obtained through gravitational coalescing would be condensing, contracting not linearly in one dimension but three-dimensionally in an omnidirectional manner due to gravity's ever-decreasing field density. This would increase its natural frequency, causing the rate of its periodic motion, like the rotation rate of a heavenly body or the operation of an atomic clock or the innate oscillations of a particle, to increase. So any correlation between a change in time's rate and motion can only be the product of a change in field density, not a product of mathematical predictions of the effect of light constancy for moving objects.

Since the condensing of any celestial body subject to gravity is essentially omnidirectional, which increases its natural frequency and accelerates the rotation rate of its periodic motion, and if we were at that celestial body or used Earth as an example where our time is established by its rotation and orbit, we could never perceive a change in the rate of the passage of our time. Time would always appear the same to us. We may feel like it's passing more quickly, but we could never attain any direct evidence of it with the exception that the periodic motions of other celestial objects in denser fields might seem to be unduly slowing.

41. Einstein, *Relativity*, 32-46.
42. "Time dilation," Wikipedia, last modified Apr 18, 2023, https://en.wikipedia.org/wiki/Time_dilation.

An outsider in a different reference frame observing our rotation rate but not experiencing our contraction could record an increase in our periodicity. So they could infer a quickening in the rate of our established time. This might be perceived as time dilation. But it would not be the product of motion and light's incorrectly assumed constancy.

It might appear to the observer that their rate of periodic motion was decreasing, running slower if they assumed ours to be constant. If they could somehow measure our relative field density, this would verify our increasing natural frequency. They could then establish whose designated rate of time is changing.

Defending Einstein's assertion that time slows for moving clocks, some may try to argue that our increasing periodicity would be a slowing rate of time for us. A unit of our time, let's say one of our seconds, would be equal to a fraction of one of the observer's seconds. Or we could say that one of the observer's seconds is equal to 1.5 of our seconds. Either way, a rate of time with a quicker or smaller second that corresponds to its clock's contraction is a faster running time and would be incorrectly interpreted as running slower or being larger or dilating.

For a clock in motion to maintain light's slower velocity at 186,000mi/s, it would have to run faster. It'd have a shorter second, a smaller rate of time that coincided with light's decreasing velocity and the contraction of the clock's reference frame in the direction of motion. A smaller second cannot be viewed as a slower running or larger time. This is another elemental flaw in relativity that no one wants to acknowledge. Moving particles subject to fields that appear to have a slowing rate of time are actually just experiencing a decrease in their natural frequency that's the product of a charge induced from their passage through the field and/or the field's increased density.

With time not actually being a property of the universe but in our real world established only by the periodicity of the object we use to set its rate, a practical alternative exists to Einstein's famous "twin paradox." One twin rockets off through outer space at a high rate of speed only to return much younger than his sibling. His faster relative motion presumably causes his time to run slower.[43]

Time's rate can only change with that object's changing periodicity that can only change with a change in the object's field density, not a difference in subjectively decided relative motion. A difference in field density means a difference in the object's size, which means a difference in the rate of its natural frequency or periodic motion. Periodic motion used as reference is what establishes time's rate.

43. "Twin paradox," Wikipedia, last modified Mar 1, 2023, http://en.wikipedia.org/wiki/Twin_paradox.

So, if one person were experiencing an increase in velocity resulting from gravitation's ever-decreasing field density, which would cause the omnidirectional contraction of all objects in their reference frame, and another person was theoretically stationary or experiencing nongravitational velocity or a different increase in velocity from a different gravity field, then there would be a difference in the rate of each person's assigned rate of time. But the person moving faster would be experiencing a quicker rate of time, not slower. Their decreasing size would be increasing their natural frequency that their rate of time is based on. (See Diagram **4** Time's Rate 7a)

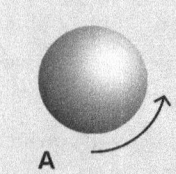

1. HOMOGENEOUS FIELD

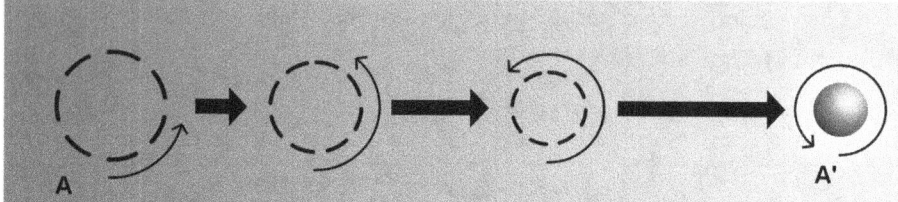

2. GRAVITY FIELD DECREASES IN DENSITY EXPONENTIALLY

TIME'S CHANGING RATE

If we could view two identical heavenly bodies, one in a field that was theoretically uniform in density and the other in a gravity field that naturally decreases in density exponentially per the inverse square law, as represented in section by the diffusing background, the body at **1.A** in the homogeneous field would remain constant, the same without motion, or contraction, or increase in its rate of rotation. This is how we perceive our environment.

Reality, though, is the condition in row **2**. The body in the gravity field would be mechanically seeking equilibrium in the diffusing density, moving in the direction of the least resistance, which is always directly toward other objects, or more accurately a common center of mass, increasing in velocity from position **A** to **A'**, as indicated by the heavy arrows, while contracting in size in a near omnidirectional manner due to the field's decreasing density. As it condenses, its natural frequency increases, increasing its rate of rotation, as implied by the increasing length of the radiused arrows.

If time's rate were established by the rate of the periodic motion of a celestial body's rotation, as ours is, then as outside observers we could see that the assigned rate of time quickens for objects traversing fields of lesser density due to their condensing which increases their rate of rotation. But for anyone inhabiting this heavenly body, there'd be no perceptible change. Without an outside reference, the length of a day would still appear to be the same despite its perpetual shortening. Those inhabitants might look out at other heavenly bodies in regions of space with higher field densities and it'd appear as if the rotation rate for those bodies was slowing.

Altering the naturally frequency of the object we've selected to establish time's rate by either condensing its mass through the natural coalescing process of gravitation or by increasing its mass by subjecting it to a charge is the only way time's assigned rate can change with or without motion.

(4 Time's Rate 7a)

The conclusion drawn from the well-known, airborne clock experiment that it proves that motion causes time dilation are incorrect. In 1971, researchers Hafele and Keating flew four cesium atomic clocks aboard commercial airliners in opposite directions around the Earth while others remained stationary on the ground. (Joseph Hafele is an American physicist, Richard Keating an American astronomer.) The results of the experiment showed that as compared to the ground-based clocks there was a negative 59-nanosecond difference in the clocks traveling in the eastward direction and a positive 273-nanosecond difference in the ones traveling westward.[44]

What should come to mind first is the apparent increase in time for the westward traveling clocks. How is it that the clocks run faster in one direction and slower in the other? Every clock in motion is supposed to always run slower.

These contradictory results are rationalized by calculating time dilation in two ways, kinematic time dilation and gravitational time dilation. But both of which are fundamentally flawed. Kinematic time dilation is time's (presumed) slowing forced by light's fixed velocity when a clock is in motion, as prescribed by special relativity. But since "time" does not actually exist and light's velocity actually varies and compounds with motion (its velocity is never constant), it doesn't work.

But even if time did exist and light's velocity was fixed, which is not possible, its self-contradictions still invalidate it. It only works in the one abstract dimension of linear motion. One dimension is the location of a line. In our real three-dimensional environment, time's (presumed) slowing (that has to apply over the entire reference frame) that (theoretically) results from motion (that's somehow subjectively decided by each observer) would cause light's velocity in the perpendicular direction (or at any angle to its motion) to conflict with its velocity in the forward direction. And it'd exceed 186,000mi/s. So it still has no possibility of ever working.

Also, if time did actually exist, its rate wouldn't be slowing with motion. As covered earlier, it'd actually have to be running faster to maintain light's fixed velocity. To coordinate with light's slower velocity and an object's contraction in the direction of its motion, time's rate would have to contract correspondingly. A contracted rate of time is a faster running time, not slower. So the entire notion that time slows with motion has no validity.

Originally proposed by Einstein in his 1907 paper, *On the Relativity Principle and the Conclusions Drawn from It*, gravitational time dilation is time's slowing

44. "Hafele–Keating experiment," Wikipedia, last modified Aug 24, 2022, https://en.wikipedia.org/wiki/Hafele-Keating_experiment.

in gravity fields. The stronger the gravity, the slower time's rate. Or the higher the altitude, the faster clocks run.[45]

It's justified by imagining a hypothetical scenario that has an inertial reference system accelerating from free fall in a gravity field. As it first falls past a higher position then a lower, both arbitrarily set at rest in the gravity field, its rate of time would be slower at the lower position because of relativistic effects. A moving reference frame's time is forced to slow to maintain light's fixed velocity. Its increasing velocity would cause its time to further slow or dilate.

Because his "principle of equivalence" has acceleration (along with braking and rotation) creating gravity fields that are the same as natural, mass-created gravity fields, their rate of time has to be equal as well. So a clock in the gravity field at the lower position has to also be running slower than one at the higher position just like the clock in the accelerating reference frame.

According to Einstein: "There exist 'clocks' that are present at locations of different gravitational potentials [in the gravity field] and whose rates can be controlled with great precision; these are the producers of spectral lines." The clock at lower position will be redshifted as compared to the higher because its slower rate of time will produce a lower frequency.[46] (Redshift in general is the displacement of the spectral lines of an object toward longer wavelengths in the direction of the red end of the electromagnetic spectrum. An object's particular spectral lines are the hallmark of the atoms and molecules of which it is composed.)

Where to begin. There's so much wrong it's hard to know. It should be noted again up front that in this same paper that he's arguing for gravitational time dilation that's contradictorily based on light's constancy, in gravity fields (pages 302-307), he's also asserting light's variability, in gravity fields (page 310). We covered this earlier in the chapter on the speed of light. So fundamentally, even in the same paper no less, he's inherently conflicted. Light can't be both constant and variable, either in or out of gravity fields.

But Let's start with the inertial reference frame's acceleration. Because it's accelerating, its motion is (theoretically) producing relativistic effects, the same effects as kinematic time dilation. So it has the same invalidating problems: no such thing as time, its slowing when it'd have to be increasing, light's impossible constancy, and the self-conflicted velocities of light in different directions that also exceeds 186,000mi/s if its constancy were possible. So right from the beginning, it has no validity either.

45. "Gravitational time dilation," Wikipedia, last modified Nov 7, 2022, https://en.wikipedia.org/wiki/Gravitational_time_dilation.
46. Einstein, ON THE RELATIVITY PRINCIPLE AND THE CONCLUSIONS DRAWN FROM IT, 302-307.

We could just stop here. There's no need to go any further. The point is sufficiently made. But we're going to continue anyway. The explanation presents us with a purely hypothetical situation. It's theoretical. The reference frame doesn't exist, and neither do the positions in the gravity field. They're just made up.

If we assume the positions are physical reference frames, what is it that's keeping them at rest? "At rest" usually means sitting on the surface of a larger body. There's nothing there to prevent their continued free fall. At best, their relationship with the inertial reference frame is a contrived "what if" situation. So the whole narrative doesn't and can't produce tangible real-world results.

What's disturbing is that we don't seem to able to comprehend the difference between what's real and what's imagined. And we don't even seem the least bit concerned one way or the other if there's a difference. We're more than ready to accept, even determined to believe this nonsense as if it were reality.

If the inertial reference frame is free falling in the gravity field, where is the necessary reaction that produces its acceleration-created gravity field? Free fall has no reaction. If the reference frame's acceleration-created gravity field doesn't exist, there's no way to equate it through Einstein's "principle of equivalence" to the natural, mass-created gravity field. So there'd be no way to infer that the reference frame's slowing rate of time can be applied to the gravity field's "at rest" positions either.

But remember, there is no such thing as "time." It's not a property of the universe. So, as already mentioned, it can't be slowing with the reference frame's acceleration. But also, if it doesn't exist, its presumed slower rate can't be equated to the natural, mass-created gravity field through his "principle of equivalence."

And his "principle of equivalence" is completely untenable. Nothing about it works. Acceleration (or braking or rotation) does not in any way produce a gravity field. And even if its (homogeneous) reaction could somehow be qualified as a field, it could never be equated to the (inhomogeneous) field of natural, mass-created gravity. They're not the same in any respect. This is covered extensively in the chapter on equivalence. You may want to jump ahead and review it now if you have doubts, which you should.

How is it exactly that an element's spectral lines can be regarded as a clock? If time doesn't exist, how can they convey its (nonexistent) rate? Even if time did exist, why would a change in its rate affect the frequency of spectral lines? It'd still be an ungrounded assumption. It'd be the same for his assertion that its slowing would shift them toward red end of the spectrum.

Also, some actual physical objects composed of the same elements would have to be present at both "at rest" positions in the gravity field to record a difference in the frequency of their spectral lines to be able to infer a difference in their rate of (nonexistent) time. They don't exist either. All of this is nothing more than delusive ravings that do not in any way legitimately explain Hafele and Keating's results.

There is another way Einstein "reasons" gravitational time dilation, but it doesn't apply to the their experiment. It's an inherent aspect of his version of gravitational redshift (the shifting of an object's spectral lines toward the red end of the electromagnetic spectrum in a gravity field[47]). He contends that the increasing rotation of a massive body creates more centrifugal gravity and because the increasing motion of higher rotation rates relativistically dilates time, time slows with increasing centrifugal-created gravity. But because centrifugal-created gravity is the same as natural, mass-created gravity, again the product of his "principle of equivalence," time must slow with increasing mass-created gravity as well. The stronger its gravity, the slower its rate of time.[48]

But it too is still based on relativistic effects and his fictitious "principle of equivalence." So it has all the same invalidating problems as kinematic time dilation and his other type of gravitational time dilation. The only difference is rotation instead of acceleration is equated to natural, mass-created gravity, which is even more preposterous. We'll cover rotating objects, equivalency, and gravitational redshift thoroughly in upcoming chapters.

Hafele and Keating's airborne clocks did apparently experience a change in their rate of operation, that is if you believe in the accuracy of the experiment's results, which have been legitimately called into question. They may have modified their figures after the fact to better fit relativistic predictions.[49] Since our rate of time is established by the cyclic motion of the Earth's rotation and the clocks' rates were synchronized to that cyclic motion, their rate of time, the same as the Earth's, could not have changed on its own accord because of subjectively assigned relative motion. It's just their rate of operation that's changed.

In reality, it's the combined effect of the varying density of Earth's magnetic and gravitational fields that decrease with the planes' increasing altitude along with a charge that's induced to the clocks from their passage them that altered their rate of operation. This is the real source for the clocks' diverging readings. It's impossible for their motion to somehow magically change (nonexistent) time's arbitrary rate to enforce light's impossible constancy.

47. "Gravitational Redshift," Wikipedia, last modified Dec 19, 2022, https://en.wikipedia.org/wiki/Gravitational_ redshift.
48. Einstein, *Relativity*, 89-90, 149-150.
49. A. G. Kelly, "Hafele & Keating Test; Did They Prove Anything?" Cartesio Episteme, www.cartesio-episteme.net/H%26KPaper.htm.

The charge acquired from their motion through the Earth's magnetic and gravitational fields, both of which remember are the same radiant energy, just different magnitudes with different directions of dissipation, would have altered the size and mass of the clocks' cesium atoms, affecting their natural oscillation frequency, which is what determines their rate of operation. This, compounded with their increase in altitude would have also altered their size and further affected their natural frequency, causing more change in their rate of operation.

Because the varying density of a gravitational field, which increases with altitude, is so much larger than the decreasing density of our magnetic field, there's very little relative change in gradient in our gravitational field's density over the planes' increased altitude. So, for the condensing-contracting effect on the clocks that causes them to run faster, it's the decreasing density of our magnetic field that becomes the determining factor. (See Diagram **5** Density 10a)

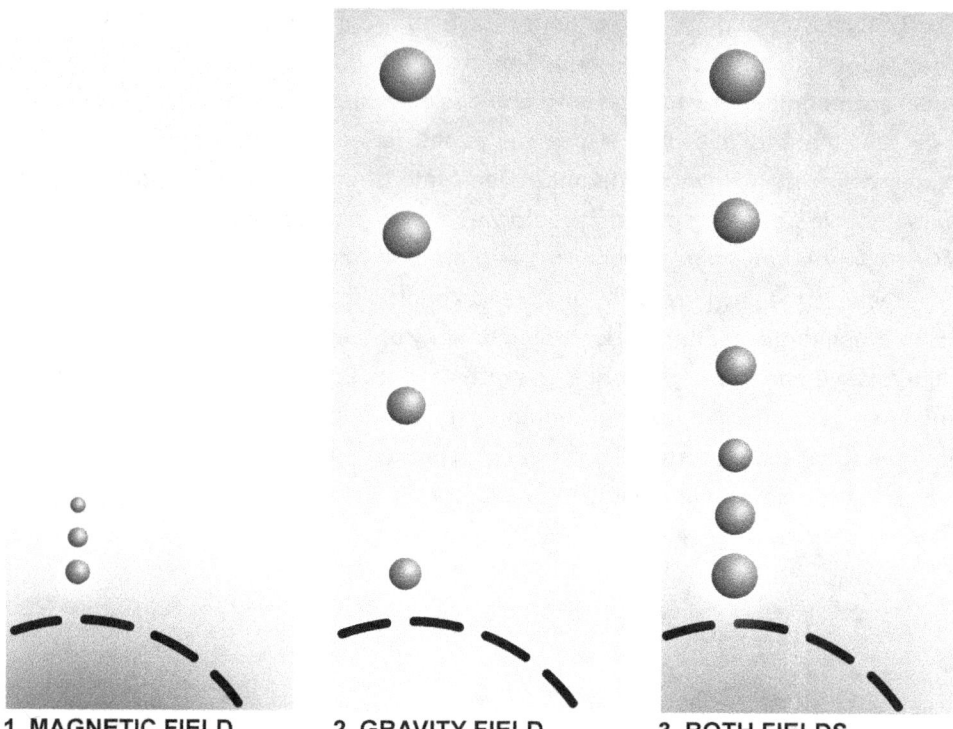

1. MAGNETIC FIELD   **2. GRAVITY FIELD**   **3. BOTH FIELDS**

FIELD'S EFFECT ON SIZE

Subatomic particles are congealed out of the universal field of radiant energy. There are no particles per se, or the objects they compose. Ultimately, there are only condensed fields that are an inseparable extension of the infinitely continuous universal field from which they arose. So when its density decreases or increases like when it expresses as a magnetic or gravity field, the size of any nearby object in it has to vary correspondingly.

In diagram **1**, imagine the Earth, omitted for clarity but where its surface is represented by the curving dashed line, without a gravity field but left with only its magnetic field. Its density dissipates from its center out exponentially, as depicted in section by the diffusing background. So all objects, including the cesium atoms of an atomic clock, would have to decrease in size correspondingly with altitude, as indicated by the sequence of spheres, which in turn causes their natural frequency to increase, making the clock run faster.

In diagram **2**, now imagine the Earth without its magnetic field but left with only its gravity field. Its density increases with altitude exponentially. So any object, or again the cesium atoms of an atomic clock, would have to increase in size correspondingly as they move farther away, causing their natural frequency to decrease and the clock to run slower.

In **3**, the compounded effect of both fields is portrayed. Objects first contract then slowly begin to enlarge as they move farther away. The gradient in the magnetic field is greater over a shorter distance because of its much smaller size while the gradient is much smaller for the gravity field because of its much larger size, which yields little change over the same distance. The sizes and effects suggested have been greatly exaggerated for clarity.

(5 Density 10a)

The motion of the clocks through the Earth's magnetic and gravitational fields when traveling eastward in the same direction as the Earth's rotation would have induced a charge in the cesium atoms, which would have increased their size and mass and decreased their natural frequency, causing the eastward airborne clocks to run slower. This effect would have been significantly stronger than the increase in frequency caused by their contraction due to their altitude in the decreasing density of our magnetic field. The result being that the eastward airborne clocks ran slower than the ground-based clocks.

When traveling in the westward direction, the clocks' altitude in our magnetic field's decreased density that reduced in the size of the clocks' cesium atoms increasing their natural frequency would have been greater than the decrease in natural frequency caused by the same charge induced from their motion through our magnetic field and lesser motion through our gravitational field in the opposite direction. This time the result was that the westward airborne clocks ran faster than the ground-based clocks. (See Diagram **6** Clocks 10a)

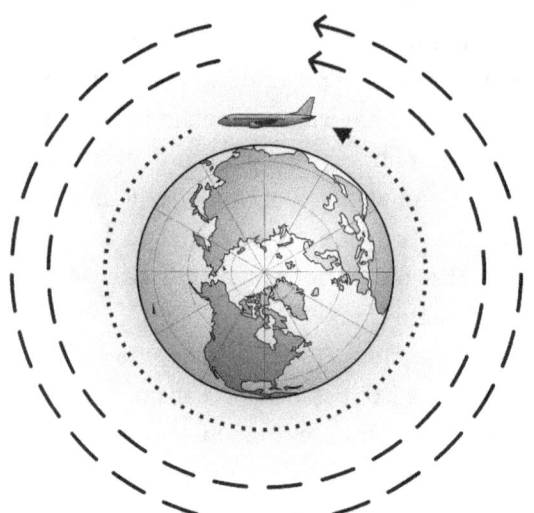

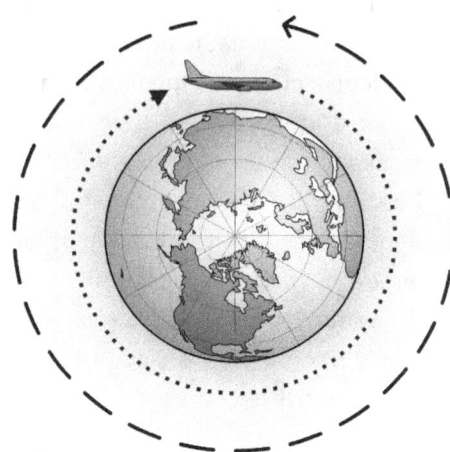

1. EASTWARD (-59ns)
THE CLOCKS' MOTION THROUGH OUR
GRAVITATIONAL & MAGNETIC FIELDS
DECREASED THEIR CESIUM ATOMS' NATURAL
FREQUENCY MORE THAN THEIR ALTITUDE IN
OUR MAGNETIC FIELD INCREASED IT

2. WESTWARD (+273ns)
THE CLOCKS' ALTITUDE IN OUR
MAGNETIC FIELD INCREASED THEIR
CESIUM ATOMS' NATURAL FREQUENCY
MORE THAN THEIR MOTION THROUGH
OUR GRAVITATIONAL & MAGNETIC
FIELDS DECREASED IT

1971 HAFELE AND KEATING AIRBORNE CLOCK EXPERIMENT

There are practical commonsense explanations for the results of all those airborne clock experiments that don't rely on special relativity's self-conflicted, metaphysical effects, length's one-dimensional contraction that's bound to nonexistent time's three-dimensional dilation that impossibly enforces light's presumed constancy.

If we assume for convenience that the speed of the jet airliners carrying the clocks is roughly 500mph and about half the speed of the Earth's rotation, we can then see how when traveling in the eastward direction, with the Earth's rotation, the airliner would complete two revolutions in the time it takes to fly one. This would induce a charge to the clocks' cesium atoms from one revolution through the Earth's magnetic field and two revolutions through its gravitational field, which would increase the cesium atoms' size and mass that would slow their natural frequency, making their clocks run slower.

In the westward direction, the clocks' motion through our magnetic field would remain the same, one revolution. But because they're traveling in the opposite direction of the Earth's rotation, they're only traveling half the distance at half the speed through its gravitational field. So their cesium atoms' acquired charge would be much less than in the eastward direction, still slowing their clocks' rate, but not nearly as much.

When the effect is compounded with the increase in the cesium atoms' natural frequency due to the aircrafts' altitude in our magnetic field where the atoms contract because of the decrease in its density, coupled with only a very slight increase in density from our gravitational field, the eastward clocks end up with a greater mass and slower natural frequency that causes them to run slightly slower than the ground-based clocks.

Conversely, for the westward clocks, not having acquired nearly the same charge, the decrease in their cesium atoms' natural frequency is less than the increase resulting from their altitude. This leaves them with a higher natural frequency than the ground-based clocks that causes them to run faster. Any effect from our orbital motion through our Sun and galaxy's gravitational field can be excluded. It's essentially the same either way.

(6 Clocks 10a)

Another observation that's mistakenly taken to affirm that motion produces time dilation and contraction in the direction of motion is that of decaying muons (subatomic particles that rapidly decay to form electrons) produced in the upper atmosphere by COSMIC rays (high energy radiation) that exist for a longer time then those produced in the laboratory.[50] Their extended life is just the effect of the charge acquired from plowing through the Earth's magnetic field that increases its mass that prolongs its existence.

Other experiments also mistakenly attribute an apparent change in time's rate to motion when it's really the varying density of our gravitational and magnetic fields or motion through them that's actually affecting the clock's cesium atoms' natural frequency. An atomic clock was also flown aboard an airplane but in a circular pattern while being tracked by a laser. Another precise clock was launched on a rocket and continually compared to a ground-based clock through radio contact. Both outcomes seemingly proved to be highly accurate as compared with calculations for gravitational time dilation.[51]

This is not unexpected. The result is correct, and probably the math too. It's just that the explanation is wrong. It's the dissipating field density inherent in the geometry of a sphere that expresses as the inverse square law that's contracting the clocks' cesium atoms, causing their natural frequency to increase. We just refuse to attribute their changing rate to practical, commonsense physical effects. Instead, we rush to confirm our fantasies, forcing everything to conform to big bang-relativity dogma.

The inverse square law is a function of the three-dimensional geometry of a sphere where the intensity, which is the same as density, of an effect like that of an electromagnetic or gravitational field or any expanding sphere with a fixed amount of material changes in inverse proportion to the square of the distance from its source or center (Density, or Intensity, α $1/r^2$).[52] (See Diagram **7.1** Inverse Sq Sphere 13a & **7.2** Inverse Sq Fields 8a)

50. "Experimental testing of time dilation," Wikipedia, last modified Dec 3, 2022, https://en.wikipedia.org/wiki/Experimental_testing_of_time_dilation.
51. "Gravitational Time Dilation," Hyperphysics, last access Dec 28, 2022, http://hyperphysics.phy-astr.gsu.edu/hbase/Relativ/gratim.html.
52. "Inverse-square law," Wikipedia, last modified Dec 13, 2022, https://en.wikipedia.org/wiki/Inverse-square_law.

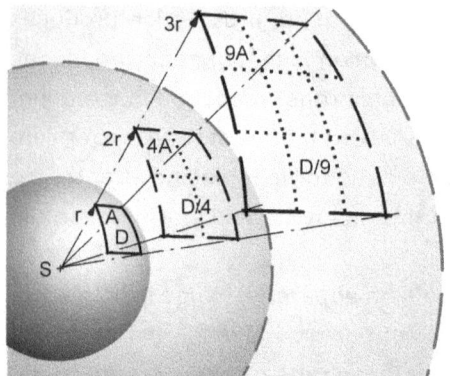

1. SPHERE'S SURFACE AREA & DENSITY
INTENSITY OR DENSITY OF AN EXPANDING SPHERE'S
SURFACE AREA, $D \alpha$ (IS PROPORTIONAL TO) $1/r^2$

INVERSE SQUARE LAW, SPHERE
The surface area of a sphere is $A = 4\pi r^2$. It increases
exponentially as it expands. When twice the radius
from its center, its area increases four times. Three
times the distance, nine times, and so on.

So the intensity or density (they're the same
thing) of any assumed flux (a quantity at the surface)
will dissipate exponentially following the inverse
square law where D (the density at a chosen radius)
= S (the original source density) / $4\pi r^2$ (the area of a
sphere). So its density twice the radius from its
center will be one-fourth its original. Three times the
radius, one-ninth, and so on, indicated in diagram **1**.

This same principle would have to apply to our
big bang universe that's assumed to be finite and
expanding. The material at each consecutive radius
throughout its interior would be simultaneously
diffusing exponentially per the inverse square law,
as represented graphically and qualified numerically
in diagram **2**.

But we observe a homogeneous and isotropic
universe. So it can't be finite, or expanding. This
simple indisputable fact of three-dimensional
spherical geometry by itself completely undermines
the big bang.

To preserve it, we've employed Einstein's belief
that it actually expresses two-dimensionally, like a
sphere's surface. Its homogeneousness is
maintained by confining galaxies to the surface of a
spherical universe that can then dissipate uniformly
with expansion. But there's no existence in two
dimensions. So the big bang can't be real, which
reduces it to nothing more than theoretical whimsy.

A more pragmatic direction would begin with an
infinite (non-expanding) universe that suggested
realistic alternatives for astronomical redshift.

(7.1 Inverse Sq Sphere 13a)

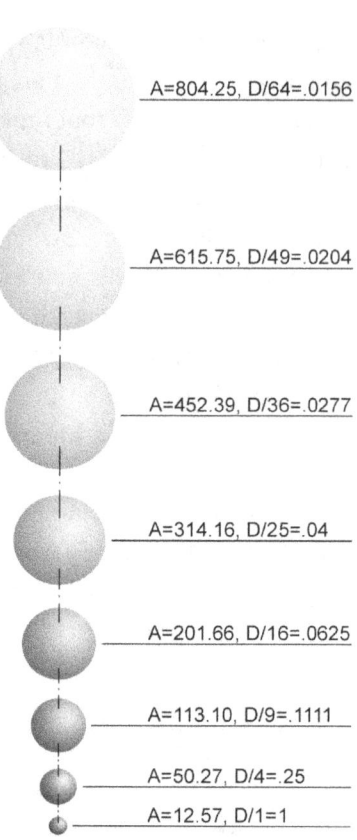

A=804.25, D/64=.0156

A=615.75, D/49=.0204

A=452.39, D/36=.0277

A=314.16, D/25=.04

A=201.66, D/16=.0625

A=113.10, D/9=.1111

A=50.27, D/4=.25

A=12.57, D/1=1

LATERAL VIEW OF AN EXPANDING
SPHERE DEPICTING THE
EXPONENTIAL DIFFUSION THAT
OCCURS AT EACH RADIUS
INCREASES FROM, r = 1, 2, 3,... 8

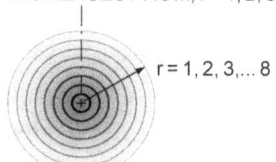

r = 1, 2, 3,... 8

SECTION VIEW THROUGH AN
EXPANDING SPHERICAL VOLUME
PORTRAYING THE EXPONENTIAL
DIFFUSION THAT OCCURS AT EACH
RADIUS SIMULTANEOUSLY AS IT
INCREASES FROM, r = 1, 2, 3,... 8

2. SPHERICAL EXPANSION
SURFACE AREA: $A = 4\pi r^2$
SURFACE DENSITY: $D \alpha 1/r^2$

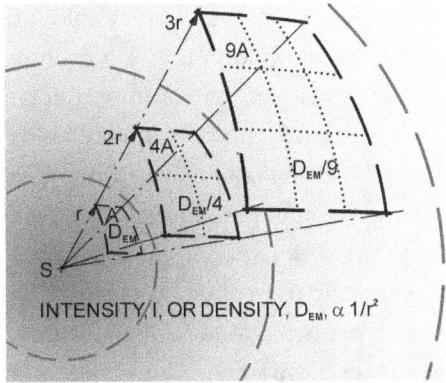

INTENSITY, I, OR DENSITY, D_{EM}, α $1/r^2$

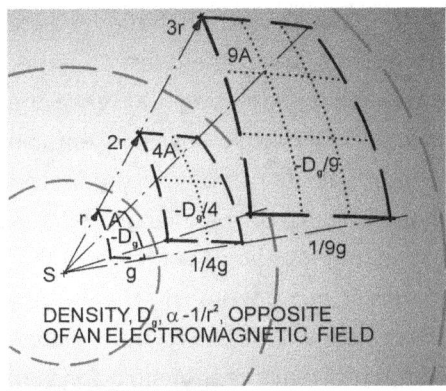

DENSITY, D_g, α $-1/r^2$, OPPOSITE
OF AN ELECTROMAGNETIC FIELD

1. ELECTROMAGNETIC FIELD
ELECTROMAGNETIC FIELD INTENSITY, WHICH IS DENSITY, DECREASES PROPORTIONAL TO THE INVERSE OF THE SQUARE OF THE RADIUS. THIS DIFFUSES ITS OUTWARD ACTING PRESSURE GRADIENT EXPONENTIALLY.

2. GRAVITY FIELD
GRAVITY FIELD INTENSITY, OR DENSITY, ALSO DECREASES PROPORTIONAL TO THE INVERSE OF THE RADIUS SQUARED. BUT ITS DIFFUSION IS INWARD. THIS INCREASES ITS OPPOSITE, NEGATIVE PRESSURE GRADIENT EXPONENTIALLY PER GRAVITY'S FORCE, g.

INVERSE SQUARE LAW, FIELD
An electromagnetic field (EM), depicted in section view as the diffusing background in diagram **1**, is subject to the inverse square law that's the product of the three-dimensional geometry of a sphere. So the field's intensity, which is the same as density that produces pressure which is force, twice the distance from its source is diluted by four times the area. This reduces its density to 1/4 the original. At three times the distance, it's spread over nine times the area, which reduces the density to 1/9 the original, and so on where D_{EM} (the density at a given radius) = S (the original density) / $4\pi r^2$ (the area of a sphere).

The big bang's radiant energy, correctly interpreted as a finite universal field, could never express uniformly as we presume. It's subject to the same physical geometry as every other field. It too would have to obey the inverse square law that'd have it diffusing exponentially with its expansion. The homogeneous isotropic expression that we observe is not physically possible for a finite, expanding universe. So in our actual three-dimensional, nontheoretical reality, the big bang is untenable.

The tangible, radiant, EM energy of our real universe that particles condense out of is all-pervasive, continuous, inseparable, and it varies in density. So the remaining ambient radiation that's not been drawn into a congealed particle has to thin inward, diffusing exponentially toward its center. This is what constitutes their, or collectively the bodies they compose, gravity field, portrayed in section view as the diffusing background in diagram **2**.

It's the opposite of an EM field. Its lowest density is reciprocal to an EM field's highest. Still bound to a sphere's inverse square law, its density, which is still intensity, which still equates to pressure and force, still has to dissipate exponentially. The gradient remains the same. It just expresses the opposite direction, diffusing inward instead of outward where D_g (the density at a given radius) = -S (the original point source strength or negative density established by a body's mass) / $4\pi r^2$ (the area of a sphere).

So at twice the distance from the center, its original negative density is diffused over four times the area, which is 1/4 the original that reduces the inward acting pressure by the same amount, decreasing gravity's force to 1/4g. At three times the distance, its negative density is spread over nine times the area, which is 1/9 as dense as its original that decreases the inward acting pressure the same, reducing gravity to 1/9g, and so on.

(7.2 Inverse Sq Field 8a)

For clocks with much higher altitudes, like those aboard spacecraft exploring the solar system, they can be expected to exhibit a slowing in the rate of their operation as they begin to leave the decreasing density of the Earth and solar system's magnetic field and enter the increasing density of their gravity fields where their atoms begin to gradually expand, which slows their natural frequency. In the end, it's only a change in the atoms' natural oscillation frequency that's the product of a change in their size that's the product of the varying density of our magnetic and gravitational fields, or the charge induced from their motion through these fields that alters their size and mass that also changes their natural frequency that's responsible for their changing rate of operation. It cannot be a supernatural change in the rate of (nonexistent) time by itself due to some arbitrarily assigned motion so that light's impossible constancy can be (mathematically) maintained.

A simple experiment that'd confirm that time's apparent dilation is really the result of an atomic clock's changing rate of operation in the varying density of the Earth's magnetic and gravitational fields would be to synchronize two atomic clocks, or any number needed for averaging, and place one on the ground floor of a tall building and the other on the top floor for some extended period of time. At the end of the time period bring the clocks together and compare the readings.

The higher clock should have run faster due to the atoms' increased natural frequency due to their contracted size in the overriding decrease in the density of the Earth's magnetic field. We should also gather that this effect would be stronger than the decrease in natural frequency resulting from the higher clock's slightly faster motion through the Earth's gravity field from its rotation.

It would be interesting to then switch the clocks for the same time period again without resynchronizing them to see how close they read when brought back together a second time. Because of the ongoing decrease in the Earth's field density due to our slow inward migration toward our galaxy's center, there might be a very slight increase in the upper clock's rate due to the difference in their rates of contraction. This difference probably wouldn't be detectable. It'd be exceedingly small.

In any case, I'm sure it'd still be vehemently argued, even though irrationally so, that any difference in (theoretical) clock rates is the result of gravitational time dilation's relativistic motion of a (theoretical) inertial reference frame that causes time's rate to dilate due to light's fixed velocity in a (theoretical) natural, mass-created gravity field whose (theoretical) acceleration (theoretically) creates gravity that's equated to the (theoretical) mass-created gravity field through Einstein's (fictional) "principle of equivalence" so that its dilated rate of (nonexistent) time can be transferred to (theoretical) locations in the (theoretical) mass-created gravity field that (theoretically) remain at rest.

Another aspect of the experiment that'd be interesting to explore is light's velocity at both locations. See if there's any difference. If it is variable as Einstein contends it is in a gravity field then it should read differently. If that variability is confirmed, then gravitational time dilation can't be the reason for the clocks' different rates. It's rooted in light's constancy.

With the effects of kinematic and gravitational time dilation seemingly correct but the explanation being totally absurd, it only takes a little common sense to rationally resolve these elemental inconsistencies. But you have to ask, why would anyone even choose to believe in light's irrational and unevidenced constancy over accepting the natural simplicity of its compounding and variability, especially when it's so conspicuous?

The answer partially lies with the human psyche. It's easily seduced and spellbound by relativity's mysticism. Our minds much prefer the excitement and intrigue of indulging in its metaphysical fantasies and perplexing puzzles, light's constancy, length contraction, time dilation, time travel, the twin paradox, and so on, over naturally intuiting the mundane, commonsense truth of light's compounding and variability along with the elemental affect that fields have on a particle's size and mass that alters their natural frequency, not their rate of time.

Einstein even intimates as much in his Preface. "May the book bring some one a few happy hours of suggestive thought!"[53] It's not, I hope my speculations aid in your own search for reality. Or, I hope my investigation broadens your horizon. Or, may this work help bring you closer to the truth. Or, may this study brighten your path. Or, may these considerations further your enlightenment. No, it's nothing of that sort. He has no idea that life's singular purpose is consciousness. To him, it's just happy hours of amusement, playing theoretical mind games just to pass the time.

53. Einstein, *Relativity*, vi.

Gravitational Fields

Einstein contends that gravitation is facilitated by a force similar to electromagnetic radiation and acts upon objects at the speed of light via waves. But he also suggests, and it's generally inferred, that it's facilitated by the curvature of two-dimensional space that dents underneath (three-dimensional) massive bodies that causes them to fall like they're rolling downhill toward one another.[54] So which is it, a force similar to electromagnetism or space's curvature? It can't be both. They're different processes.

Also, if space's two-dimensional curvature really did compel objects to fall toward one another, that'd be a mechanical reaction. Mechanical reactions essentially act instantaneously, not at the speed of light.

Since space by definition doesn't actually exist, There's nothing there to be curving. Space is the three-dimensional expanse in which objects and events occur. It's the nothingness around objects or the void area in between them that has measurable distance whether we're contemplating subatomic particles, common everyday items, or the region between stars and galaxies that occupy the endlessly vast condition beyond the Earth's atmosphere.[55]

And if it were something with an actual existence, it'd have to be three-dimensional. Two dimensions only define the location of a plane.[56] Without a third dimension, there's still nothing there. And a three-dimensional volume of something can never curve. It's a physical impossibility. It can only vary in density, an innate property of three dimensions. Curvature is a property limited to a one-dimensional line or a two-dimensional plane.

So, since space doesn't exist and couldn't curve even if it did, gravitation must be caused by something else. The universal field of radiant electromagnetic energy that comprises the entire universe does exist. It's everywhere. It extends indefinitely. If it could be separated from matter, it could be said to correspond to all (three-dimensional) space. It's continuous. Its continuity is unbreakable. And it can't be uninterrupted. But its intensity, which is the same as density, can and does vary. None of this is the least bit controversial.

Few would also disagree that something is never created from nothing. Particles don't just pop into existence out of nowhere. They condense into being out of the universal field itself. There's no such thing as matter per se. Particles aren't composed of any actual material. Nor do they have a surface. They're just

54. Einstein, *Relativity*, 71-73, 92-107, 112-116; "Gravity," Wikipedia, last modified Jan 1, 2023, https://en.wikipedia.org/wiki/Gravity.
55. "Space," Merriam-Webster, last modified Mar 26, 2023, https://www.merriam-webster.com/dictionary/space.
56. "Plane (mathematics)," Wikipedia, last modified May,25, 2023, https://en.wikipedia.org/wiki/Plane_(mathematics).

small condensed spherical fields of radiant energy that have an increasing density that reaches some maximum concentration at their center. There's no separation between what's defined as a particle, or the objects they comprise, and the field they originate from and now reside. They're one and the same.

Einstein contends that, "Physical objects are not *in space*, but these objects are *spatially extended*. In this way the concept of 'empty space' loses its meaning."[57] Is he actually saying objects are space? Sounds like it. He also says, "There is no such thing as an empty space, *i.e.* a space without a field. Space-time does not claim existence on its own, but only as a structural quality of the field."[58] Try making sense out of all that.

In reality, all objects are *electromagnetically extended*. When a particle "spontaneously" congeals into existence, the universal field is drawn together from all directions, radially, condensing toward a single point. This puts the entire universal field at further loss. It thins commensurate with the particle's condensing. But that thinning isn't spread evenly throughout the entire universe. What's not taken up and drawn into the particle has to dissipate inward immediately around it, spherically.

Because of a sphere's innate three-dimensional geometry, the field's inward diffusion has to dissipate exponentially per the inverse square law. Whether the diffusion dissipates inward or outward doesn't matter. It's still bound to the exponential gradient inherent to spherical geometry.

It's the inward diffusion of the universal field's ambient electromagnetic energy that's not been drawn into the particle but remains outside it that defines its, and the bodies they compose, gravity field. It's the innate compounding of those gravity fields that causes their density to always be at their least directly in between bodies, toward their common center of mass, regardless of the distance.

Naturally compelled to seek equilibrium in the ever-decreasing density of their ever-compounding fields, all bodies from particles to galaxies are constantly pushed mechanically by the highest density toward the lowest. This causes their unrelenting coalescing toward one another in the ever-decreasing density of their ever-combining fields. Runaway coalescing/condensing naturally ensues.

This continues until enough material accumulates that the resultant pressure triggers fusion reactions that ultimately transmute every particle back into the radiant electromagnetic energy from which it arose. If you believe that the fundamental constituent of the universe is radiant electromagnetic energy, that it's continuous and uninterruptible, and that subatomic particles spawn from it then you'll have a difficult time denying this elemental unifying conclusion.

57. Einstein, *Relativity*, vii.
58. Einstein, *Relativity*, 176.

Imagine that it was possible to reach out in front of you with your hands and gather together some radiation and compress it into a tight sphere of matter like you were forming a snowball. The leftover hollowed out region couldn't form a void as it would with the snow.

The universal field that you gathered the radiation from would still remain. It's innately uninterruptible. Its continuity can't be broken. It's impossible for it to be disconnected from itself. But it can and does vary in density. It has to diffuse inward exponentially with the same gradient as any electromagnetic field but in the opposite direction. It's still bound by a sphere's inverse square law.

It's tempting to view buoyancy as analogous to gravitation. When underwater, the density beneath an object is greater than above it. That difference in gradient tends to push it upward or outward toward the surface with increasing velocity. As it does so, the decreasing pressure on the object allows it to expand.

The water's increasing density with depth is a product of the decreasing density of the Earth's gravity field. That decreasing density is a function of a sphere's inherent geometry that creates an exponentially increasing pressure gradient that pushes downward or inward on coalescing objects. Because the decreasing field density permeates those objects, they're compelled to condense as they approach one another.

The very fact that a field of electromagnetic energy causes motion and resistance to motion confirms that it is a tangible something having physical properties and that its interactions are mechanical in nature. And since all mechanical reactions are instantaneous, or practically so, then gravitation must also propagate instantaneously.

Physical mechanisms for either particles (gravitons) or waves acting at the speed of light that can pull, or even push, objects together simply don't exist. It would not be a stretch to equate real gravitation with what's described as the Casimir effect.[59] (Hendrik Casimir was a Dutch physicist, 1909-2000.) It has the field density between any two parallel plates innately decreasing, causing their apparent attraction. (See Diagram **17** Curvature 6a, **27** Fields 12a & **28** Gravitation 12.1a)

59. "Casimir effect," Wikipedia, last modified Mar 31, 2023, https://en.wikipedia.org/wiki/Casimir_effect.

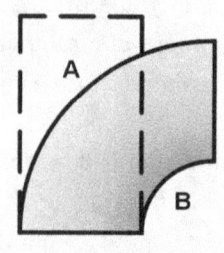
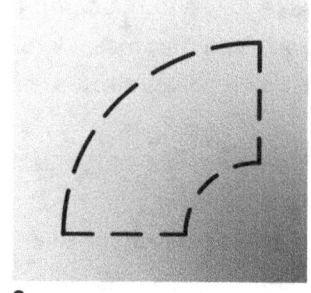
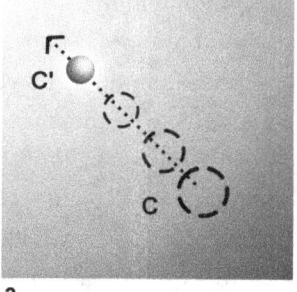

1   2   3

CURVATURE & DENSITY

If we were to envision the shape in diagram **1** as three-dimensional, giving it the same depth as its width, we can see how if it were bent over where its two-dimensional surfaces, at **A** & **B**, now curved, its interior would have to vary in density, as represented in section view by the diffused shading that dissipates toward the outer curved side with a greater radius, **A**. Note how the density along both the outer and inner curved surfaces would remain constant as **A** diffused and **B** condensed when bent.

If we were to imagine that shape as part of a larger volume of the same material, as depicted in diagram **2**, we can more easily see how any three-dimensional volume of something can never curve but can only vary in density. Curvature is a property of only one and two dimensions, which don't actually exist in reality. Without the third dimension, they only define a location of a line or a plane.

These principles of curvature and density would also have to hold true for the three-dimensional volume that's commonly refer to as space. It could never curve. It could only vary in density. But space is the nothingness between objects. It's nonexistent by definition, and it would also be nonexistent if it were a curving two-dimensional plane. So there's nothing there to vary or curve. But there is something that does correspond exactly to all space: radiant electromagnetic energy. Everywhere that space is, so is the all-pervasive universal field of radiation that comprises the electromagnetic spectrum. This field is continuous, its continuity uninterruptible, and it's average volume is fixed per any given quantity. And its density varies.

Because of its innate continuity, the density of the radiation that's been drawn together to form a particle has to diffuse. But it diffuses inward toward the particle's center, still dissipating exponentially per the inverse square law because of the three-dimensional geometry of a sphere. It's the compounding of those inward diffusing fields of the particles themselves, and the objects they compose, that's responsible for gravitation.

Referring to diagram **3**, if we were to place a small object into the decreasing density of a gravity field of a much larger object, it would begin moving with acceleration in the direction of least resistence directly toward the larger object as it mechanically sought to find equilibrium in the field's decreasing density. It'd naturally move from the highest density toward the lowest. But it would also be contracting in size. Its atoms also naturally coalesce and condense within the field's decreasing density, as indicated by the sequence of dashed circles representing the object's motion from **C** to **C'**. In reality, each object would move toward the other in proportion to the size of their gravity fields toward a common center of mass because of the compounding effect of their gradient.

(17 Curvature 6a)

78

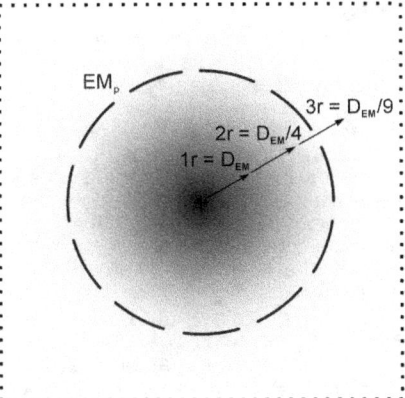

1. ELECTROMAGNETIC FIELD
POSITIVE DENSITY DIMINISHES, $D_{EM} \propto 1/r^2$

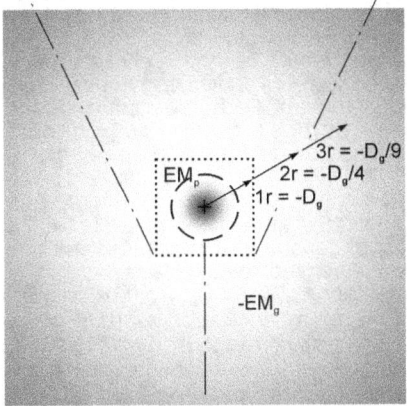

2. GRAVITY FIELD
NEGATIVE DENSITY DIMINISHES, $D_g \propto -1/r^2$

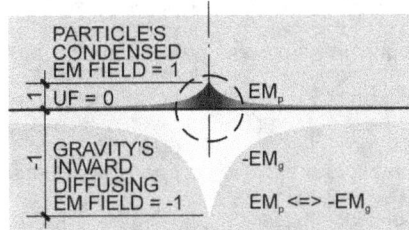

3. RELATIVE FIELD DENSITY & SIZE
A RECIPROCAL EM GRAVITY FIELD (-EM$_g$) WITH NEGATIVE DENSITY & INWARD DIFFUSION IS INNATELY CREATED BY THE RESULTING DEFICIT WHEN A PARTICLE (EM$_p$) CONGEALS OUT OF THE UNIVERSAL FIELD (UF).

(27 Fields 12a)

FIELDS

The fundamental constituent of the universe is radiant electromagnetic energy that manifests as a universal field (UF). It expresses in two ways, as an electromagnetic field (EM) and its inverse, a gravity field (-EM$_g$). They're the same UF but with opposite diffusion and different densities that are the intrinsic byproduct of a particle's inception. They coexist and coincide. But they're reciprocal. And one cannot exist without the other because of their uninterruptable continuity.

When the UF "spontaneously" congeals into a subatomic particle (EM$_p$), it's defined as matter with an assigned amount of mass. But at its essence, it remains radiant EM energy. It's portrayed in an expanded section view by the diffusing background in diagram **1**. The dashed circle represents its theoretical limit or that of any individual body.

The radiation drawn into and composing the particle is the radiation from the UF. That deficit is what constitutes the particle's gravity field (-EM$_g$). They're equivalent because they're the same. They're inverse because of the particle's innate condensing, which is also why it's much smaller, as represented in **2**.

If the UF were assigned the neutral value of zero (0) and the condensed particle (EM$_p$) a value of one (1), the resultant inward diffusion of the UF, the particle's gravity field (-EM$_g$), would have a corresponding negative quantity equivalent to negative one (-1). They naturally reciprocate despite their divergent size and opposite diffusion, as implied in diagram **3**.

It makes no difference whether the EM field's diffusion dissipates inward or outward. The gradient still has to diminish exponentially. Both are subject to the inherent geometry of a sphere that's bound to the inverse square law (Intensity or Density $\propto 1/r^2$).

The exponential outward diffusion of a particle's EM field from higher inner density to lower outer density creates outward acting radial pressure. This should be interpreted as a positive charge, having a male or originative quality.

The exponential inward diffusion of gravity's EM field from higher outer density to lower inner density creates inward acting radial pressure, which should be viewed as a negative charge, having a female or receptive quality.

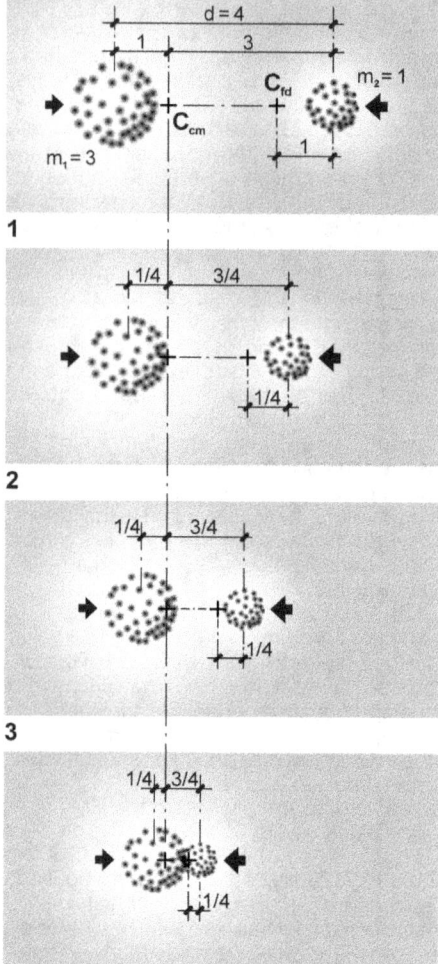

1

2

3

4

COALESCING BODIES

BODIES AREN'T ATTRACTED TO ONE ANOTHER. THEY'RE CONSTANTLY PRESSED TOGETHER BY HIGHER FIELD DENSITY TOWARD LOWER FIELD DENSITY AS THEY MECHANICALLY REACTIVELY SEARCH FOR EQUILIBRIUM IN THE EVER-DECREASINGDENSITY OF THEIR EVER-MERGING GRAVITY FIELDS.

(28 Gravitation 12.1a)

GRAVITATION

A natural consequence of a particle's emergence, gravity fields, depicted in section view by the diffused background, necessarily diffuse inward exponentially because of basic spherical geometry and the uninterruptible continuity of radiant electromagnetic energy.

Gravity fields' innate compounding causes that inward diffusion to always be at its least directly in between the particles and the bodies they surround at their common center of mass, C_{cm}, which is the same as their common center of gravity.

Mechanically pursuing equilibrium in the ever-decreasing density of their ever-compounding gravity fields, all bodies, be it particles or galaxies, are constantly pushed by the highest field density toward the lowest. This inexorably leads to runaway coalescing that ultimately ends with fusion reactions transmuting all matter back into the radiant energy it originated from.

Because gravity fields not only surround but also permeate all bodies, including atoms, depicted as the small spheres comprising the spherical bodies, their compounding simultaneously causes both coalescing and condensing at all scales consistent with Newton's law of gravitation: $F = G(m_1 m_2) / d^2$, where F is the "attractive" force, G is the gravitational constant, m the mass, and d is the distance between their centers.

The distance to their C_{cm} from m_1 is $d_{cm} = m_1 d_1 + m_2 d_2 / m_1 + m_2$, where $d_{cm} = 3(0) + 1(4) / 3+1$ or 1. From m_2, it'd be $1(0) + 3(4) / 3+1$ or 3.

C_{fd} indicates the location in between them where they share a common field density. The distance to their C_{fd} is opposite of or naturally reciprocal to their C_{cm}. Both their C_{cm} and C_{fd} could be interpreted as non-centrifugal Lagrange points where the gravitational influence remains in equilibrium.

Actual Lagrange points incorporate orbital motion's centrifugal force. It's not included in this example for clarity. If it were, their C_{fd} would become the L_1 Lagrange point that'd have to be closer to m_1 to compensate for the outward centrifugal force.

The distance to their C_{cm} and C_{fd}, their relative rate of motion toward each other, and their relative condensing, all remain proportional to their masses as they relentlessly gravitate in the ever-thinning density of their ever-compounding gravity fields, conceptually portrayed in the sequence of diagrams **1-4**.

The coalescing/condensing effect of self-gravitation is always the product of a reduction in the density of the universal electromagnetic field that innately decreases toward the center of every single body, which defines its gravity field. It could also be said that the universal field diffuses inward toward the common center of mass of all the bodies that form a larger body, defining their collective gravity field. Either way, it always dissipates radially, inward three-dimensionally.

The process of gravitation is the same regardless of the number of bodies. The only difference is how they're qualified. It could be considered that they're separate individual bodies or the components that comprise a larger single body like the particles of an atom, or the atoms of an object, or the stars that make up a galaxy, or the galaxies that form our (presumed) finite universe.

All bodies still move toward one another, or toward their common center of mass, whether they're considered separate or constituents of a larger body because of their reactive, mechanical search for equilibrium in the ever-decreasing density of their ever-combining gravity fields that's always less in between them toward their common center of mass.

With the density of their ever-combining gravity fields constantly decreasing as they coalesce, objects can never stretch as they gravitate toward one another. They can only continue to compress, spherically, in an omnidirectional manner. But their condensing will always produce a minor elongated distortion that leaves them slightly egg-shaped. Their more pointed, more condensed end will always point directly toward the other body, or more precisely, toward the common center of mass of any number of bodies due to the exponential decrease in density of their combined gravity fields across both or all of their diameters radially. (See Diagram **11.1**, **11.2**, **11.3**, **11.4** Shape 13a)

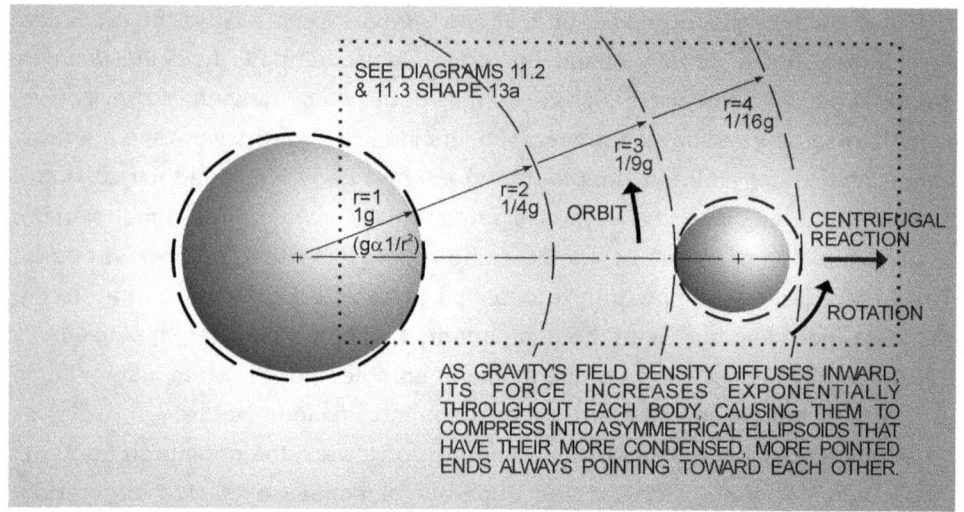

SEE DIAGRAMS 11.2
& 11.3 SHAPE 13a

r=4
1/16g

r=3
1/9g

r=2
1/4g

ORBIT

r=1
1g

$(g \propto 1/r^2)$

CENTRIFUGAL
REACTION

ROTATION

AS GRAVITY'S FIELD DENSITY DIFFUSES INWARD, ITS FORCE INCREASES EXPONENTIALLY THROUGHOUT EACH BODY, CAUSING THEM TO COMPRESS INTO ASYMMETRICAL ELLIPSOIDS THAT HAVE THEIR MORE CONDENSED, MORE POINTED ENDS ALWAYS POINTING TOWARD EACH OTHER.

GRAVITY'S CONDENSING (TOP-DOWN VIEW)

THE SHAPE OF GRAVITATING BODIES - 1

Bodies don't stretch or "spaghettify" as they gravitate. They continue to condense, contracting omnidirectionally into ellipsoidal shapes that are slightly asymmetrical similar to an egg. This is due to the exponential decrease in density of their compounding gravity fields that permeate each other's bodies, portrayed in section view by the diffusing background.

For simplicity, if we set the smaller body's diameter equal to the larger's radius and locate it three radiuses out then the larger's force of gravity, defined as 1g at its surface, would radially affect the smaller, sweeping across/through its entire body, exponentially decreasing from 1/9g at its closest point to 1/16g at its farthest, causing more condensing/compression at the closer end.

The smaller's gravity field would affect the larger in the same way but much less, creating a slight asymmetry in their condensing that has their more pointed, more compressed ends always pointing toward one another. Or more precisely, they point toward their common center of mass. This applies for any number of objects.

If the smaller body had a decreasing orbit or none at all, the asymmetry of its deformation would increase while it continued to condense/compress until they merged. If it had an increasing orbital and/or rotation rate where its outward centrifugal force began to exceed gravity's inward condensing, its material would begin to loosen, become weightless, and start dispersing.

But that dispersion would begin first from its backside, from its outermost point where the centrifugal force would be the greatest and gravity's compounded force would be at its weakest. This is routinely observed as the fanned dust tails of comets that always diffuse to the outside of their elliptical orbits opposite the Sun.

An obvious example of a body's asymmetrical ellipsoidal deformation is the Moon's, and to a lesser degree the Sun's, effect on the Earth's oceans. Water's pliability causes it to more readily deform than the rocky crust below, making its distortion easily perceivable. Tides are simultaneously at their highest both facing and opposite the Moon where they're slightly lower.

This distortion is not the product of the "pull" of the Moon's gravity as many believe. If it were, there'd be a gravity source on the opposite side pulling those oceans into their high tide. Those tides are often explained as the result of no pull, or sometimes more rationally but still incorrectly, the product of the Earth-Moon system's centrifugal force.

(11.1 Shape 13a)

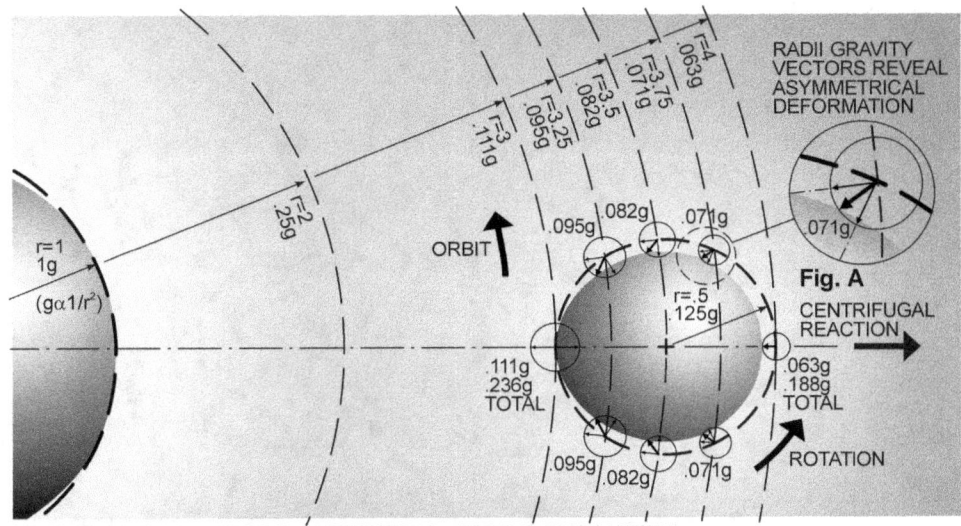

RADII GRAVITY
VECTORS REVEAL
ASYMMETRICAL
DEFORMATION

Fig. A

CENTRIFUGAL
REACTION

ROTATION

PLOTTING GRAVITY'S CONDENSING (TOP-DOWN VIEW)

THE SHAPE OF GRAVITATING BODIES - 2

The inward exponential diffusion of compounding gravity fields permeates gravitating bodies causing their ongoing condensing into evermore compact asymmetrical ellipsoids. This effect can be approximated by bisecting the angle established by the larger body's gravity expressed as proportional radii vectors to both bodies, as indicated in **Fig. A**. The bodies' uniform compression from their own self-gravity is assumed, represented by the heavier dashed circle.

Each body's volume ($V = 4/3\pi r^3$) can be used to establish their relative gravity. But any valve could be adopted. Setting the larger's radius at 1.0 (the unit of measure doesn't matter), its volume will be 4.19. For the smaller whose radius is half of the larger's, its volume will be .524. So the smaller body's volume will be about one eighth of the larger's. That's the ratio we'll use for their gravity: the larger, 1g, the smaller, .125g.

At 3.0 radiuses out, the bodies' closest point, the larger's gravity ($g \propto 1/r^2$), will be .111g ($1/(3)^2$). With gravity's radii vectors (exaggerated for clarity) pointing in opposite directions, they counteract. So no distortion is produced. The total gravity here would be .236g (.111g + the smaller's .125g).

At 3.25 radiuses out, the larger's gravity decreases to .095g ($1/(3.25)^2$). An approximation of the smaller's distortion at this location can be plotted by applying a gravity vector from the smaller body's surface that bisects the angle set by each body's radii.

Using the same method at 3.5 radiuses where the larger's gravity diffuses to .082g ($1/(3.5)^2$) and at 3.75 radiuses where it's weakened to .071g ($1/(3.75)^2$), the smaller's deformation at those locations can be charted.

At 4.0 radiuses out, the farthest point from the larger, its gravity diminishes to .063g ($1/(4)^2$). A radii vector proportional to that gravity defines the outer limit of the smaller's condensing. The total gravity here will be only .188g (.063g + .125g).

With the total gravity at the outermost location always being less than the closest, the material of any body with an increasing rotational and/or orbital velocity will always begin to become weightless, dislodge, and start dispersing from their outermost point first as their increasing centrifugal force's outward dispersal begins to exceed their gravity's inward coalescing/condensing.

This simplified representation reveals how gravitating bodies distort into asymmetrical ellipsoids that continue condensing until they merge unless subject to high enough centrifugal forces that cause them to begin shedding material, which always occurs from their backside first.

(11.2 Shape 13a)

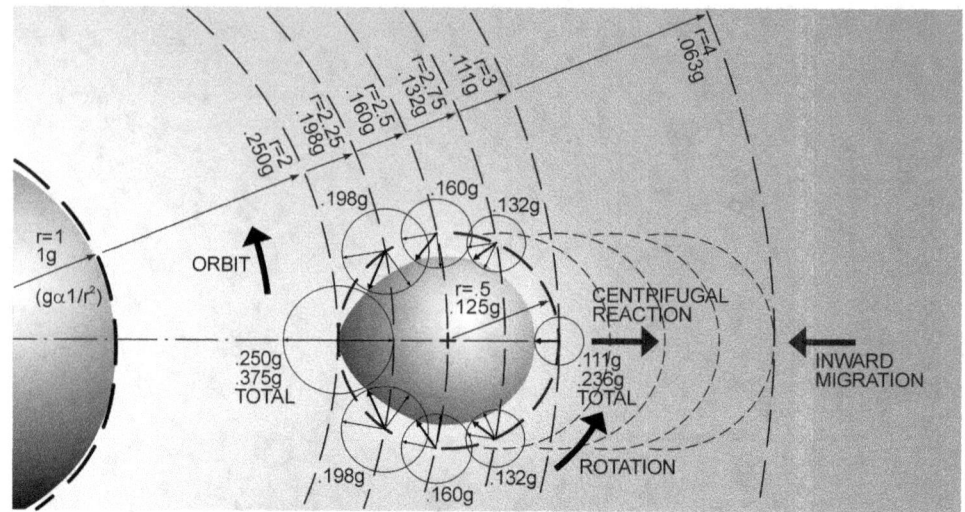

PLOTTING GRAVITY'S INCREASING CONDENSING (TOP-DOWN VIEW)

THE SHAPE OF GRAVITATING BODIES - 3

The ongoing contraction and increasing distortion of gravitating bodies can be conceptually demonstrated by graphically charting gravity's effect at the closer distance of 2.0 radiuses in the same manner that was done at the 3.0 radius distance (also exaggerated for clarity). The conclusion is self-evident: The increasing pressure from the exponentially increasing gradient that's produced by the inward diffusion of gravity's field causes ever-increasing asymmetrical ellipsoidal condensing. No stretching. No spaghettifying.

The relentless coalescing of gravitating bodies, that's a natural byproduct of gravity's inherent runaway nature, has to continue unabated until they merge unless increasing outward centrifugal forces begin to exceed gravity's inward coalescing/condensing. This causes the smaller body's material to begin to loosen, dislodge, and disperse, which is always initiated from the farthest point on its backside.

When the merging gravity fields of coalescing bodies create enough inward pressure, fusion reactions are triggered that begin converting their matter back into the radiant electromagnetic energy it originated from.

(11.3 Shape 13a)

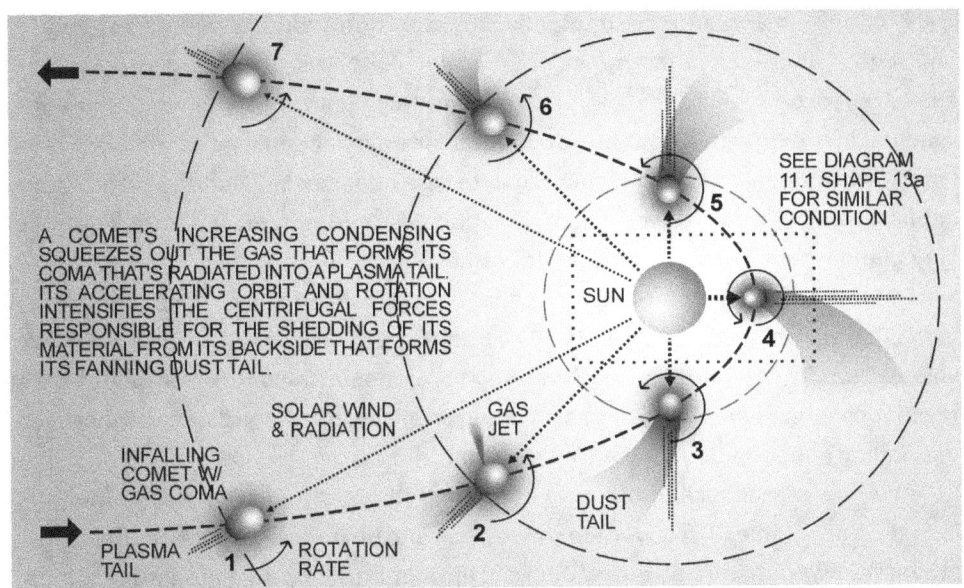

The following labels appear within the diagram:

7

6

SEE DIAGRAM 11.1 SHAPE 13a FOR SIMILAR CONDITION

5

A COMET'S INCREASING CONDENSING SQUEEZES OUT THE GAS THAT FORMS ITS COMA THAT'S RADIATED INTO A PLASMA TAIL. ITS ACCELERATING ORBIT AND ROTATION INTENSIFIES THE CENTRIFUGAL FORCES RESPONSIBLE FOR THE SHEDDING OF ITS MATERIAL FROM ITS BACKSIDE THAT FORMS ITS FANNING DUST TAIL.

SUN

4

SOLAR WIND & RADIATION

GAS JET

INFALLING COMET W/ GAS COMA

3

PLASMA TAIL

1

ROTATION RATE

2

DUST TAIL

GRAVITY'S EFFECT ON COMETS (TOP-DOWN VIEW)

THE SHAPE OF GRAVITATING BODIES - 4

1. A comet that was theoretically uniform and pliable would assume the shape of an asymmetrical ellipsoid that was continually condensing while its smaller, more compressed end always pointed toward the Sun due to the exponential decrease in the density of the Sun's gravity field, portrayed in section view as the diffusing background. That decreasing density in its gravity field at first gently squeezes out the comet's gas, mostly hydrogen, to form its coma, that may or may not have been evaporated/sublimated from internal material by the intensifying pressure and heat from its increasing compression. The Sun's radiant energy then begins to ionize the gas and blow it straight back to form the comet's plasma tail.
2. As its condensing continues, the gas is sometimes seen jetting out at high velocities. This confirms its internal origin that's more likely caused by the pressure originating from the increasing compression than the external heat from increasing sunlight.
3. The comet's increasing condensing also increases its rate of rotation, indicated by the increasing length of the radiused arrows. When its outward acting rotational and orbital centrifugal forces begin to exceed gravity's inward acting condensing, its material begins to dislodge, fall away, and disperse into an arcing fan shape to form its dust tail. This always occurs from the comet's backside opposite the Sun where the combined centrifugal forces are the strongest and gravity's compounded condensing is the weakest.
4. The comet's coma along with its plasma and dust tails continue to increase until it reaches its closest point to the Sun, perihelion, where its condensing and centrifugal forces and the Sun's radiant energy are all at their maximum.
5. As it begins to leave the Sun's vicinity, the now increasing density of the Sun's gravity field begins to reverse the comet's condensing/contraction that in turn slows its rate of rotation. Together with its slowing orbital velocity, its rotational and orbital centrifugal forces weaken, curbing its loss of material, which reduces the size of its dust tail.
6. As it continues to move farther away, the compression responsible for its outgassing also eases while the solar wind and radiation diminish. This reduces the size of its plasma tail as well.
7. The comet's decompression proceeds while its rotation and orbital velocity continue to decrease all the way to its aphelion, its farthest point from the Sun, where the sequence begins again.

(11.4 Shape 13a)

With the universal field supplanting (nonexistent) three-dimensional space, any further decrease in the density of a gravity field, from either the presence of a new object or an increase in an object's matter, would impart an instantaneous reaction in the gravity field that would act over an infinite distance because of its innate mechanical nature. It would not act at the speed of light via waves. Einstein implies that this would have to be concluded if it were not for, "the fact velocity c plays fundamental role in this theory [of relativity]."[60]

But he still contradictorily asserts that light's constancy can't occur in gravity fields. It's variable in gravity fields.[61] And gravity fields are everywhere. So light's velocity has to be variable everywhere. It's constant nowhere. Without light's constancy, none of relativity works. Without relativity, the theoretical prohibition on gravitation's mechanical, instantaneous nature no longer exists.

Imagine you're holding one end of a board about 8' long where the other end is resting on a table. If you were to move your end up and down, the effects of that movement would be felt instantaneously throughout the board's entire length, at least for all practical purposes it would. It's exactly the same effect for gravitation. An increase in the amount of merging material composing an object would instantaneously further decrease the density of its ambient radiation, its gravity field. This would instantaneously increase its inward acting pressure, the gravity field's strength.

So circumventing altogether any discussion concerning the existence of an aether, no matter its defined and whether it's at rest or not, this demonstrates how the radiant energy that corresponds to all (three-dimensional) space has to be an instantaneous material carrier of information. This basic understanding offers practical explanations for some of the instantaneous phenomena associated with quantum theory known as entanglement where particles seem to be interconnected over distance.

The very notion that the universal field and matter are inseparable by itself subverts the theory of gravitational waves. Just as with our moving board example, the universal field's innate interconnection with objects prevents it from having an independent ongoing reaction, a wave motion instigated by the object that continues to propagate separately on its own.

There can only be a change in the universal field when there's a change in the quantity or location of an object's mass. If we could somehow repeatedly add and then remove a portion of its material, then its gravity field's density would decrease and increase correspondingly. But this would not be or create a wave function.

60. Einstein, *Relativity*, 53.
61. Einstein, *Relativity*, 84-85.

The recurring periodicity in the amount or location of mass that produces an associated oscillation in field density only simulates wave motion.

This principle would hold true for the extreme conditions of rapidly orbiting dense bodies whose orbits around a common center of mass produce a back and forth motion relative to the Earth. The closer then farther away their positions in their orbits, the stronger then weaker their gravity fields would be at our location, that's if we were actually able to detect them.

This could in no way be considered a gravitational wave. It only reflects the instantaneous physical change in the field's density that repetitively approaches then recedes from us. There simply can be no emanating gravitational wave produced by the presence of an object that propagates independently no matter how extreme the motion or violent the merging of bodies. Regrettably, we've spent untold millions on experiments to falsely validate these nonexistent waves.

Coordinate Systems

Einstein seems to be arguing that because gravity fields warp spacetime, meaning that space curves, along with the objects in it, while time's rate (and light's velocity) varies, the use of Euclidean geometry is limited. Normal Cartesian coordinate systems only work for conventional conditions.

To accommodate space's (presumed) non-Euclidean nature, and time and light's variability, he proposes a curving, extra-dimensional Gaussian coordinate system. He contends that it's more correct to make physical descriptions by referring to coinciding spacetime events assigned with four Gaussian coordinates, "which have not the least direct physical significance, but only serve the purpose of numbering the points of a continuum in a definite but arbitrary manner." He suggests that it can accommodate an infinite number of dimensions.[62] Sounds purely theoretical to me. Any number beyond the three of our real physical world is wholly impractical.

(Euclidean geometry is an elementary geometry of straight and parallel lines of two and three dimensions. Non-Euclidean is a geometry of curving, converging or diverging, nonparallel lines. Euclid of Alexandria was a Greek mathematician that lived around 300 BC. A Gaussian coordinate system is essentially a Cartesian coordinate system whose grid is not rectilinear but curves to accommodate spherical applications. Carl Friedrich Gauss was a German mathematician, 1777-1855.)

Einstein appears to be implying that a curving, "Gaussian system is a logical generalization of [a rectilinear] Cartesian coordinate system." It's better suited for larger scales. But even then, any deformation would still be minor. (It might also be inferred that it only works for outside observers.) The normal Euclidean properties of a Cartesian system with three rectilinear spatial and one time coordinate would remain applicable at the relatively smaller scales we encounter in our everyday lives.[63]

So from a practical standpoint, how would that work exactly? If a Gaussian system only applies to theoretical conditions, how is that useful? If it were functional but only applies to larger scales, what are they exactly? When does the Cartesian system stop and the Gaussian begin? And how would they interface? Wouldn't they have to be overlapping at some point? So they'd always be commingled, at least to some degree. How would you ever know which one to use and when to use it?

Also, how could calculating the exact shape of distorted space or the distorted objects in it ever be possible when the effect of gravity is felt over an infinite distance?

---

62. Einstein, *Relativity*, 92-107.
63. Einstein, *Relativity*, 99-100.

Every object is in an infinite number of gravity fields, deformed in an infinite number of ways by the gravity fields of every other object in the universe. Given real world conditions, the use of a Gaussian system has no potential benefits.

But wait a minute. Let's back up. Assuming again "time" actually exists, why would it even be allocated a coordinate, in either in a Cartesian or Gaussian system? It's rate would not be variable. It can only vary when light's velocity is fixed. Light's constancy and time are conceptually interrelated and inseparable. A three-dimensional coordinate system with a fourth time coordinate only makes sense if light's constancy forces time's rate to change (that is if you accept that any reference frame can have an infinite number of rates of time).

If its velocity is not fixed but varies in gravity fields, as Einstein contends (which means it varies everywhere), then it doesn't.[64] No apparent reason, cause, or mechanism exists that induces a change in time's rate. So how can its rate be variable? And if it's not variable, why would it need a coordinate?

Gravitational time dilation can't cause time's variability in a coordinate system. Its changing rate is not a result of location but a product of relativistic effects due to hypothetical motion associated with special relativity's effects despite that light's velocity is variable in gravity fields.

It can't be spacetime's inherent geometry. Light's variability in it would always maintain time's constancy everywhere. If this were the case, it'd eliminate the dilemma of having an infinite number of rates of time for any reference frame or body due to light's constancy. That problem would be resolved.

If time's rate did actually vary in spacetime, it'd be varying everywhere and in every direction. So light's constancy would be maintained everywhere. The inherent conflict with its variability and its velocity in the perpendicular direction would be resolved. But a reference frame or body with an infinite number of rates of time is not possible. Despite Einstein's contradictory assertions, it'd be hard for him to disagree.[65]

With light's variability, time's variability is not possible. It'd have no impetus. So the time coordinate of his four-dimensional Gaussian coordinate system has no purpose. It's a "dunsil." There's no reason for it. Which also means that the time component of his "space-time" is pointless.

What we're left with is just normal, three-dimensional space. But just like with time, space is innately nonexistent. It's the nothingness between objects. So the whole notion of a "space-time" is baseless. And that's in addition to the fact that its four-dimensionality is an inconceivable concept. There can be no such thing as "space-time."

64. Einstein, *Relativity*, 85.
65. Einstein, *Relativity*, 31.

Time and light's variability has huge implications for general relativity. They invalidate all of it. But we ignore this and all of its other crippling flaws. Insisting instead that it all makes perfect sense while parroting its delusive canon as if it were a rational, factually verified reality.

Aside from (nonexistent) time's impossible variability and light's contradictory constancy and variability and the impossible curvature of (nonexistent) space, what Einstein fundamentally failed to comprehend is that any measuring device in his distorted spacetime, whether it was measuring distance or angles, or time's assigned rate, would be condensing and deforming in the same manner as the object it was attempting to measure. So everything would always appear normal regardless of its distortion. So our Euclidean geometry would be maintained.

To any outside observer, the deformation of any single reference frame or body from self-gravity would always manifest as omnidirectional condensing as it was compounding with the self-gravity of every other nearby body. So its ongoing condensing and contraction would appear as if it were receding. So for both observers, a normal Euclidean geometry would still be preserved, even if it's only apparent. (See Diagram **9** Parallel 6a)

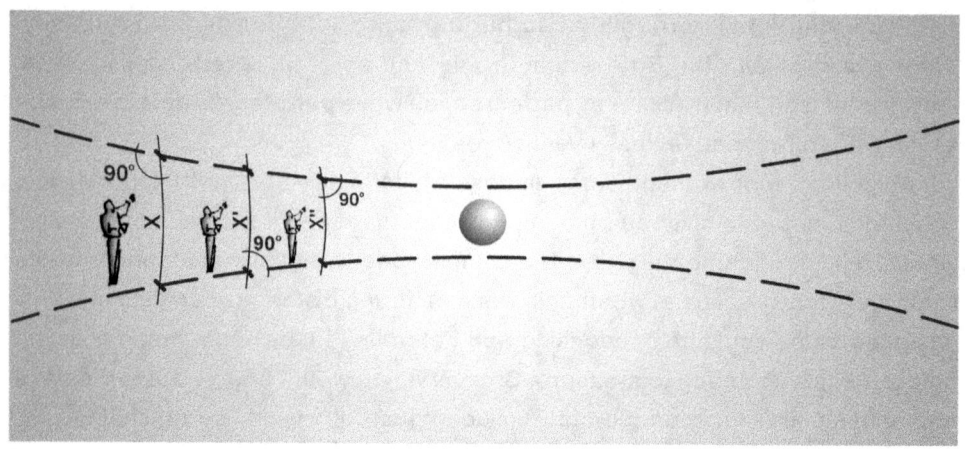

PARALLEL LINES

A gravity field's decreasing density, represented in section view by the diffusing background, causes the omnidirectional condensing of the objects in it, light's decreasing velocity, along with time's increasing rate, which we establish by selecting an object with periodic motion that increases with contraction. Theoretically, to an outside observer, any parallel lines near a massive body and perpendicular to their line of sight of the body would appear to converge. This makes the body appear farther away than it actually is, suggesting distance.

For someone at the lines, they'd continue to record right angles and measure a consistent distance between them as they moved toward the body. **X**, **X'** & **X"** would all read the same. Light's slowing velocity, time's increasing rate, and the omnidirectional contraction of all objects, including any measuring devices along with the measurer, would all occur in unison, which would render any change imperceptible while maintaining our ordinary Euclidean geometry under all circumstances.

(9 Parallel 6a)

If space could actually stretch/curve in three dimensions as the mass of a body increases, the distance between grid lines of Einstein's theoretical, extradimensional, Gaussian coordinate system centered on that body would be increasing. In contrast, grid lines correlating to the decreasing density of a real, three-dimensional gravity field centered on a body would be condensing as its mass increased. The grid would be ever-converging but never merge because the field, no matter its thinness, can never cease to exist.

With light's slowing velocity in a gravity field's decreasing density that's synchronized to time's increasing assigned rate, the body's condensing would essentially be imperceptible. The size and shape of any objects on it, including any measuring devices, would be condensing in unison with it in an omnidirectional manner. So a rectilinear Cartesian coordinate system and Euclidean geometry would always be preserved, despite any distortions and their decreasing size. (See Diagram **10** Coordinates 7a)

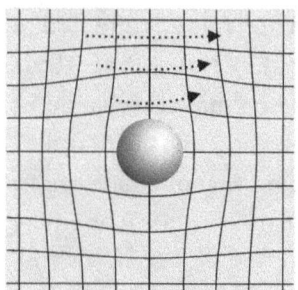 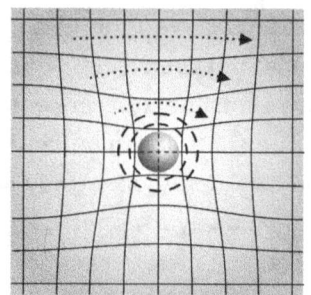 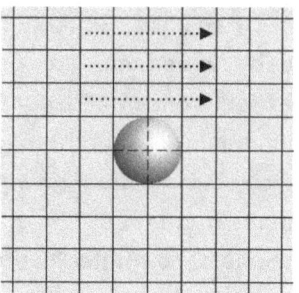

1. GAUSSIAN EINSTEIN	**2. CARTESIAN REALITY**	**3. CARTESIAN PERCEIVED**
LIGHT VARIES IN GRAVITY	LIGHT VARIES EVERYWHERE	LIGHT APPEARS CONSTANT
TIME VARIES EVERYWHERE	TIME IS NONEXISTENT	TIME APPEARS CONSTANT
SPACE CURVES BUT IN 3 DIM.	FIELD DENSITY DECREASES	FIELD LOOKS UNIFORM
OBJECTS DISTORT	OBJECTS CONDENSE	OBJECTS SEEM STATIC
(ALL CONTRADICTORY)	(ALL PRACTICAL)	(ALL APPARENT)

COORDINATE SYSTEMS

Diagram **1** attempts to conceptually portray all the inherent contradictions of Einstein's curving, extradimensional, Gaussian coordinate system as applied to his (inconceivable, nonexistent) four-dimensional "space-time." The dotted lines indicate how "light varies with position," but only in gravity fields and despite (nonexistent) time's variability, in (nonexistent) space that somehow stretches and curves two-dimensionally under three-dimensional massive bodies while its density remains constant, represented by the uniform background. The grid suggests how his stretching (nonexistent) space would actually behave in three dimensions. It's always incorrectly shown two-dimensionally with its length contradictorily stretching while its width condenses, "spaghettifying" all infalling objects.

His recognition of light's variability, in gravity fields, means he's lost light's constancy, everywhere. There are no places where gravity fields aren't. They're everywhere. So light's velocity has to vary everywhere. Without light's constancy, the impetus for time's changing rate no longer exists (not to mention the premise that all of relativity is based on).

It is light's fixed velocity that (mathematically) forces time's changing rate. Without time's changing rate, there's no need for the fourth coordinate of his Gaussian (or a Cartesian) coordinate system. He unknowingly reduced his (inconceivable, nonexistent) four-dimensional "space-time" to just (nonexistent) three-dimensional space.

In diagram **2**, light's velocity is also variable. It slows in a gravity field's decreasing density, depicted in section view as diffusing background. "Time" is not a factor. It doesn't exist. It's not a property of the universe. So its rate can't be changing even if light were constant. Its assigned rate can only change when the size and/or mass of the object or body we've selected for reference changes, which alters any periodic motion like its rotation.

If that body was condensing in the decreasing density of its and another's compounding gravity fields, the periodicity of its natural frequency would be increasing, which increases its rate of rotation. So if its rate of time is based on that rotation, it could be interpreted as increasing. But it'd be difficult for anyone located on that body to be aware of it.

All measuring devices, along with the measurer, would be contracting in unison with the body while light's velocity decreased and time's assigned corresponding rate increased in the fields' decreasing density. So effectively, any change in size, shape, or passage of time could never be perceived. Everything would always appear normal.

It's the same for any outside observers. Its spherical, omnidirectional condensing would appear as recessional velocity despite that the distance to its center would remain unchanged. So with time's actual nonexistence, or even if it did exist, with light's factual variability everywhere, there can be no "time" variable. So there's no need for a Gaussian coordinate system's (theoretical) extra dimension or unworkable Lorentz transformations. Our normal Euclidean geometry and rectilinear Cartesian coordinate system are preserved essentially in all cases, even if it's only apparent, as implied in diagram **3**.

(10 Coordinates 7a)

The space around so-called "singularities" at the center of (theorized) black holes is always misrepresented with a contradictory and impossible two-dimensional, funnel-shape. A superimposed grid somehow endlessly elongates in the direction of infall while condensing in the perpendicular direction. Objects are wrongly shown as stretching and elongating as they are being endlessly "pulled" in toward the black hole's center. This is referred to as "spaghettification," supposedly the product of the difference in gravity's force from the top to the bottom of the infalling object.

In reality, the object would be condensing inward, spherically, while free falling toward a point in (three-dimensional) space. It's impossible for it to be sucked down the two-dimensional surface of a funnel of (nonexistent) space. That environment does not and cannot physically exist.

But even if it did, it'd require the existence of an opposing gravitational source that somehow tugged at it in the opposite direction, resisting its infall, causing its elongation and (nonexistent) space's stretching. (A black hole is a region of space, or an object, with so much mass that any nearby object, and supposedly even light, cannot escape its gravitational pull. A singularity is a theoretical point thought to reside at the black hole's center where gravitational forces presumably cause spacetime to become infinitely distorted and matter to be infinitely compressed denoting its ultimate fate.)

The notion of objects stretching in free fall is at conflict with the central concept of gravitation. Objects coalesce and condense in a runaway process that never ceases. They can only further compact in the ever-decreasing density of their ever-compounding gravity fields, whether that occurs at a presumed black hole or not.

Also, an object's stretching in the direction of free fall is in conflict with the relativistic effect of their contraction in the direction of their motion to maintain light's constancy. How can they be stretching in the direction of free fall while simultaneously contracting in the same direction? That's not possible. (See Diagram **12** 3-D Space 13a)

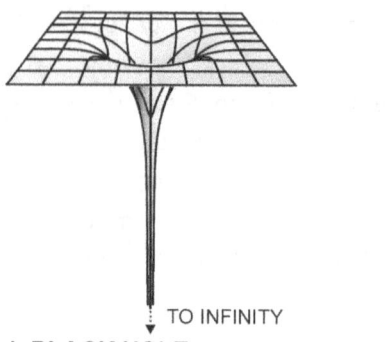

1. BLACK HOLE

TO INFINITY

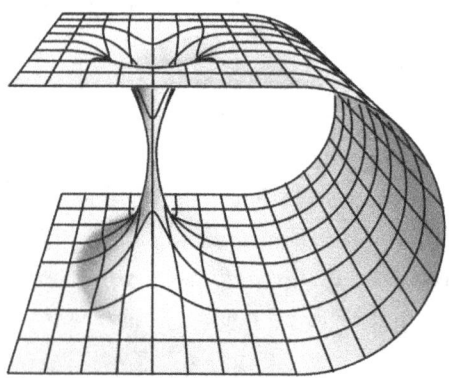

2. WORMHOLE

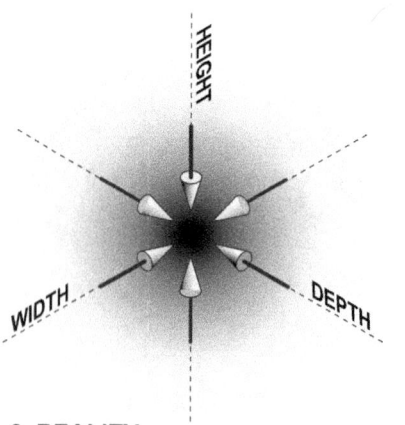

3. REALITY

(12 3-D Space 13a)

THREE-DIMENSIONAL SPACE

The notion that space is something and that it expresses two-dimensionally has led us to mistakenly reason black holes. Their gravity is presumed to be so strong they create a two-dimensional funnel-shaped curvature in their inconceivable four-dimensional space-time, usually portrayed in diagrams similar to **1**. It's imagined that any object, or light, that strays too close will be drawn over a theorized event horizon, never to be heard from again, where it endlessly infalls while stretching, or "spaghettifying" as it's called, as it accelerates without end toward the infinite (mathematical) condition of a singularity.

The same mistaken logic is applied to so-called wormholes. Space's fictional two-dimensional composition, commonly depicted in diagrams similar to **2**, would supposedly allow us to shortcut through a black hole that's not infinite but tied to another black hole at another location in our curving two-dimensional universe.

In reality, space is the nothingness between objects. There's nothing there. But if it really were something, it'd still have to be three-dimensional, as indicated in diagram **3**. It'd have to have width, height, and depth. That's it. Additional dimensions are inconceivable. And two-dimensions, despite being incorrectly justified as merely an analogy, only describe the location of a plane. Without the third dimension of height, existence isn't possible.

Whether you insist that space is actually something or accept the universe as a field of radiant electromagnetic energy that corresponds to all space, in our reality of three actual dimensions, coalescing infalling material, represented by the condensing grey three-dimensional sphere in **3**, would quickly arrive at a common center of mass. There's nowhere else for it to go. It cannot slip over the edge of a nonexistent two-dimensional funnel-shaped space on an endless journey to infinity. That's a make-believe science fiction fantasy that's justified through two-dimensional mathematical gimmickery.

Gravity's exponentially decreasing field density continuously coalesces and condenses matter together in a runaway recycling process where it's increasingly compressed until fusion reactions are triggered that perpetually convert it back to its original state of radiant energy.

Imagine that objects actually could contract in the direction of their motion while their (nonexistent) time slows. For simplicity, disregard motion's supposed changeableness that's subjectively decided by each observer. Because all objects already have some relative motion due to their ongoing gravitation, we'd then be faced with the complexity of having to calculate Lorentz transformations for its linear contraction and its changing rate of time in the direction of motion for each the four spacetime variables of its curving Gaussian coordinate system that would be continuously changing due to spacetime's curvature/deformation. But we'd also have to simultaneously calculate them for the motion resulting from gravitation that's be commingled. Would that even be possible?

Einstein argues after the fact in an attempt to salvage special relativity that its relativistic effects (time dilation with motion, the contraction of objects in the direction of motion, an object's increasing mass with increasing velocity, all the product of his assumption of light's fixed velocity) can only be considered outside the influence of gravitational fields where light's velocity is fixed.[66] This isn't possible. Everything exists under the influence of gravitational fields.

Every object whether it's at the subatomic level or our entire (presumed) finite universe has self-gravity. It could also be argued that because those fields propagate over an infinite distance, they affect every other object. So where can we find light with a velocity that can be fixed so that special relativity can be employed? The answer obviously is nowhere.

Again, with the loss of its underlying premise, relativity becomes incoherent, all of it. It no longer works. Some might argue an exception could be made for his "principle of equivalence" and his "finite and yet unbounded" universe. But they're totally fallacious for other reasons that are just as subverting. They're addressed in upcoming chapters.

Einstein also gives undue importance, and in fact credits the success of relativity, to Hermann Minkowski's (Russian mathematician, 1864-1909) idea of a "four-dimensional space-time continuum," whose phrasing long ago degenerated into meaningless, colloquial rhetoric. But there's nothing cryptic or mystical about what a warping of his "space-time" should mean.

Space is a three-dimensional, physical something that distorts that in turn causes time's rate to change to maintain light's constancy.[67] But he concedes light's variability, which has to vary everywhere, which undermines the whole notion. Light's variability and time's variability together become meaningless. They don't work when they exist together. One or the other has to be fixed.

66. Einstein, *Relativity*, 85, 109, 171.
67. Einstein, *Relativity*, 61-64.

The primary problem, though, is that space by definition doesn't actually exist. There isn't anything there. So there's nothing there to distort. He also portrays it as a two-dimensional plane so it can curve to facilitate gravitation and even maintain the uniformity of our entire (presumed) finite universe. But two dimensions can only define the location of a plane. So just like space, it doesn't exist either.

But if we accept that electromagnetic radiation is something real, is omnipresent, inhabits all of (three-dimensional) space, and that its continuity cannot be disrupted but does vary in density/intensity then we'd have something that can distort three-dimensionally, not impossibly curve, where light's velocity can naturally vary through refraction in its varying density. That's simple enough.

Also, as touched on earlier, the theoretical four-dimensional distortion resulting from spacetime's curvature would fundamentally conflict with the one-dimensional distortion resulting from special relativity's effects. And this nullifying conflict is exacerbated by his assertion that all motion can be interpreted as relative motion, established by the arbitrary and subjective choices of each observer. This suggests that length contraction, time dilation, and increasing mass along with gravitational interactions are all determined by and can be changed at our whim without any regard at all to reality, or anybody else's choice.

Just as problematic, though, is Einstein's inability to see that it's not possible for any single object or reference frame to have varying or multiple rates of time or different rates of time in different directions. They'd all be in conflict. He seems to recognize this at one point.[68] But it's just one more in a long list of inherent contradictions.

This sabotaging error is similar in concept and magnitude to his inability to see that time's "slowing" for a moving object to maintain light's constancy in the direction of motion creates an unresolvable conflict with light's velocity in the other two dimensions of our three-dimensional universe that aren't in motion but still have time's slower rate. Light's velocity in those directions would exceed 186,000mi/s.

It demonstrates light's compounding with motion, which voids his assumption of light's constancy that in turn invalidates all of relativity. But he and almost everyone else can't, or more accurately, refuse to see it. We much prefer our delusions over reality.

68. Einstein, *Relativity*, 31.

Equivalence

Einstein contends that, "The gravitational mass of a body is equal to its inertial mass... that this important law had hitherto been recorded in mechanics, but it had not been interpreted. A satisfactory interpretation can be obtained only if we recognize the following fact: The same quality of a body manifests itself according to circumstances as 'inertia' or as 'weight'."[69] So it follows that a person experiencing a one g acceleration in an enclosed reference frame, theoretically free of "preexisting" gravitational fields, could not distinguish the difference between that acceleration's inertia and the free fall effect of gravity resisted by the Earth's surface. Because the difference can't be perceived, they must be equivalent.

He wholly believes that acceleration's inertia (which for him also includes braking as if a train were coming to a stop) actually creates a real gravity field for the person. This embodies "the natural law of the equality of inertial and gravitational mass" that he in effect restates as a "principle of equivalence."

What he does is first adopt the supposition that inertial mass and gravitational mass are separate and individual types of masses. Then he suggests that they've already been found to be, "what comes to the same thing." Then he qualifies that equality as a significant preestablished law of nature that's a, "fact... strongly confirmed empirically... We have thus good grounds for extending the [general] principle of relativity to include bodies of reference which are accelerated with respect to each other [the natural law of the equality of inertial and gravitational mass] and as a result we have gained a powerful argument for a generalized postulate [principle] of relativity." And in circular reasoning, he also contends that the, "extension of the [general] principle of relativity implies the *necessity* of the law of the equality of inertial and gravitational mass."[70] But it's all manufactured. And none of it is even remotely workable.

If "the law of the equality of inertial and gravitational mass" is really his "principle of equivalence," and if "the general principle of relativity" ("[where] all bodies... are equivalent for... the general laws of nature [mechanics and light's constancy], whatever may be their state of motion"[71]) encompasses "the law of the equality of inertial and gravitational mass," doesn't that make both his "principle of equivalence" and "the law of the equality of inertial and gravitational mass" redundant? If his general principle of relativity was actually realizable, they'd essentially be useless.

69. Einstein, *Relativity*, 73-74.
70. Einstein, *Relativity*, 75-79, 172.
71. Einstein, *Relativity*, 69, 108.

It can easily be shown that acceleration does not produce a gravity field, and that its reaction is not even close to natural, mass-created gravity's. Only in the most generic nontechnical way can acceleration for the person in the enclosed reference frame be interpreted as gravity here on Earth. To draw that correlation, you have to intentionally ignore an excess of conspicuous, indisputable, disqualifying facts.

Acceleration's reaction is uniform throughout all locations of the reference frame. It acts only in one dimension, opposite the direction of motion. It doesn't coalesce and condense objects. And it requires motion.

But remember an object's rate of motion, or whether it even has any motion, is supposed to be a subjective choice of each observer. So all motion is relative and discretionary.[72] If that were actually true, how could acceleration-created gravity be real? It could be manifested at the whim of every observer. Also, it'd conflict with the choices of every other observer. And it'd wreak havoc with gravitational interactions.

Natural, mass-created gravity's reaction is nonuniform. It increases exponentially. And it acts three-dimensionally, spherically, radially inward toward the center of every mass or common center of multiple masses. The three-dimensional distortion it produces in objects is completely different from acceleration's one-dimensional distortion.

Mass-created gravity doesn't require motion. It creates motion. It perpetually coalesces, and condenses, in a runaway process that never ceases. Objects are mechanically pushed toward one another as they seek equilibrium in the ever-decreasing density of their compounding fields.

Acceleration's reaction is mechanical, which is essentially instantaneous. Convention has gravitational attraction acting at the speed of light via waves by a force similar to electromagnetism. Contradictorily, Einstein believes the same.[73] But that attraction is also somehow simultaneously mitigated by unobservable massless graviton particles that somehow physically exist without mass. Which if they actually were particles, wouldn't be able to act at the speed of light. They'd relativistically become infinitely large.

How could acceleration-created gravity ever coexist with real, mass-created gravity? For any accelerating object, there'd be two types that inherently conflict. Also, his misinterpretation of his mass-energy relation that has the mass of an accelerating object metaphysically increasing from its kinetic energy would

72. Einstein, *Relativity*, 16-18, 30, 67, 68...
73. Einstein, *Relativity*, 72.

exacerbate the conflict.[74] The increase in mass would be increasing its real gravity while its acceleration-created gravity would also be increasing.

If the acceleration of a particle through an electromagnetic field increases its charge, which increases its mass that in turn increases its natural gravity, as is commonly accepted, then this would also further conflict with relativistic-created mass and acceleration-created gravity.

And according to Einstein, the relativistic effects of special relativity can't occur in (natural, mass-created) gravity fields. Light's velocity is not fixed in gravity fields. It's variable: "the velocity of propagation of light varies with position [in gravity fields]... [Special relativity's] results hold only so long as we are able to disregard the influences of gravitational fields... The special theory of relativity has reference to Galileian domains, *i.e.* to those in which no gravitational field exists... The principle of inertia and the principle of the constancy of the velocity of light are only valid with respect to an *inertial system*."[75] So relativistic effects have no possibility of ever working in gravity fields.

If all gravity fields really were the same, and if relativistic effects can't occur in gravity fields, how could they ever occur for any accelerating object? Subatomic particles in particular come to mind. They'd have gravity fields from their acceleration. So how could they ever demonstrate the increasing mass of "Einstein's" celebrated mass-energy equation, $E=mc^2$, along with their time's presumed dilation.[76]

But how can relativistic effects ever work anywhere under any conditions? Gravity fields are everywhere. They extend indefinitely. And every object has its own self-gravity whether it's a subatomic particle or our (presumed) finite universe. So there's no place where gravity fields don't exist. So there's no place where light's velocity can be fixed. So there's no place where relativistic effects can occur.

Beyond light's nullifying variability in gravity fields, its constancy is conceptually impossible. It's unworkable in three dimensions. In our real world, it also compounds with all relative motion.

Time is nonexistent as well. It's not an inherent property of the universe. We create time and define its rate. So all relativistic effects do not actually exist in reality. They're abstract contrivances. Any inquiry into them has to remain a strictly theoretical endeavor.

It's not difficult to come to the realization that acceleration's inertia and gravitational inertia are not and cannot be the same effect. So "the law of the equality of inertia and gravitational mass" or its concocted twin "the principle of equivalence" have to be false precepts. (See Diagram **13.1** Acceleration 8a)

74. Einstein, *Relativity*, 49-54.
75. Einstein, *Relativity*, 85, 109, 171.
76. Einstein, *Relativity*, 49-54.

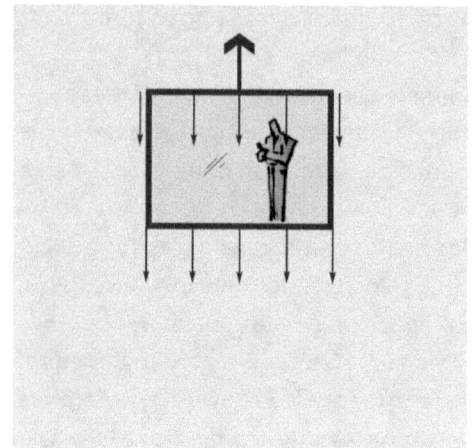

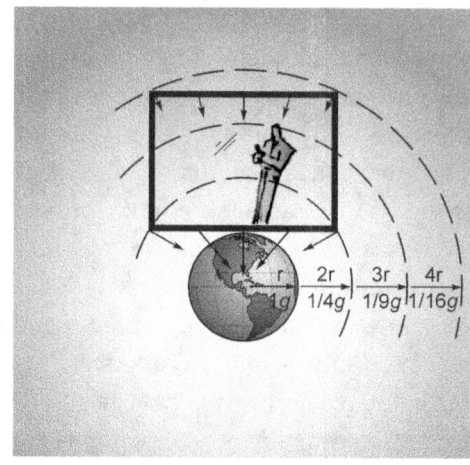

1. ACCELERATION **2. GRAVITY**

ACCELERATION & GRAVITY

Einstein asserts that acceleration, which for him also includes braking as if a train were coming to a stop, creates an actual gravity field. But it can easily be shown that acceleration and gravity's reaction are not the same. Using an example similar to his, let's imagine that we've placed someone inside a large crate. But let's make ours transparent. We'll assume it's an independent reference frame.

For diagram **1**, when the crate is being pulled upward with a one g acceleration out in space somewhere theoretically free of gravitational fields, we can see how the reactive force, indicated with the smaller arrows that we'll say corresponds to weights on springs, has to be essentially instantaneous. It's mechanical in nature. This is contrary to Einstein's claim that gravity acts at the speed of light via waves by a force similar to electromagnetic radiation.

It's also equal everywhere throughout the crate and acts in only one dimension, opposite the direction of motion. This might initially appear or feel like gravity to the person experiencing the acceleration until they look outside or acquire the proper equipment to detect the uniform distribution of the reaction.

Now for diagram **2**, let's exaggerate our condition here on Earth and imagine that it's been compressed down to a few feet in diameter but where gravity's force remains the same at the surface. For clarity, let's keep our crate the same size and shape, compensating for any distortion that would also reveal that gravity is entirely different from acceleration.

Because the strength of a gravity field always dissipates exponentially per the inverse square law and because it always radiates spherically, three-dimensionally, from the center of a mass, gravity's force, as indicated by the varying length and direction of the smaller arrows that still corresponds to weights on springs, has to vary in strength and direction at every location within the crate as it rests on the Earth's surface.

Gravity coalesces. Acceleration doesn't. Gravity acts inward three-dimensionally and increases/decreases exponentially. Acceleration acts only one-dimensionally and uniformly. A cursory comparison easily reveals how gravity can never be created by acceleration, which completely undermines Einstein's principle of equivalence.

(13.1 Acceleration 8a)

Equivalency's disqualifying contradictions don't end with inertial and gravitational masses' innate incongruity. The supposition completely collapses with the realization that there's no such thing as "inertial mass" or "gravitational mass" per se. They're manufactured abstractions as well. They don't actually exist. In reality, there's only mass, the amount of material an object contains.

Whether that mass is in motion or affected by gravity does not bestow it with some unique and separate quality. When it's in motion, it has inertia. When it's "at rest" resisting the gravitational influence of a more massive body, it has weight. First separating mass into different types, inertial and gravitational, that don't actually exist, then turning around declaring that they're not actually different but the same nonexistent thing through an invented, nonexistent "principle of equivalence" is as nonsensical as it gets.

More concerns arise from this artificial precept. It's generally accepted that since a photon has momentum (because it's mistakenly assumed to be a particle), it must have inertial mass. And because of the "principle of equivalence," that inertial mass must also be gravitational mass. So a photon has gravitational mass. Which means, it should be affected by gravity like any other object.

This errant reasoning has led to the misconceptions of a black hole's event horizon, the threshold where the pull of gravity has become so strong or spacetime so curved that not even a photon can escape; the notion of a gravitational lens, a photon's path is altered by gravity's pull because of its mass; and the more conventional explanation for the idea of gravitational redshift, a photon's impeded escape from a gravitational well due to gravity's pull on its mass that decreases its energy, which reduces its frequency. All of these notions are premised on the supposition that a photon is a real particle that actually has "gravitational mass" even though that mass has never been observed.[77]

Max Planck (German physicist, 1858-1947) first advanced the hypothetical notion of a photon (a massless quantum of radiant energy) in an attempt to explain the frequency distribution of radiation. Unfortunately, the concept has evolved and become fixed in quantum theory as a real particle.[78]

Still, energy's wave-particle duality is heavily weighted toward being exclusively wavelike in nature. But the whole issue of a photon's particle nature is undermined by simply acknowledging that radiation by definition is not matter. So it cannot have mass, whether it is quantized or not. With the acceptance of this elemental fact, the concepts of a black hole or at the very least its event horizon, a photon based

77. "Principle of Equivalence," HyperPhysics, last access Dec 28, 2023, http://hyperphysics.phy-astr.gsu.edu/hbase/Relativ/grel.html.
78. "Max Planck," Wikipedia, last modified Dec 11, 2022, https://en.wikipedia.org/wiki/Max_Planck.

gravitational lens, and the orthodox explanation for gravitational redshift all evaporate prior to invoking Einstein's unfounded "principle of equivalence."

He continues to deceptively argue that because a "ray of light" projected perpendicular by someone experiencing acceleration curves downward as it traverses their acceleration-created gravity field, his general principle of relativity (the laws of nature must hold true for all reference frames regardless of their motion) allows us to, "conclude, *that, in general, rays of light are propagated curvilinear in* [all] *gravitational fields.*"[79] What he's really arguing for in a backhanded way is equivalence.

Not unlike selling snake oil, he intentionally misdirects us away from the question of how, or if, light would actually curve in an acceleration-created gravity field. He wants us to instead blindly accept that he's providing us the factual evidence that's already been established.

With more sleight of hand, he attempts to correlate two completely different conditions. Neither of which produce light's actual curvature. Its apparent curvature in the gravity fields of massive bodies is the product of its refracted slowing through the field's decreasing density, not because it's following the geodesic (the shortest possible line between two points on a curved surface) of nonexistent spacetime's impossible two-dimensional curvature or because its photons are being "pulled" from their otherwise straight path by gravity. He'd have a hard time disagreeing with that. He (contradictorily) affirms himself that light refracts: "A curvature of rays of light can only take place when the velocity of propagation of light varies with position [in gravity fields]."[80]

His assertion that light curves when its source is under acceleration is just as wrong. He (deliberately) misrepresents how a perpendicular "ray of light" would behave under acceleration, implying that its end would be dragged upward as if it were attached to its moving source that causes its increased bending with increasing velocity. He wants us to envision it as if the light was arcing downward as it departs, appearing like water streaming out from a garden hose.[81]

But the path of a "ray of light" would always propagate perpendicular in a straight line from its source, assuming as he does that it's theoretically free of "existing" gravity fields. This can be easily shown if we quantify it into a series of projected photons.

Each consecutive photon would be seen as defining an inverted arc with decreasing curvature if viewed in a series of stop-action photos taken at equal intervals.

79. Einstein, *Relativity*, 84.
80. Einstein, *Relativity*, 85.
81. Einstein, *Relativity*, 83-85.

But each of their individual paths would have to remain straight and perpendicular relative to their point of origin. His misrepresentation is readily seen when comparing it to what light's behavior would be under constant velocity. (See Diagram **14** Ray 4a)

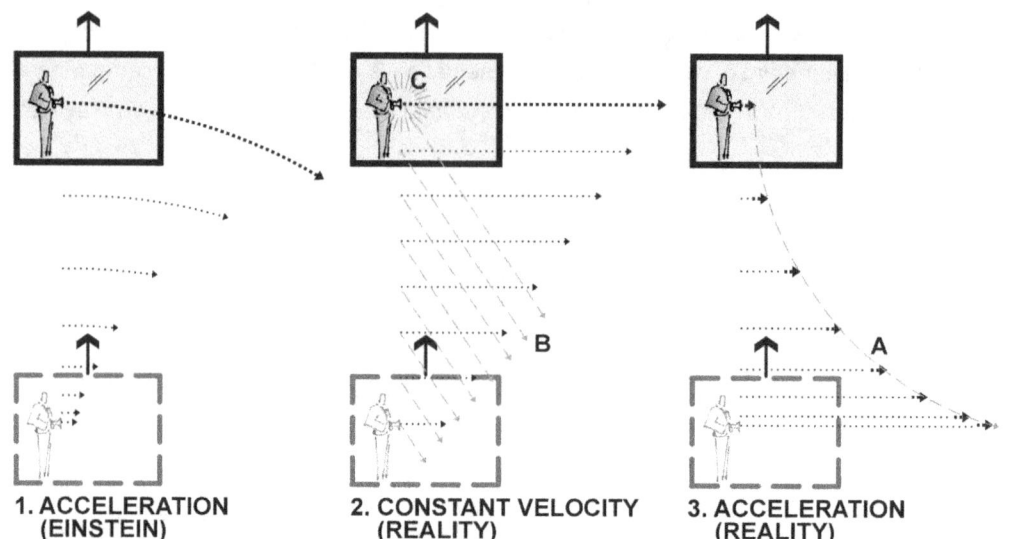

1. ACCELERATION (EINSTEIN)

2. CONSTANT VELOCITY (REALITY)

3. ACCELERATION (REALITY)

LIGHT UNDER PERPENDICULAR ACCELERATION

Using our example that's similar to Einstein's, let's again imagine that our transparent crate is being pulled upwards with a one g acceleration somewhere out in space theoretically free of any gravity fields so that the person standing inside would not be able to perceive the difference between the crate resting on the ground or its upward acceleration.

Einstein contends that because the conditions appear the same they must be the same. He argues that the crate's uniform acceleration must actually produce a real gravitational field. In a backhanded way, he indirectly correlates how a perpendicular light ray transmitted by the upward accelerating person, as depicted in diagram **1**, would curve in a manner no different from the curving light rays passing through the gravity field of a massive body like the Sun.

But to make the argument, he misrepresents how a perpendicular ray of light propagates under acceleration. He implies that it moves upward in unison with the crate as it would under constant velocity, as shown in diagram **2**. And then he wants us to infer that it should be bending downward as if being "pulled" by gravity similar to how water arcs downward as it leaves a garden hose.

In reality, an upward accelerating source would produce a new quantum of light each moment, leaving behind the previous, each propagating parallel in the perpendicular direction with an upward velocity that would remain constant at the rate it was emitted, as indicated in diagram **3**. Once the light separates from its source, there's no force causing its upward velocity to increase.

Inferring an inverted, downward curving path, indicated by the dashed line at **A**, as a ray of light would not be correct either. It would be connecting different quanta of light propagating separately in parallel perpendicular paths. The angled dashed lines at **B** though could be considered a single ray of light.

Light emits radially in all directions from its source, not just in the direction the flashlight is pointing, as implied at **C**. Under constant velocity, it would still be moving in unison with its source. But it's not bent or in any fashion being pulled downward or following the presumed geodesic of nonexistent spacetime's impossible curvature.

Just as Einstein uses light's nonexistent curving under acceleration to infer a gravity field, we have to use its actual perpendicular propagation under acceleration as evidence of the absence of a gravity field. This has us concluding again that acceleration and gravitation are not the same. So his principle of equivalence cannot be valid.

(14 Ray 4a)

The invalidating contradictions don't stop here either. As if aware of the flaws but desperate to rationalize them anyway, Einstein later, when off topic, covertly slips in an ad hoc qualifier, as if acknowledging its disqualifying difference makes it okay. He suggests that acceleration-created gravity fields are "homogeneous."

"[For a] system of reference... in uniform acceleration... there exists a state which, at least to a first approximation, cannot be distinguished from a gravitational field. The following concept is thus compatible with the observable facts: [The system of reference] is also equivalent to an 'inertial system'; but with respect to [that system of reference] a (homogeneous) gravitational field is present (about the origin of which one does not worry in this connection)."[82]

He doesn't want us worrying about the origin because he knows (maybe mostly subconsciously) that a homogeneous gravity field is impossible and nonexistent and contradictory. If it were homogeneous, it wouldn't be, and can't be, a real gravity field. Their inherent nature is to vary everywhere. It's their innate variability that causes the gravitational coalescing and condensing of objects.

Their accelerating free fall toward one another is in reality due to their mechanical reactive search for equilibrium in the ever-decreasing density of the universal field's electromagnetic energy that corresponds to all (three-dimensional) space. Its density innately decreases exponentially at every object because of the inherent geometry of a sphere that's bound to the inverse square law. The greater pressure of its higher density constantly pushes all objects toward its lowest density that always lies directly between them toward their common center of mass.

The "homogeneous" stipulation of acceleration-created gravity fields is sometimes found in online descriptions as well, also hastily slipped in and ignored without qualification because of the obvious contradiction. Acceleration's presumed gravity field cannot be inhomogeneous. And real gravity's cannot be homogeneous.

In his 1907 paper, *On the Relativity Principle and the Conclusions Drawn from It*, Einstein states, "According to §17, equation (30) is also applicable to a coordinate system in which a homogeneous gravitational field is acting... While assuming that equation (30a) holds for an inhomogeneous gravitational field as well."[83] He's referring to his reasoning behind gravitational time dilation, which is dependent on equivalence and contradictorily on light's constancy in a gravity field.

---

82. Einstein, *Relativity*, 172.
83. Einstein, ON THE RELATIVITY PRINCIPLE AND THE CONCLUSIONS DRAWN FROM IT, 306. Also on page 307, he footnotes (1) a reference to an equation in Chapter §18 (Space and time in a uniformly accelerated reference system).

What's significant is that it seemingly reveals his awareness of the inherent difference between accelerated-created, "homogeneous" gravity fields and natural, mass-created, "inhomogeneous" gravity fields. He apparently realizes the problem for equivalence, but accepting the invalidating consequences is just not an option for him.

So he proceeds anyway, working it both ways, moving back and forth, intentionally obscuring the nullifying conflict, hoping it won't be noticed. Despite his relentless obfuscation, acceleration's theoretical, nonexistent, homogeneous gravity fields can never equal real, mass-created, inhomogeneous gravity fields. It just doesn't work.

How can Einstein argue with a straight face that acceleration (and braking and rotation) create gravity fields because their reactions are the same as natural, mass-created gravity's, but then come back later after the fact and concede that they're not actually the same? They're different (homogeneous, inhomogeneous). Yet still maintain they're gravity fields. It was their sameness that was the whole reason why he decided they had to be gravity fields in the first place. If that sameness is lost, the reasoning doesn't hold. Only Einstein is brash enough to propose such spurious reasoning and expect us to believe it. It's inexplicable (almost) that most do.

The other invalidating contradiction is light's curving in gravity fields. If the, "curvature of rays of light [through gravity fields] can only take place when the velocity of propagation of light varies with position," as Einstein contends, a consequence of its refraction through the field's "inhomogeneous" density,[84] how could it ever curve through a field with "homogeneous" density? It wouldn't be able to. So if acceleration's gravity fields are made homogeneous to correspond to acceleration's homogeneous reaction then a ray of light's curving becomes impossible and the argument again falls apart.

If light doesn't actually curve for the accelerated person then either light doesn't curve in all gravity fields, which violates his general principle of relativity and nullifies Eddington's 1919 (presumed) observational confirmation of it[85] or acceleration doesn't create a gravity field, which invalidates his "principle of equivalence" (Sir Arthur Eddington was an English astronomer, physicist, and mathematician, 1882-1944). At best, the whole assertion, like so much else of relativity, is permanently relegated to the theoretical realm with no possibility of ever having any practical relevance to reality. (See Diagram 15.1.1, 15.1.2 Refraction 5a. Non-bold reference numbers indicate repeat diagrams duplicated for clarity.)

84. Einstein, *Relativity*, 85, 172.
85. Einstein, *Relativity*, 116, 145-147.

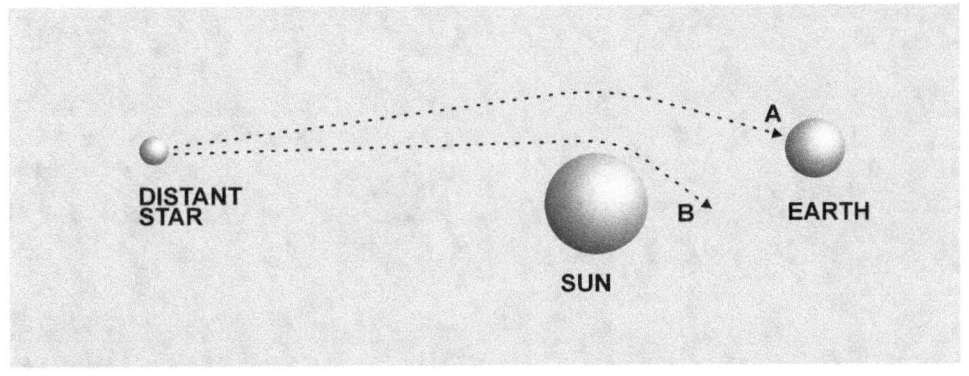

DISTANT STAR

A

B

EARTH

SUN

LIGHT'S BENDING

Our current belief is that a "ray" of light from a star or any distant object passing near a massive body like our sun is being pulled by gravity, that it's being bent from its otherwise straight path in the direction of **B** as it follows space's geodesic that somehow curves two-dimensionally in the vicinity of mass. And when viewed from Earth, its position is distorted in the direction of **A** due to an optical illusion.

Even though Einstein contends that light's distortion is actually due to its slowing through gravity fields, which is nothing more than refraction, which contradicts relativity's founding premise, light's fixed velocity, we reject his explanation. Instead, we hold to our belief that a photon, which remember is only a hypothetical quantum of massless energy, is subject to gravity's influence. We first mistakenly assumed that a photon is a particle. And then we incorrectly reason that because it's in motion it must have momentum. If it has momentum, it must have inertial mass. And then because of relativity's principle of equivalence, if it has inertial mass it must also have gravitational mass. And if it has gravitational mass, it must then be affected by gravity.

We're highly motivated to retain this convoluted logic because if we use light's refracted slowing like Einstein, we're abruptly confronted with the total collapse of relativity, which is wholly dependent on light's constancy. Incredibly, Einstein actually agrees that relativity would completely unravel if it were found that light's velocity was not fixed but variable.

(15.1.1 Refraction 5a)

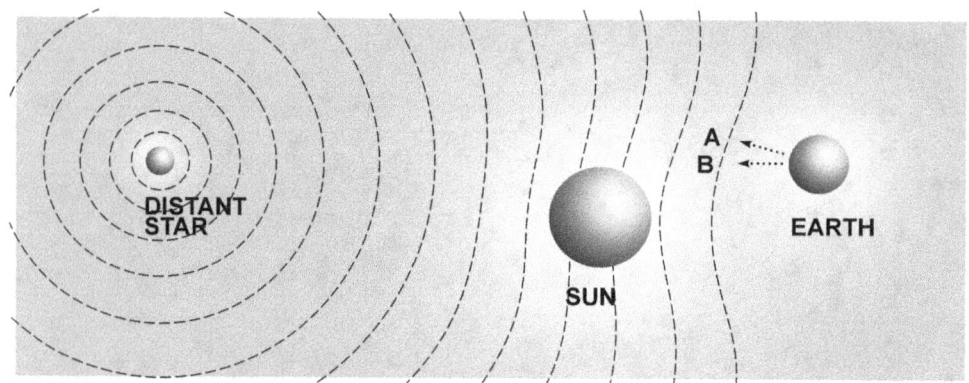

LIGHT'S REFRACTION

Light refracts through gravity fields. The distant star appears displaced in the direction of **A** not because light rays follow the impossible curvature of two-dimensional space or a resultant optical illusion but because the light in that direction reaches us slightly before the light coming directly straight from the star in the direction of **B**. Light's velocity slows through the decreasing density of the Sun's gravitational field, depicted in section as the diffusing background, just as any wave travels slower through a less dense medium, as portrayed by the series of circular and waving dashed lines that indicate the varying velocity of the incoming light emanating from the distant star.

It's also light's refracted slowing that's responsible for the gravitational lensing of distant galaxies or quasars that are split into two or more images that are assumed to be the product of the mass of some unseen foreground galaxy that's closer but fainter. But more often than not, it's just the common center of mass of any number of galaxies or galaxy clusters that is located between us and the object along its line of sight that's responsible for the lensing effect, which is why the refracting mass is so often never identified.

(15.1.2 Refraction 5a)

Rotating Bodies

Still constrained by relativity's foundational premise, light's constancy, which is conceptually impossible but according to Einstein forces time dilation, length contraction, and the increasing mass of accelerating objects through the relativistic effect of special relativity,[86] he reasons that for any rotating body, like a flat circular disk similar in shape to a phonograph record or the base of a merry-go-round, its perimeter would contract while time's rate at its perimeter would slow and the disk's radius would remain constant. Time's rate at its center would remain unchanged. But its rate would slow incrementally from its center out.[87]

He contends that its fixed radius would yield a larger value for π (pi, the ratio of a circle's circumference to its diameter, which is always 3.14): "If, then, the observer first measures the circumference of the disc... then the diameter... on dividing the one by the other, he will not obtain as quotient the familiar number π = 3.14..., but a larger number."[88] Your initial reaction might be to question whether he's really serious. Maybe he's just toying with us, seeing if we're paying attention. Mine certainly was.

A disk with a circumference that contracted with rotation but somehow maintained a constant radius is of course conceptually impossible. It can't work. Also, it wouldn't produce a larger value for π. It'd be smaller. A circumference of 3.14 divided by a diameter of 1 equals 3.14. A smaller circumference of let's say 2.14 divided by a diameter that theoretically remained unchanged at 1 would equal 2.14.

But wait. Isn't special relativity's time dilation and length contraction excluded from all rotating objects? "[His] basal principle, which was the pivot of all our previous considerations, was the *special* principle of relativity."[89] He qualifies it as, "the *principle of relativity* (in the restricted sense)." It applies only to Galilean reference frames in uniform motion: "natural phenomena run their course... according to exactly the same general laws... [for] uniformly moving co-ordinate system[s] devoid of rotation."[90] "With reference to... (Galileian reference-bod[ies]) the laws of nature... should be exactly equivalent... provided that they are in a state of *uniform rectilinear and non-rotary motion*... In this sense we speak of the *special* principle of relativity."[91]

So whether it's for particles, everyday objects like a rotating phonograph record, massive bodies like planets or stars or galaxies, or anything else that's rotating, he's asserting that special relativity can't apply. But he employs it anyway.

86. Einstein, *Relativity*, 32-46, 49-54.
87. Einstein, *Relativity*, 90-91.
88. Einstein, *Relativity*, 91.
89. Einstein, *Relativity*, 67.
90. Einstein, *Relativity*, 16.
91. Einstein, *Relativity*, 68-69.

But what's even more fundamental, remember light's velocity cannot be fixed. It's conceptually impossible. It compounds with motion. And it's variable. Also, "time" itself does not exist. It's not an independent property of the universe. Nor can it change with motion. That's only possible in the theoretical domain of mathematics. So all concepts or outcomes rooted in relativistic effects are invalid from the start.

And even if time could change with motion, its rate would not be slowing. It'd be increasing. A time that corresponds to light's reduced velocity and length's contraction so that its constancy is maintained when its source is in motion would yield a contracted rate of time. A contracted rate of time would be a faster running time. So again, the whole notion is conceptually flawed from the beginning, just like the rest of relativity.

But for the purpose of discussion, let's go ahead and assume for now that light's velocity can remain constant and that time does exist and can slow with motion. The disk is always going to be one continuous reference frame. It doesn't matter whether you're an outside observer or not. It'd be the same single reference frame for both.

So time's slower rate at its edge would cause an unresolvable conflict with light's unchanging velocity in the radial direction perpendicular to its tangent, or at any angle, where there'd be no or less contraction and motion. This would cause light to exceed 186,000mi/s in that direction.

Also because the disk's rotation progressively slows toward its center, light's tangential velocity at the edge would conflict with its tangential velocity at every other location on the disk. Or time's rate would have to increase correspondingly toward the disk's center to maintain light's fixed velocity. Einstein actually acknowledges this: "Thus on our circular disc... a clock will go more quickly or less quickly, according to the position in which the clock is situated (at rest)."[92]

But a single reference frame with multiple rates of time is conceptually impossible. He knows this: "Every reference-body (co-ordinate system) has its own particular time."[93] A reference frame with a different rate of time in different directions is just as unfeasible. But he refuses to address it. He recognizes the difficulty, "but I do not wish to go any further into this question."[94] He brushes the issue aside as if it's not pertinent but an unnecessary diversion.

These same conflicting conditions would apply to a rotating sphere. Every location between the equator and the poles is of the same reference frame. But there's no rotation at the poles. This would cause light's velocity to impossibly

92. Einstein, *Relativity*, 90.
93. Einstein, *Relativity*, 31.
94. Einstein, *Relativity*, 90.

vary over the entire sphere while exceeding 186,000 mi/s. Or time's rate would have to impossibly vary at every location. Neither is a viable option.

And if the sphere were our Earth, that varying rate of time would have to somehow be compounded with time's slowing from its orbital motion along with the orbital motion of our solar system in our galaxy and our galaxy's motion through the universe. Try imagining the consequences of all that.

A simple experiment that would confirm the impossibility of time's changing rate with rotation, or any motion, due to light's constancy, would be to synchronize two atomic clocks. Then place one at one of the Earth's poles and the other at the equator. Position both at the same elevation in the same environment. Then wait and see if there's any difference after a period of time. We can be certain they'd read the same.

The only complication would be compensating for the additional charge imparted to the clock's cesium atoms. The longer distance traveled at the equator through the Earth's magnetic and gravity fields would increase the atoms' size and mass that would in turn slightly reduce their natural frequency, causing a minute slowing the clock's rate of operation.

Invoking his "principle of equivalence" again, Einstein also contends that a person standing at the edge of the rotating disk (or on the surface of a rotation sphere) would feel the outward pull of its centrifugal force as if it were gravity. So it must be gravity. Like with acceleration/braking, he actually believes that rotation creates a real gravity field.[95] But the centrifugal force's outward thrust also only mimics gravity's apparent attraction to someone unable to perceive or measure the actual effect. Also just like with acceleration, it's not anywhere near the same.

Rotation's centrifugal force acts outward. And it acts in only two dimensions, perpendicular to the rotation's axis. Natural, mass-created gravity acts inward, three-dimensionally, spherically, radially, toward the center or common center of every mass.

A centrifugal force becomes stronger with distance. Natural gravity becomes exponentially weaker with distance. A centrifugal force disperses objects outward. Real gravity continuously coalesces and condenses objects inward toward a center of mass or common center of mass in a ceaseless runaway process that perpetually recycles them back into their primordial state of radiant or plasma energy (electrically charged particles in a quasi gaseous state).

A centrifugal force's reaction is mechanical, essentially instantaneous. Convention and Einstein hold that gravity acts via waves at the speed of light by

95. Einstein, *Relativity*, 88-89.

some kind of force similar to electromagnetism.[96] It's also thought to be mitigated somehow by graviton particles that are believed to exist physically without mass.[97] (If they actually did exist, they'd have mass. So they wouldn't be able to act at the speed of light. Relativistically, they'd become infinite.)

The centrifugal force doesn't require mass. Real gravity does. The centrifugal force requires rotation. Natural gravity doesn't require rotation or any other motion. The centrifugal force becomes stronger as rotation increases. Gravity doesn't. How can rotation-created gravity be real when according to Einstein the rate of an object's rotation or whether it even has any rotation is a subjective choice of each observer? They'd be imparting it with or withdrawing its gravity.

The centrifugal force of a rotating body would vary from zero at its poles where there's no rotation to its maximum at its equator where rotation would be the fastest. So centrifugal-created gravity would vary at the surface over the entire body while natural, mass-created gravity is the same at the surface over the entire body.

The distortion that each imparts to a body is completely different. Rotation's centrifugal force acts two-dimensionally while dispersing outward. Gravity's distortion acts three-dimensionally and condenses inward.

If the relativistic effects of special relativity, time's slowing, length contraction, the increasing mass of accelerating objects, can only manifest outside of gravity fields where light's velocity is no longer variable but constant, as Einstein insists,[98] how can they occur for any rotating body? They'd have a centrifugal-created gravity field. And according to his "principle of equivalence," it's the same as natural, mass-created gravity. So it'd have to be causing the same variability in light's velocity that nullifies relativistic effects.

Einstein never suggests that the accelerating rotation of a flat disk, or of a celestial body, or of a particle would generate a gravity field opposite the direction of its rotation or spin. If it did, that'd mean, according to his principle of relativity, that for any spherical body with linear acceleration that also had increasing rotation, it'd be experiencing a slowing in its overall rate of time and a decrease in its diameter in the direction of its linear motion. It'd have acceleration-created gravity that acted opposite the direction of its linear motion along with a relativistic increase in its mass that'd be increasing its real gravity.

Its increasing rotation would also be decreasing its circumference while slowing its rate of time that would vary from zero at its poles to its maximum at its equator. Time's variable rate from increasing rotation would conflict with time's slowing rate from linear acceleration.

96. Einstein, *Relativity*, 72.
97. "Graviton," Wikipedia, last modified Nov 29, 2022, https://en.wikipedia.org/wiki/Graviton.
98. Einstein, *Relativity*, 85, 104, 109, 171.

The body would also experience increasing centrifugal-created gravity along with acceleration-created gravity that acted tangentially opposite its increasing rotation that would vary from zero at its poles to its maximum at its equator. Both of which would conflict with its natural, mass-created gravity and the gravity created by its linear acceleration and that would also be relativistically increasing its mass. Rotation's increasing rate would be relativistically increasing its mass as well.

So what we'd end up with is a huge mess, six conflicting types of gravity:

- natural gravity from the body/particle's innate mass
- more natural gravity from relativistically-created mass from linear acceleration
- equivalency's acceleration-created gravity from linear acceleration
- equivalency's increasing centrifugal-created gravity from increasing rotation
- increasing gravity from relativistically-created mass from increasing rotation that varies from its center out
- equivalency's acceleration-created gravity from increasing rotation that varies from its center out

It's not difficult to understand why he may have been motivated to overlook these invalidating contradictions.

But what about gravitational time dilation? If it were actually real, how could it ever be accurately determined for massive bodies that also had any of the other conflicting types of gravity fields, particularly equivalency's acceleration-created or rotation-created gravity. Or what if they had relativistic time dilation from linear acceleration or relativistically increasing mass from linear acceleration that increased their natural, mass-created gravity? We could never know how much time dilation to attribute to each type of gravity.

The same question applies to gravitational redshift. It could never be accurately determined for massive bodies that also had increasing gravity from relativistically acquired mass from linear acceleration, or equivalency's acceleration-created or its rotation-created gravity.

All this begins with and results from Einstein's assumption that centrifugal mass (and inertial mass) and gravitational mass are all the same. For anyone willing to take an objective look, there can be no unique mass qualified as centrifugal or gravitational, or inertial. The effect of rotation's centrifugal force is also only inertia, an object's reaction to its outward thrust that doesn't at all appear or behave like gravitation. With rotation's centrifugal force also incapable of being or creating a gravity field, just like with acceleration we're again left with an erroneous "principle of equivalence." (See Diagram 13.1 Acceleration 8a, **13.2** Centrifugal 8a, **13.3** Distortion 8a & **48.1**, **48.2** SR&G 6a)

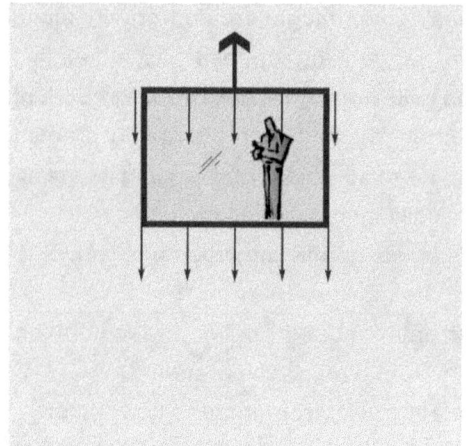 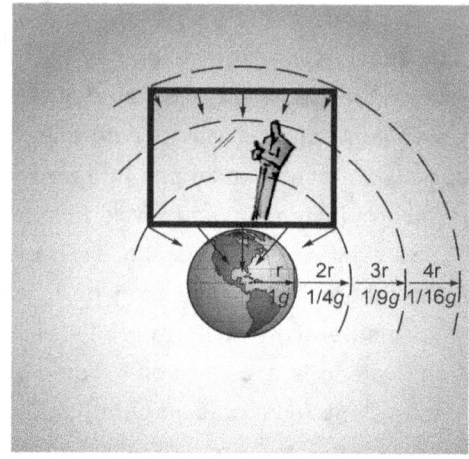

1. ACCELERATION   **2. GRAVITY**

ACCELERATION & GRAVITY

Einstein asserts that acceleration, which for him also includes braking as if a train were coming to a stop, creates an actual gravity field. But it can easily be shown that acceleration and gravity's reaction are not the same. Using an example similar to his, let's imagine that we've placed someone inside a large crate. But let's make ours transparent. We'll assume it's an independent reference frame.

For diagram **1**, when the crate is being pulled upward with a one g acceleration out in space somewhere theoretically free of gravitational fields, we can see how the reactive force, indicated with the smaller arrows that we'll say corresponds to weights on springs, has to be essentially instantaneous. It's mechanical in nature. This is contrary to Einstein's claim that gravity acts at the speed of light via waves by a force similar to electromagnetic radiation.

It's also equal everywhere throughout the crate and acts in only one dimension, opposite the direction of motion. This might initially appear or feel like gravity to the person experiencing the acceleration until they look outside or acquire the proper equipment to detect the uniform distribution of the reaction.

Now for diagram **2**, let's exaggerate our condition here on Earth and imagine that it's been compressed down to a few feet in diameter but where gravity's force remains the same at the surface. For clarity, let's keep our crate the same size and shape, compensating for any distortion that would also reveal that gravity is entirely different from acceleration.

Because the strength of a gravity field always dissipates exponentially per the inverse square law and because it always radiates spherically, three-dimensionally, from the center of a mass, gravity's force, as indicated by the varying length and direction of the smaller arrows that still corresponds to weights on springs, has to vary in strength and direction at every location within the crate as it rests on the Earth's surface.

Gravity coalesces. Acceleration doesn't. Gravity acts inward three-dimensionally and increases/decreases exponentially. Acceleration acts only one-dimensionally and uniformly. A cursory comparison easily reveals how gravity can never be created by acceleration, which completely undermines Einstein's principle of equivalence.

(13.1 Acceleration 8a)

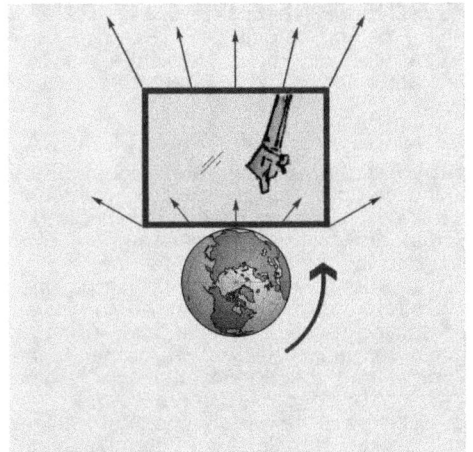 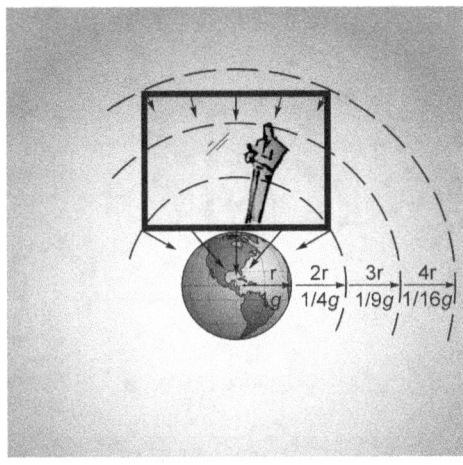

1. CENTRIFUGAL FORCE **2. GRAVITATION**

ROTATION & GRAVITY

Einstein's inference of equivalency with gravity isn't limited to acceleration and braking. He actually believes that rotation creates gravity as well. But this too can easily be shown to be false. Rotation's centrifugal force and gravity are not anywhere near the same. Neither are rotation and acceleration. For our discussion, let's continue to use our reference frame that's similar to his and imagine that we've put someone inside a large transparent crate. But let's say that it's attached to the Earth that we've again compressed down to a few feet in diameter.

For diagram **1**, if we were to eliminate the Earth's gravity and set its rotation rate to where the centrifugal reaction approximated gravity's, we can see how that reaction, as indicated with the smaller arrows that could still correspond to weights on springs, would again be mechanical, essentially instantaneous. Which is still contrary to Einstein's assertion that gravity acts at the speed of light via waves.

Its reaction would vary throughout the crate like gravity's but only in two dimensions, perpendicular to the axis defined by its rotation. It would also act outward from the axis where it'd be nonexistent, becoming increasingly stronger farther out. It might appear like gravity to the crated person unless they could see outside or had the equipment to detect the reaction's opposite direction and two-dimensional dispersal.

For diagram **2**, let's again keep gravity's force the same for our shrunken Earth and let's eliminate its rotation for clarity. Let's also keep the crate's same size and shape as it rests on the surface, omitting any distortion for now for simplicity. The difference in the rotating crate's distortion, just like for the accelerating crate, would also reveal that they don't create gravity.

Because the strength of gravity always dissipates exponentially, radially, three-dimensionally, from the center of any mass, gravity's force, as indicated by the varying length and direction of the smaller arrows that could also represent weights on springs, would vary in strength and direction but at every location within the crate, three-dimensionally, not two-dimensionally. And its reaction would be inward, opposite of the centrifugal reaction, and weaken from the center out.

Gravity coalesces and increases inward, three-dimensionally. Centrifugal forces disperse and increase outward, two-dimensionally. Acceleration neither coalesces nor disperses and it acts uniformly in only one dimension. None are the same. This simple analysis clearly shows that rotation and acceleration do not create gravity, which again invalidates Einstein's equivalence principle.

(13.2 Centrifugal 8a)

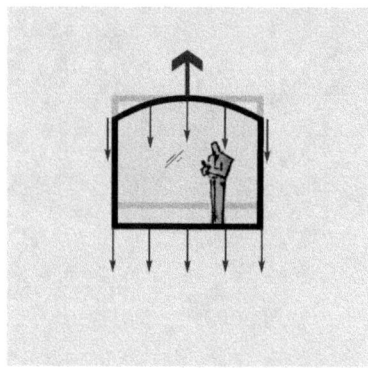

1. ACCELERATION/BRAKING

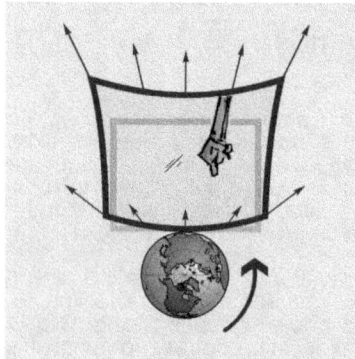

2. CENTRIFUGAL REACTION

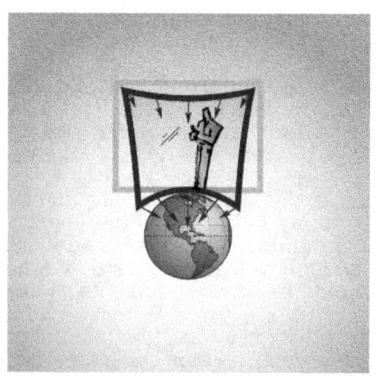

3. GRAVITY

(13.3 Distortion 8a)

REFERENCE FRAME DISTORTION

To continue the argument, let's examine our reference frame's distortion conceptually when accelerating and rotating and compare the results to gravity's.

Imagine that our original crate is now theoretically no longer rigid but made out of some pliable rubber-like material that consistently stretches and compresses to some degree when subjected to external forces. The result indicated by the darker distorted crate that's superimposed over the original rectilinear crate that's lighter gray.

Diagram **1** shows acceleration/braking's one-dimensional stretching. If our crate was being pulled uniformly from the top, not from a single point as indicated, it'd remain rectilinear. If it was being pushed evenly from the bottom, it'd also remain rectilinear. But it'd be uniformly compressing.

Diagram **2** portrays the two-dimensional outward expanding diffusion of rotation's centrifugal reaction. Diagram **3** depicts the three-dimensional inward compressing of gravity's innate coalescing. It's impossible for them to produce the same distortion. This indicates that they are not at all experiencing the same force.

Try to imagine the result if you had a rotating reference frame that was subject to linear acceleration. Not only would its distortion from acceleration and rotation impossibly conflict with each other, but they'd also conflict with that of real gravity, which is always present for any quantity of mass, including that of subatomic particles.

They innately have mass and spin. So they have gravity and centrifugal reactions. And they're routinely accelerated. If equivalency were actually real, what would the physical effect be of commingling their compaction, diffusion, and stretching all at the same time?

This again easily establishes that acceleration and rotation do not and cannot create gravity fields. So Einstein's principle of equivalence is a fallacy.

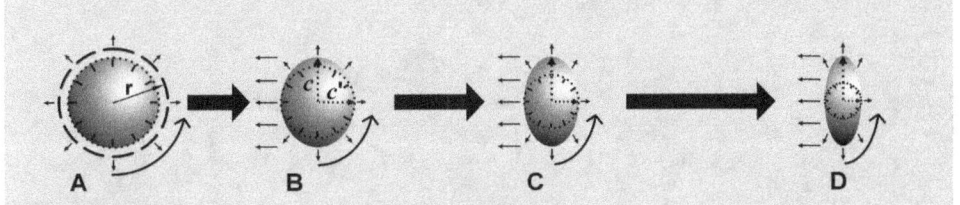

1. RELATIVITY

SPECIAL RELATIVITY & GRAVITY

Let's say we could place an object like a heavenly body somewhere out in Einstein's two-dimensional curving but uniform space and let's superimpose a clock face on it that allows us to monitor time's changing rate and then let's give the object some rotation, as implied by the radiused arrows. He asserts that because light's velocity is fixed, the object's perimeter contracts as its time slows while its radius remains the same, depicted in top view at diagram **1.A** where the dashed circle indicates the object's original size.

If the radius of a circle is always proportional to its circumference, just how does that work exactly? Also, the velocity of its rotating perimeter varies from its equator to zero at its poles. This would yield a varying rate of time over the object's entire surface. And what about the velocity of light perpendicular or at any angle to a tangent at its surface? That would also give it an endless number of conflicting rates of time.

Einstein also claims that in addition to its self-gravity, the centrifugal force from the object's rotation creates another gravity field that acts outward, as denoted by the smaller outward pointing arrows. Now let's imagine we could give the object an increasing push that imparts some constant acceleration. He also contends that this would create another gravity field that acts opposite the direction of travel, as indicated by the array of small arrows at **1.B**, **1.C** & **1.D**, despite that its reaction is uniform and gravity's is nonuniform. So we have three types of conflicting gravity fields, one from self-gravity that coalesces three-dimensionally, one from rotation that disperses two-dimensionally, and one from acceleration that acts uniformly in only one dimension that neither coalesces nor disperses.

He also maintains that the moving object's light has to slow in the forward direction to maintain its constancy, as indicated by the dotted arrow at c', that becomes infinitely slow at 186,000mi/s, and the object's time has to slow or dilate concurrently, as portrayed by the shrinking clock-like circle that represents its changing rate, that also becomes infinitely slow at the speed of light, and the object has to contract correspondingly in the direction of motion, as depicted at **1.B**, **1.C** & **1.D**, that becomes infinitely small at 186,000mi/s as well. But how can time be slowing when light's slower forward velocity and the object's contracted size would require a smaller, faster running time to maintain light's constancy? And what about the light perpendicular to the object's linear motion, at c, or in any direction except directly forward? It's not required to slow to maintain light's constancy but is still subject to time's slower rate. This creates a conflicting velocity that exceeds 186,000mi/s.

Also because of light's fixed velocity, Einstein asserts through his mass-energy relation, $E = mc^2$, that an object's mass increases with its acceleration, becoming infinitely large at the speed of light. How can that be if light's fixed velocity also requires its contraction in the direction of motion to become infinitely small at the speed of light?

Beyond that, he actually avows that because of light's variability in gravity fields all of special relativity, light's constancy, time's slowing, an object's length contraction, and its ever-increasing mass, is only valid outside of gravity fields. How can any of it ever be valid when every object generates its own self-gravity field, and every rotating or accelerating object generates another gravity field? And let's not forget the gravity fields from every other object in the universe. They extend indefinitely. So there's no place where light's velocity can be fixed. But his assertion of light's variability itself undermines nearly all of relativity. It's founding premise is light's constancy.

(48.1 SR&G 6a)

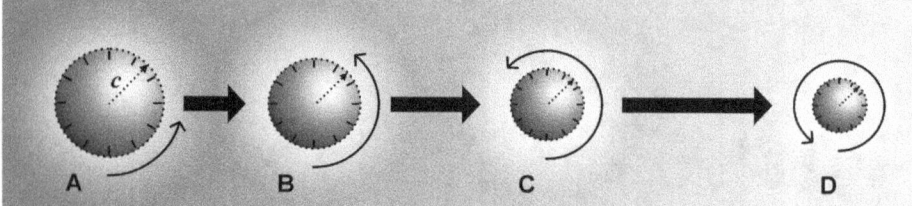

A   B   C   D

2. REALITY

SPECIAL RELATIVITY & GRAVITY - CONTINUED

You have to also remember that Einstein asserts that it's a completely subjective choice by each observer as to which object is in motion and their relative rate of motion. That even includes heavenly bodies subject to gravity. So right now I can choose that the Moon is no longer orbiting around the Earth but that the Earth is orbiting the Moon. And this is perfectly acceptable to him despite that my choice conflicts with the laws of gravitation and everyone else's choice, which is also deciding light's velocity, time's rate, an object's contraction in the direction of motion, and the amount of mass it has.

But there is a real-world way that an object in motion can contract and its rate of time can change while light's velocity appears to remain constant. Let's take our same object and instead of placing it in Einstein's nonexistent space that impossibly curves two-dimensionally, let's put it in a real gravity field, the universal field of radiant energy of the electromagnetic spectrum that innately diffuses inward exponentially at each object, as represented in section view by the varying background in diagram **2**. Objects naturally gravitate toward one another as they mechanically seek equilibrium in its decreasing density that's always at its least directly in between them or at their common center of mass for multiple objects.

Pushed by the higher density, the rotating object begins accelerating toward the lessor density while contracting in size in a nearly uniform omnidirectional manner, as portrayed in top view by the sequence **2.A**, **2.B**, **2.C** & **2.D**. Its contraction increases its natural frequency, which increases its rate of rotation, as indicated by the increasing length of the radiused arrows. If its rate of rotation is used to establish its rate of time like the Earth's is for us then it could be said that its rate of time is increasing with motion, as the contracting clock face implies. Light's velocity also slows correspondingly in the gravity field's diffusing density, as suggested by the decreasing length of the dotted arrows at c. This renders its variability along with the object's contraction imperceptible for an experiencer, which maintains a normal Euclidean environment.

(48.2 SR&G 6a)

Gravitational Redshift

Einstein is generally credited with being the first to describe gravitational redshift. But his version is completely different from what's generally accepted now. The most common explanation begins with the assumption that a photon's escape from a massive body is impeded by gravity. If it's affected by gravity then it must have gravitational mass.

A photon doesn't innately have mass. But it's thought that its velocity gives it inertia. So it must have inertial mass. Then because of Einstein's "principle of equivalence," its inertial mass is gravitational mass. They're the same. That's how photons can be affected by gravity. (Note the reliance on his debunked "principle of equivalence.")

Since they have to travel at the speed of light (because light's velocity is presumed to be fixed), their impeded escape does not decrease their velocity. Instead, they lose energy that expresses as a decrease in light's frequency, which is a longer, redder wavelength.[99] Those explanations that accept light's varying velocity in gravity fields simply reason that as it emerges from a gravity well, it is its slower velocity that reduces or redshifts its frequency.

Even though Einstein realizes that light varies in gravity fields, the stronger the gravity the slower light's velocity, his gravitational redshift is still based on nonexistent relativistic time dilation that's produced by a body's rotation that would also have nonexistent centrifugal-created gravity. That fictitious gravity is then impossibly correlated with real natural, mass-created gravity through his trumped-up "principle of equivalence."

He "reasons" that since time for, "an observer [with a] clock at the edge of [a rotating] disk... goes at a rate permanently slower than that of [a] clock at the centre [and because they're aware of] a force which acts outwards in a radial direction, and which would be interpreted as an effect of inertia (centrifugal force)... [and because] the observer believes in the general theory of relativity... The force acting on himself, and in fact all other bodies which are at rest relative to the disc, he regards the effect of a gravitational field."[100]

The observer's motion around the disk's circumference dilates their rate of time and because they're in a centrifugal-created gravity field this would establish a relationship between time and gravitational potential. Increasing rotation relativistically slows their time's rate while increasing their centrifugal-created gravity.

99. "Gravitational Redshift," Wikipedia, last modified Dec 19, 2022, https://en.wikipedia.org/wiki/Gravitational_redshift.
100. Einstein, *Relativity*, 89-90.

Then with his "principle of equivalence" that qualifies all gravity as the same, he equates centrifugal-created gravity with natural, mass-created gravity. But he also applies time's dilation created by centrifugal-created gravity's rotating motion to natural, mass-created gravity despite that the prerequisite rotation necessary for time's dilation no longer exists. He just casually asserts without explanation that time's dilation "will hold quite generally for [all] gravitational fields. Furthermore, we can regard an atom which is emitting spectral lines as a clock, so that the following statement will hold:

"*An atom absorbs or emits light of a frequency which is dependent on the potential of the gravitational field in which it is situated.*

"The frequency of an atom situated on the surface of a heavenly body will be somewhat less than the frequency of an atom of the same element which is situated in free space (or on the surface of a smaller celestial body)... Thus a displacement towards the red ought to take place for spectral lines produced at the surface of stars as compared with the spectral lines of the same element produced at the surface of the earth..."[101]

No, none of this has been misquoted. He actually expects us to believe and accept that redshift corresponds to time's relativistic dilation caused by the motion of a rotating body that also has centrifugal-created gravity. But because of his "principle of equivalence," centrifugal-created gravity is the same as natural, mass-created gravity. So the relativistically derived time dilation of centrifugal-created gravity has to be applied to natural, mass-created gravity. Nice trick.

We should go over that again. Motion, including rotating motion, causes time dilation because light's velocity is fixed. The spectral lines of atoms can be regarded as a clock. Time's dilation shifts those spectral lines toward the red end of the spectrum.

A body's rotation causes both time dilation and centrifugal-created gravity. So a relationship exists between time's slowing, indicated by redshift, and centrifugal-created gravity. The faster the body's rotation, the more dilated its time, the farther redshifted its light, the stronger its centrifugal-created gravity.

Because of equivalence, rotation's centrifugal-created gravity and nonrotating, natural, mass-created gravity are the same thing. So rotation's centrifugal-created gravity's dilated rate of time, expressed as redshift, has to apply equally to nonrotating, natural, mass-created gravity. So higher redshifts indicate stronger nonrotating, natural, mass-created gravity.

101. Einstein, *Relativity*, 147-150.

It's difficult sometimes to rationally explain irrationality, to accurately convey nonsensical ideas and how ill-conceived they are. They're inherently hard to follow because they don't make sense. And in Einstein's case, no one wants to believe they're incoherent and fallacious. But the fabricated complexity and disqualifying absurdity of this particular logic chain that's based on delusive assumptions and correlations exemplify the convoluted, illusive reasoning and the manic, fanciful nature that permeates all of relativity.

But let's go ahead anyway and try to put his version of gravitational redshift in proper context. First, time doesn't really exist. We've gone over it before. And we'll have to do it again. It's not an actual property of the universe. Also, it's conceptually impossible for light's velocity to remain constant. It compounds with motion, and it's variable.

Einstein has already conceded light's variability in all gravity fields.[102] Its constancy can only occur outside of gravity fields. If its velocity cannot be fixed in gravity fields then any attempt to establish a relationship between redshift and the strength of a gravity field that's based on time's relativistic slowing due to light's constancy, in a gravitational field, is nonsensical.

So the relativistic foundation of his gravitational redshift subverts it from the start. If light's velocity is not fixed, rotating motion cannot cause nonexistent time to slow. And if there's no such thing as time then an atom's spectral lines cannot be regarded as a clock.

Considering spectral lines as clock has no basis anyway. He deceptively asserts it as fact and declares that it's so. If time did actually exist, there's no reason why its slowing would shift spectral line, or shift them in the red direction over the blue.

The rotation of a celestial body or of a disk varies from zero at its poles or center to its maximum at its equator or edge. This means its rotation-created gravity will vary accordingly over the entire body. With an infinite number of rotation-created gravities to choose from along with its associated redshifted light, how can it ever be rationally correlated with a singular, mass-created gravity that's the same for the entire body to establish a valid gravitational redshift?

But if a relationship does exist between the displacement in the spectral lines of an atom and its natural frequency, which is perfectly reasonable to infer, we could rationally theorize several ways that a celestial body might produce a red or blue shift in its light. The natural frequency of an atom would increase when subjected to stronger gravity, a field of decreased density. Its contraction would innately produce faster oscillation, which would tend to blue shift its light.

102. Einstein, *Relativity*, 85, 109, 171.

But it's possible that its spectra might still be redshifted if it were originating from a stronger gravity field. Light's velocity propagates slower in fields of decreased density. Also, any celestial body that was rotating faster due to its contraction would be induced with an increasing charge. This would increase the mass of its atoms, causing an increase in its gravity and a slowing in its natural frequency. Both would produce a redshift. These processes would give rise to higher redshifts for more condensed bodies. This would cause them to appear to be receding with a Doppler shift. But the actual distance between them measured from their centers would remain the same.

This offers a much more feasible explanation of the redshifted displacement of an atom's spectral lines than Einstein's contradictory assertion of nonexistent time's impossible relativistic slowing from a body's rotation that he irrationally contends creates a gravity field from its centrifugal force that he illusively correlates through contrived equivalence to natural, mass-created gravity.

A rotating mass does not create centrifugal gravity. For all the reasons previously detailed, it does not and cannot exist. So if it's nonexistent then no equivalence can be drawn between it and natural, mass-created gravity. If equivalency can't be drawn between centrifugal mass (or inertial mass) and gravitational mass then his "principle of equivalence" is a fallacy as well. But even if centrifugal gravity did exist and could be associated with dilated time, indicated by redshift, there's still no basis to infer that centrifugal gravity's redshift correlates to the strength of natural, mass-created gravity.

He never attempts to explain how that might work. He can't. The correlation cannot be rationally drawn. He again just declares that it's so. That doesn't make it reality. As if dealing three-card monte, what he does (maybe subconsciously, it's hard to say) is skillfully obfuscate redshift's cause and effect. He transfers it from rotation's relativistic time dilation to natural, mass-created gravity's strength, using its inferred association with centrifugal-created gravity as the translating intermediary. If it's not intentional misdirection, it's profound delusion.

Einstein never suggests that acceleration-created gravity would redshift an atom's spectral lines. If equivalence compels all gravity to be the same, it'd have to. The greater the acceleration, the stronger the gravity, the more relativistically redshifted they'd have to be. If this were actually true, it'd have significant consequences.

But if time actually did relativistically slow for moving objects and if an atom's spectral lines could actually be regarded as a clock then the light of any moving object, whose rate of motion remember is supposedly subjectively decided by each observer, would have to be redshifted as well. Take a moment to consider the implications of that.

But even beyond that, let's say acceleration-created gravity, rotation-created gravity, the relativistic effects of special relativity, and his or conventional gravitational redshift all existed and worked as theorized. We could never know the true gravitational redshift of any accelerating or rotating body. In addition to their natural, mass-created gravity, they'd have increasing gravity from relativistically acquired mass from linear acceleration, or increasing gravity from equivalency's acceleration-created or rotation-created gravity. But we'd have no way of ever determining the source or sources of the gravity or how much gravity was originating from each. So the body's gravitational redshift would be meaningless.

The relativistic basis and delusive, metaphysical gyrations of his inferences mount to nothing more than a theoretical game of make-believe. None of it has any chance of ever actually manifesting in our real physical world of three actual dimensions.

Even so, Einstein still declares (for his version of gravitational redshift) that, "If [it's discovered that] the displacement of spectral lines toward the red by the gravitational potential does not exist, then the general theory of relativity will be untenable."[103] Can't disagree with that at all.

It'd also mean that special relativity would have to be untenable. His version of gravitational redshift is based on special relativity's effects. If his gravitational redshift does not exist that'd have to mean that special relativity's effects do not exist. If that's true then special relativity would be just as untenable as general relativity.

But on the other hand, if his gravitational redshift was found to exist then general relativity, along with special relativity, would have to be judged sound and legitimate. If that were the case, then cosmological redshift couldn't be originating solely from the universe's stretching. It'd also have to include his gravitational redshift. It'd also have to include any additional gravitational redshift from additional gravity from his relativistically created mass from linear acceleration, and additional gravitational redshift from equivalency's acceleration-created gravity and its rotation-created gravity.

Convention used to hold that cosmological redshift was produced by a Doppler shift due to the recessional velocity of galaxies from universal expansion. It's now more widely held that cosmological redshift is a consequence of space's stretching from the big bang's expansion that's displacing or lengthening light's wavelength toward the red.[104]

But with all these other potential redshift sources, we'd have no way of ever knowing how much redshift to attribute to them or cosmological redshift. So we could never know the universe's rate of expansion, or even whether it was expanding. This would completely undermine our inference of an expanding big bang universe.

103. Einstein, *Relativity*, 151.
104. "What is cosmological redshift," How Stuff Works, last access Dec 27, 2022, https://science. howstuffworks.com/cosmological-redshift.htm.

Einstein accepts that cosmological redshift from recessional velocity (or now stretching) originates from universal expansion.[105] Which means, redshift cannot be an indication of gravitational potential. Apparently, he doesn't realize that he's again cornered himself in a basic contradiction. This one completely nullifies his gravitational redshift that in turn completely invalidates relativity.

Our big bang-relativity cosmology is fundamentally self-conflicted. To preserve the big bang, we have to decide that Einstein's version of gravitational redshift is nonexistent, which would invalidate relativity. If we embrace his gravitational redshift, we then have no consistent redshift source. Inferring universal expansion, much less its rate, would be impossible.

So the entire big bang would collapse. We're forced to choose one or the other. Relativity and the big bang can't coexist. But we continue to ignore this crucial and obvious nullifying contradiction.

But there's more. Conventional, nonrelativistic gravitational redshift can't be ignored. It has to be considered as well. So does gravitational redshift from light's slowing. Both are just as conflicted with cosmological redshift. Neither of them can coexist with it either.

And they pose an even greater threat. They're more rational, and more difficult to discount, especially gravitational redshift from light's slowing. So the big bang is in even greater jeopardy. One or the other, gravitational redshift or cosmological redshift, still has to be chosen.

Remember though, conventional gravitational redshift is dependent on Einstein's "principle of equivalence," which we totally debunked. It's not legitimate. So it can be discounted. That leaves us with light's slowing in gravity fields as the singular source of gravitational redshift that threatens to undermine the big bang.

What's even more threatening, the big bang's accepted cosmological redshift from stretching directly conflicts with any other potential source of galactic redshift, i.e., Doppler effect, tired light, relativistic time dilation, the reduced frequency of atoms due to an induced charge from their motion, the Doppler shift from the recessional velocity of each galaxy's infalling material, and so on. As many as eight seem viable. There may be more. The Doppler redshift from the recessional velocity of the continuously coalescing infalling material at each galaxy's due to gravity's runaway nature is the most rational.

If even one other galactic redshift source were found to be legitimate, we'd be facing the same difficulty. We could never know how much redshift to attribute to stretching and how much to the other source.

105. Einstein, *Relativity*, 152-154.

But even stretching itself poses a potential conflict. It innately produces recessional velocity. Recessional velocity has an associated redshift from its Doppler shift. So even if we discount all other potential redshift sources, we're still left with two types of redshift for every galaxy, one from stretching and the other from recessional velocity.

And what about the smaller number of galaxies that exhibit a blueshift? Are we to believe that the universe is contracting in the direction of those galaxies for some unknown reason? How is this question to be resolved? It would seem that one way or another relativity and the big bang are both doomed by their inherent contradictions.

Michelson-Morley

The Michelson-Morley experiment attempted to confirm the existence of an aether by comparing light's velocity in the direction of the Earth's orbital motion to the perpendicular direction. It essentially consisted of pairs of mirrors perpendicularly placed on a rotating table a fixed distance from a central beam splitter in a cross fashion such that a recombined beam of light aligned with the direction of the Earth's orbital motion would reveal any difference in velocity through an interference pattern. It unexpectedly found no difference. So it failed to establish the aether's existence.[106]

As an explanation of the result, Einstein argues that it's Lorentz and George FitzGerald's (Irish physicist, 1851-1901), "[assumption] that the motion of the body [the Earth] relative to the aether produces a contraction of the body [the Earth] in the direction of [orbital] motion, the amount of contraction being just sufficient to compensate for the difference in time... But on the basis of the theory of relativity the method of interpretation is incomparably more satisfactory."[107]

He contends that length contraction and time dilation does not occur to satisfy the notion of an aether, as they propose, but to conform to his theory of relativity. But for either case, the rate of an object's time cannot change to enforce light's incorrectly presumed constancy just because of some assumed relative motion.

There's no such thing as time. It's not a property or attribute of the universe. We ourselves establish "time" and its rate through our selection of objects with periodic motion that we then use as a reference. Any variation in its rate has to remain theoretical applied through mathematics, unless the periodicity of our selected object changes. That's the only way time's rate can change.

Since time's rate cannot change with an object's motion, they cannot be contracting in the direction of their motion and the speed of light cannot be fixed. It must compound with the motion of its source just as Michelson-Morley (and Sagnac) demonstrate.

Also, the relativistic effects of special relativity, time dilation, length contraction, and the increasing mass of accelerating objects, can only manifest outside of gravity fields where light's velocity is constant.[108] So the Earth could not be contracting in the direction of motion or its time slowing to maintain light's fixed velocity. It's very much under the influence of all kinds of gravity where light's velocity is variable, most immediately its own self-gravity. But it's also under the influence of the self-gravity of the solar system, and our galaxy, and the entire

106. "Michelson-Morley Experiment," Wikipedia, last modified Dec 26, 2022, https://en.wikipedia.org/wiki/Michelson-Morley_experiment.
107. Einstein, *Relativity*, 58-60, 167-168.
108. Einstein, *Relativity*, 85.

universe (especially if it's presumed to be finite), along with the gravity of every other object in it. Gravity's influence extends indefinitely.

The very notion of an object contracting but only one-dimensionally when in motion, especially the Earth, is absurd enough on its own to warrant its outright dismissal. But let's go ahead and carefully examine what this contraction would mean for the Michelson-Morley experiment if it really could occur.

Because the experiment was concerned with the effects resulting from the Earth's orbital motion, it's implicit that Einstein is referring to that orbital motion as its chosen system of reference that produced the results of the experiment: "Thus for a co-ordinate system [a reference frame] moving with the earth the mirror system of Michelson and Morley is not shortened, but it *is* shortened for a co-ordinate system [a reference frame] which is at rest relative to the sun [stationary with respect to the solar system]."[109]

Remember, he insists that all motion is relative. Which reference frame is moving and which isn't and the rate of that motion is a strictly personal choice, one that can be made by each observer.[110] That's the whole point of "relativity."

But wouldn't the Earth's contraction resulting from its rotation, or from its other motions, our solar system's motion through our galaxy and our galaxy's motion through the universe, have to also affect the outcome of the experiment? They'd have to. There's no way to eliminate them, unless of course you believe as Einstein does that someone could decide beforehand that they aren't applicable.

His "reasoning" taken to its logical conclusion could actually have anyone arbitrarily deciding that the Earth is no longer spinning, but that the Sun and the rest of the entire universe are now rotating around the Earth. Needless to say, that's over the top ludicrous. But Einstein wouldn't have a problem with it. He sees, or at least he presents, his nonsensical ideology as serious science.

He does seem to quickly slip in what appears to be a (subverting) qualifier that, like so much else of relativity, is inherently contradictory: "only experience can decide as to [the] correctness or incorrectness [of an object's motion.]"[111] Well, is he now actually saying that motion is not relative after all, nor is it subjectively decided by each observer, but that there exists an absolute reality for all observers? Looks like it.

If this is correct, then there can be no such thing as relative motion or an individual's ability to chose a reference frame's motion. This would fundamentally conflict with the very basis of relativity. The whole reason the theory acquired its name is because its underlying formation is the subjectivity and "relativity" of motion.

109. Einstein, *Relativity*, 60.
110. Einstein, *Relativity*, 16-18, 30, 67, 68...
111. Einstein, *Relativity*, 68.

Of course, we know that our personal preferences of motion have no validity. All motions have to be in play. And if motion actually causes contraction in the direction of motion, as Einstein insists, then the Earth would have to be contracting from all of them. But to simplify our argument, we'll limit our scrutiny to just its orbit and rotation. Both would have to be reducing the physical dimensions of the experiment's equipment, compensating for light's fixed velocity and time's slowing rate.

So, if the Earth really was contracting in the direction of motion, then in the direction of its orbital motion, at its equator and everywhere else latitudinally (horizontally), it would acquire the shape of an ellipsoid that had its shortest axis always pointing in the direction of its orbital motion. This means that the construct of the experiment would have been contracting and expanding latitudinally (horizontally) because of its rotation around the ellipsoid-shaped Earth.

But it would have also been contracting latitudinally (horizontally) without fluctuating due to the Earth's rotation. The combined effect would have produced a constant dimension in the equipment in the longitudinal direction (vertically) where there would be no contraction because there's no motion in that direction. But there would have been a continually fluctuating latitudinal (horizontally) measurement that would have been shortest at noon and midnight, and longest at dawn and dusk, along with a minor fluctuation in the height of the equipment and its distance from the ground that would be largest at noon and midnight that wouldn't affect the results.

All of this fluctuation in the measurement of the equipment would have been occurring while time's slower rate remained constant because of its constant orbital and rotational motion. This should have produced a varying, periodic interference pattern. But none was observed. Which can only mean that the Earth is not contracting in the direction of its orbital or rotational motion. And if there is no contraction then light's constancy, along with time's varying rate, must be a fallacy. Its velocity must then compound with motion. (See Diagram 1.1, 1.2, 1.3, 1.4 M-M Exp 6a & 16.1, 16.2 MM 7a)

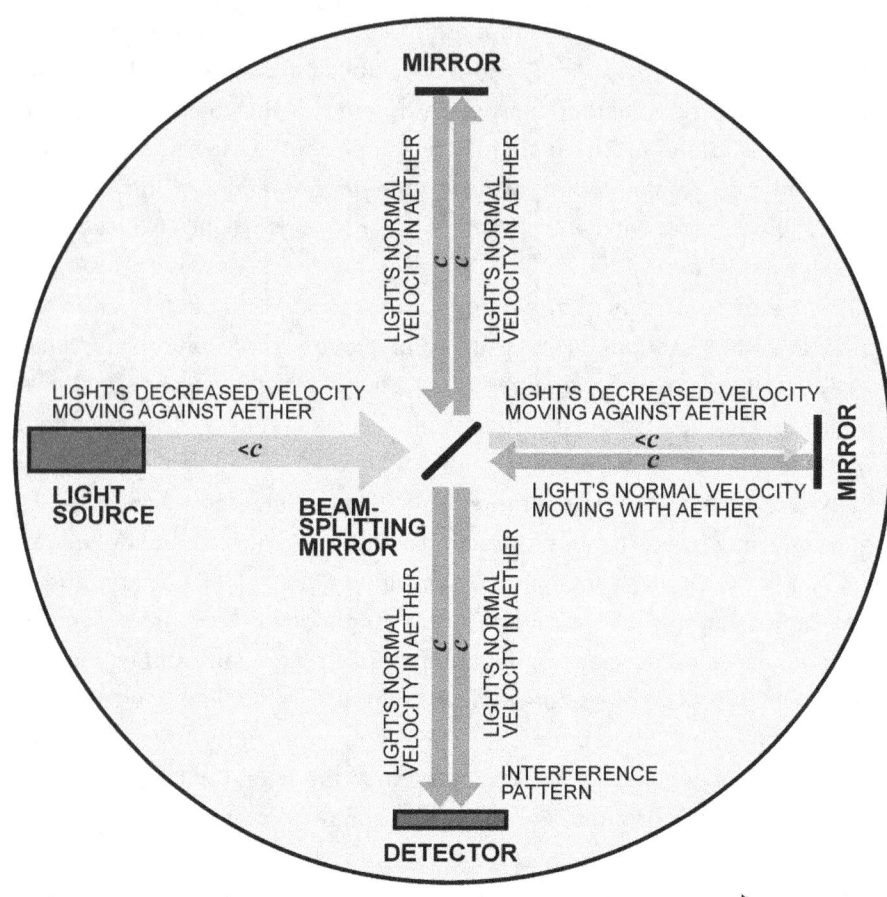

DIRECTION OF EARTH'S ROTATION & ORBITAL MOTION ➡️

MICHELSON-MORLEY - CONCEPTUAL DIAGRAM
EXPECTED RESULT
The experiment essentially consisted of a light source projected onto a series of mirrors arranged perpendicular equal distances from a central beamsplitter mounted on a rotating table oriented with one beam projected in the direction of the Earth's orbital motion and the other perpendicular. When the light was recombined, it was expected to produce an interference pattern due to its decreased velocity from the theorized aether "headwind." This would confirm the aether's existence. But no interference pattern was found.

(1.1 M-M Exp 6a)

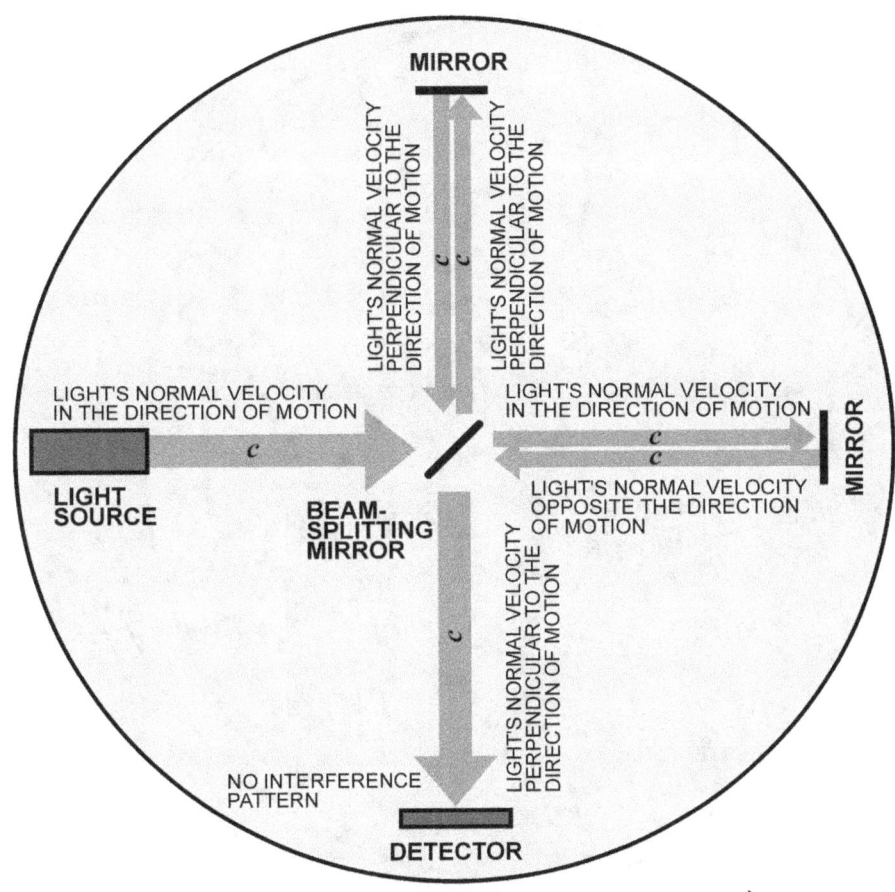

DIRECTION OF EARTH'S ROTATION & ORBITAL MOTION ➡️

MICHELSON-MORLEY - CONCEPTUAL DIAGRAM
ACTUAL RESULT

What the experiment actually showed is that light always leaves its source at 186,000mi/s in every direction at the same time as we'd naturally expect. This indicates its compounding with the motion of its source and implies its compounding with other reference frames. Which means that because everything's in motion, its velocity can never be fixed at 186,000mi/s but will always be some slower or faster rate that can be any velocity up to instantaneous.

If someone was out in space stationary with respect to the solar system, they'd be in a different reference frame recording a compounding of light's varying velocity, which is determined by the field density at their location, plus/minus the Earth's rotational and orbital velocity or some vector angle of it.

(1.2 M-M Exp 6a)

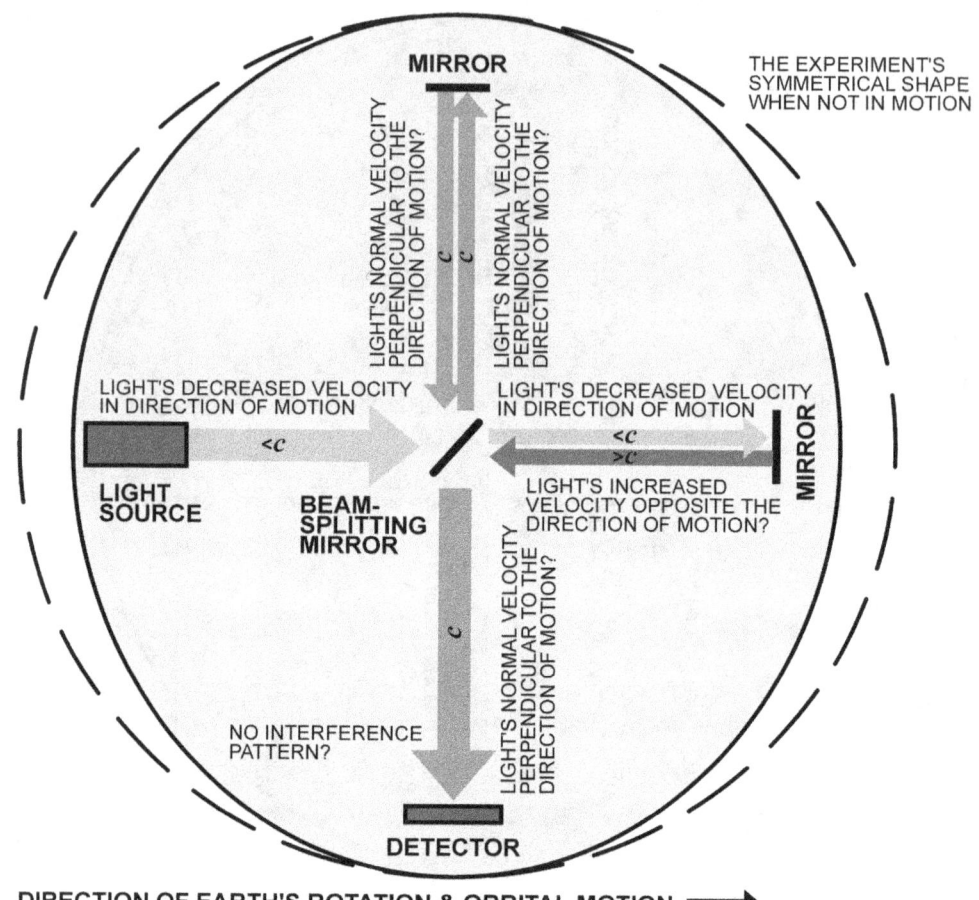

THE EXPERIMENT'S
SYMMETRICAL SHAPE
WHEN NOT IN MOTION

MIRROR

LIGHT'S NORMAL VELOCITY PERPENDICULAR TO THE DIRECTION OF MOTION?

LIGHT'S NORMAL VELOCITY PERPENDICULAR TO THE DIRECTION OF MOTION?

LIGHT'S DECREASED VELOCITY IN DIRECTION OF MOTION

LIGHT'S DECREASED VELOCITY IN DIRECTION OF MOTION

$<c$

$<c$
$>c$

LIGHT SOURCE

BEAM-SPLITTING MIRROR

MIRROR

LIGHT'S INCREASED VELOCITY OPPOSITE THE DIRECTION OF MOTION?

LIGHT'S NORMAL VELOCITY PERPENDICULAR TO THE DIRECTION OF MOTION?

NO INTERFERENCE PATTERN?

DETECTOR

DIRECTION OF EARTH'S ROTATION & ORBITAL MOTION ➡

MICHELSON-MORLEY - CONCEPTUAL DIAGRAM
LORENTZ'S EXPLANATION ADOPTED BY EINSTEIN

What Lorentz proposed to explain the result, and Einstein later adopted for relativity, was that objects contract in the direction of their motion while time slows to maintain light's fixed velocity so that an interference pattern is not produced. But what happens to the experiment for the reflected light moving opposite the direction of motion? To maintain light's constancy wouldn't it have to expand while time's rate increased? And how about in the perpendicular direction? Without any contraction wouldn't time's "slower" rate cause light's velocity to exceed 186,000mi/s? These irresolvable conflicts confirm that light's velocity cannot remain fixed in our real nontheoretical environment of three actual dimensions but must compound with motion, which invalidates any Lorentz contraction and undermines nearly all of relativity.

(1.3 M-M Exp 6a)

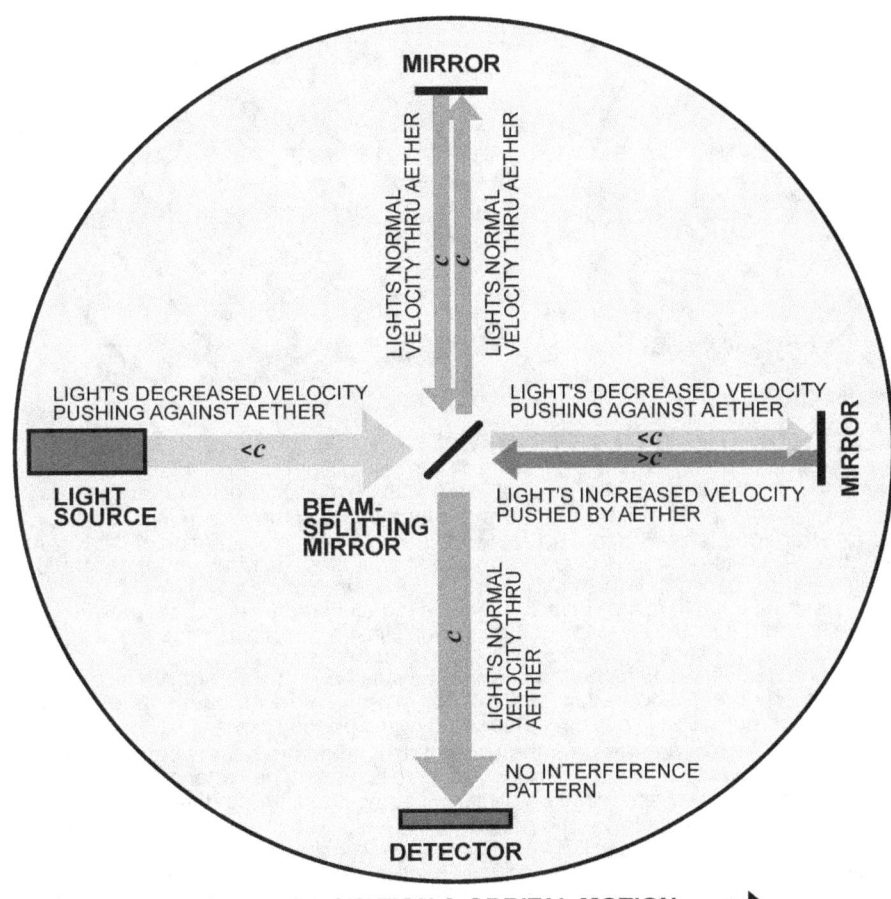

DIRECTION OF EARTH'S ROTATION & ORBITAL MOTION ➡️

MICHELSON-MORLEY - CONCEPTUAL DIAGRAM
AETHER EXPLANATION
Instead of contriving the fantastic, self-conflicted notion that objects physically contract but only in the direction of motion while time's rate slows to maintain light's constancy, why wouldn't you simply reason that light's velocity is first slowed by the aether's "headwind" then is increased by the same amount from its "tailwind" after it's been reflected backward? It's not the correct explanation. But at least it's rational.

(1.4 M-M Exp 6a)

135

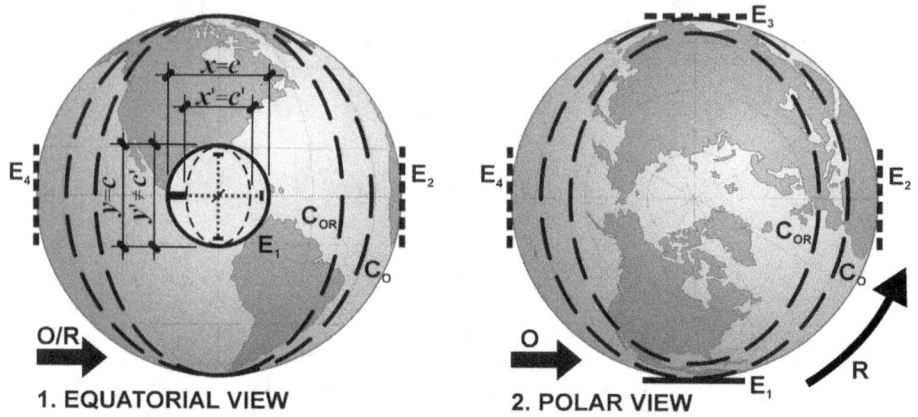

1. EQUATORIAL VIEW   **2. POLAR VIEW**

MICHELSON-MORLEY EXPERIMENT - 1

The experiment essentially consists of several mirrors arranged rectilinearly in a cross pattern around a central beamsplitter on a table that can be rotated. For simplicity, ours is at the equator, E_1. It showed that light, c, always radiates at the same rate in all directions at the same time regardless of motion. The arrow at O/R indicates the direction of the Earth's orbit & rotation. At O, its orbit. The curved arrow at R indicates the direction of its rotation. For our purposes, we can ignore the motion of our solar system and galaxy.

Einstein assumes that light's velocity is fixed. It's (special) relativity's basal premise. So objects have to contract in the direction of motion while time's rate slows correspondingly to maintain its constancy. That's how he explains the experiment's result.

So in the direction of its orbital motion, the Earth would have to contract into an ellipsoid shape, indicated by the dashed ovals at C_0. Its rotation would also cause its ellipsoidal contraction. The combination of both is represented by the dashed ovals at C_{OR}.

As the experiment revolves around the Earth, this would cause it to contract in the x direction, as implied by the dashed oval at E_1, which decreases the distance between the mirrors while time's rate slows to maintain light's constancy (assuming "time" is something that actually exists and that it slows with motion). In the y direction, there is no motion. So there's no contraction. The distance between the mirrors remains the same.

Since time's slower rate has to apply equally over the experiment (and over the entire Earth, they're of the same reference frame), it affects light's velocity, c', the same in both the x and y directions. But with contraction only in the x direction, not in the y, this creates different velocities for light. y' will always be faster than x'. This would cause different arrival times, which should have produced a negative interference pattern. But it didn't.

For the Earth's contraction from its orbital motion, the distance between the mirrors in the y direction would also remain unchanged, while the distance in the x direction would be constantly fluctuating. The Earth's rotation would pass the experiment through its orbital tangents, causing maximum contraction at E_1 & E_3 with no contraction at its perpendicular positions, E_2 & E_4, which would compound with its constant contraction from its rotation. The result would be C_{OR}. This should have also produced a negative interference pattern, but one that was expanding and contracting every twelve hours. But this didn't happen either.

Einstein also asserts that special relativity's effects can't occur within gravity fields where light's velocity varies. But the experiment rests in the Earth's gravity field, and all others. They extend indefinitely. So how can it even be considered to explain the results?

The experiment clearly demonstrates that objects do not contract in the direction of motion. Nor does (nonexistent) time slow with motion. So its velocity can't be fixed. It compounds with the motion of its source and that of other reference frames, and that's in addition to its variability. With a founding premise that's untenable, relativity has no viability.

(16.1 MM 7a)

136

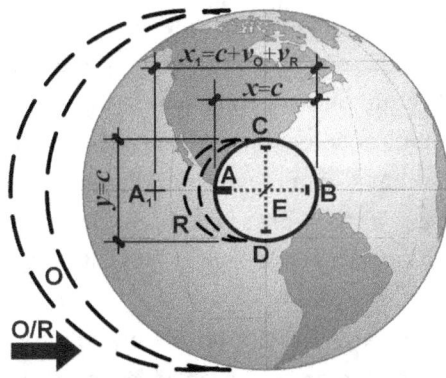

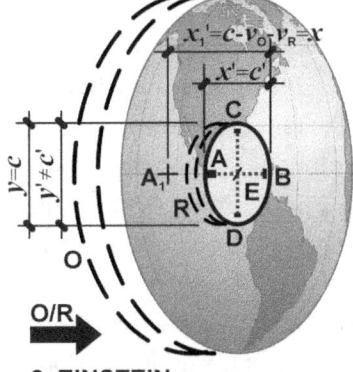

1. REALITY:
LIGHT'S COMPOUNDING
TIME IS NONEXISTENT

2. EINSTEIN:
LENGTH CONTRACTION
TIME DILATION

MICHELSON-MORLEY EXPERIMENT - 2

Light always radiates at the same rate in all directions at once regardless of the motion of its source. So someone at the experiment, **E** in diagram **1**, measuring light's velocity from **A** to **B**, distance x, and also from **C** to **D**, distance y, would get 186,000mi/s in each direction, $x=c$ and $y=c$, despite their motion. They're of the same reference frame, moving in unison with the Earth's orbit and rotation, implied by the dashed ovals at **O** and **R**.

If we positioned ourselves just outside the Earth's orbit, stationary with respect to the solar system, which is a different reference frame, we'd see that the measurement they made in the x direction actually covered a longer distance from A_1 to **B**, distance x_1. But for them, they still measured from **A** to **B**, covering the shorter distance x due to their motion.

If we measured light's velocity from **A** to **B** as the Earth sped by, our measurement would also begin at A_1 and end at **B**. But light would traverse the longer distance, x_1, in the same time it took for it to go from **A** to **B**, distance x, for the person at **E** because we're not moving with the Earth. So for us, light's velocity would have to include the Earth's orbital velocity, v_0 (66,000mph), and its rotational velocity, v_R (1,000mph). Their compounded velocities over distance $x_1=c+v_0+v_R$. That works out to be about 186,020mi/s.

With his underlying assumption that light's velocity is fixed regardless of motion, Einstein reasons that time slows for moving objects while their length contracts to preserve light's constancy. So for the Earth, light's velocity has to decrease by its orbital and rotational velocity, v_0 and v_R (excluding our solar system and galaxy's motion), while time's rate has to slow and the Earth has to contract correspondingly, along with everything else in its reference frame, in the direction of those motions, as portrayed in diagram **2**. The distance x_1' has to contract to distance x to maintain light's constancy where $c'=c-v_0-v_R$.

But "time" doesn't actually exist. It's not a property of the universe. So there's nothing there to slow with motion. And even if it did exist, its rate wouldn't slow. It'd increase. A unit of time, like a second, that corresponded to distance x that was theoretically contracted to the shorter distance x' would be a faster second. Compressed "time" is a faster running "time."

Einstein also "ignores" light and time's innate three-dimensionality. Time's changing rate and length's contraction only work in the one abstract dimension of linear motion. In the three actual dimensions of our real environment, they're inherently conflicted. There's no motion in the y direction. So there's no contraction. y' and x' can never be the same length. But time's "slower" rate still has to apply in the y direction. It's the same reference frame. So light's velocity can never be the same. y' 's will always exceed x' 's. This invalidates Einstein's basal premise, which undermines relativity in its entirety, and that's without even considering his contradictory but correct assertion of light's variability that does so as well.

(16.2 MM 7a)

If we were to change the experiment's location longitudinally (vertically) then the rate of the Earth's rotation would also change. The speed of light and time's rate would read differently from the Earth's equator to its poles where there would be no fluctuation in measurement, or any contraction or change in the rate of time from rotation. Differing rates of time and light's velocity across the entire face of the Earth, which is a single reference frame, presents us with another insurmountable conflict that also clearly indicates that light's velocity has to compound with motion.

More fundamental, Einstein and Lorentz and FitzGerald simply ignore the implications for the projected light oriented longitudinally (vertically), perpendicular to the direction of the Earth's orbital, and rotational, motion. Being of the same reference frame as the rest of the experiment, subject to all the same conditions arising from motion but pointed perpendicular to the direction of motion, the distance between the mirrors would not experience contraction. So they'd remain the same measurement while experiencing a slower rate of time.

Without a coinciding decrease in the distance between the mirrors but a slower rate of time, this would cause the light to be out of phase, again producing the negative interference pattern, which did not occur. For the light to remain in phase, its speed perpendicular to the direction of travel would have to have exceeded 186,000 mi/s while the light projected in the direction of motion remained constant, which, of course, didn't happen either.

Accurately interpreting the results from this real-world experiment (and others like it including Sagnac's that we covered earlier) forces us to conclude that Einstein's and the Lorentz-FitzGerald explanation is without doubt ill-conceived, and that relativity's most fundamental premise, light's constancy, is incorrect, which renders it unworkable.

It's truly remarkable how these obvious and elemental conclusions can remain completely overlooked. If there were only one insight that you could take away from this book, this would be it. Because once you've challenged relativity's bluff, exposed its deception, and broken through its barrier of misdirection, seeing the fallacy of its underlying assertions through a highly regarded experiment, understanding them to be completely baseless and amazingly incoherent, then you've freed yourself from the influence of its intimidating rhetoric and overly complicated math. With this unimpeded perception, you can now easily challenge the rest of its conceptual absurdities. What little remains will also quickly crumble.

Principles of Relativity

Einstein "defines" three illusive principles of relativity: "the special principle of relativity," "the general principle of relativity," and "the exact formulation of the general principle of relativity." He maintains that each is correct and functionally in force despite that each supersedes and replaces the previous. This by itself is a contradiction. The very definition of a principle implies a fundamental law that's fixed and unchanging.

His first is special relativity or, "the *special* principle of relativity," or "this principle of relativity," or "the principle of relativity," or, "the special theory of relativity... i.e. the principle of the physical relativity of all *uniform* motion... [where] uniformly moving co-ordinate systems devoid of rotation... [have] exactly the same general laws... called the *principle of relativity* (in the restricted sense)" They're all used interchangeably... "the idea it conveys to us, every motion must be considered only as relative motion... [where the] general laws of nature (e.g. the laws of mechanics or the law of the propagation of light *in vacuo*) have exactly the same form in both cases [when drawing comparisons]."[112]

He stipulates that, "all bodies of reference... should be exactly equivalent ... for the formation of natural laws, provided that they are in a state of *uniform rectilinear and non-rotary motion*... The whole content of the special theory of relativity is included in the postulate: The laws of Nature are invariant with respect to the Lorentz transformations."[113]

In essence, all he's really doing is restating, "the [presumed] law of the propagation of light *in vacuo*" that's thought to always be 186,000mi/s.[114] There's no new principle here or anything *special* about it. Light's (presumed) constancy is masquerading as special relativity. It has him concluding time dilation, length contraction, and his mass-energy relation.[115] It also has him concluding that it's a completely subjective choice of each observer which objects (reference frames) are actually in motion and the rate of their motion.[116]

It's not unreasonable that the laws of nature should include a consistent rate of time or an object's inability to contract one-dimensionally, don't you think? They might also include light's compounding with motion and its variability. It's ironic that the whole point of the Lorentz transformation is to alter these natural laws to accommodate the incorrectly assumed natural, "law of the propagation of light *in vacuo*."

---

112. Einstein, *Relativity*, 16, 67-68.
113. Einstein, *Relativity*, 69, 169.
114. Einstein, *Relativity*, 23.
115. Einstein, *Relativity*, 25-46, 49-54.
116. Einstein, *Relativity*, 16-18, 30, 67, 68...

Einstein cautions us that when detecting or describing the motion, "it is in principle immaterial to what reference-body we refer the motion... this is self-evident [in the special principle], but it must not be confused with the much more comprehensive statement called 'the [general] principle of relativity'."[117] Notice how he qualifies it as a "much more *comprehensive* statement." It's inclusive and he knows it. So why keep the special principle? It's redundant. And excluding rotating bodies, as special relativity does, makes no difference.

His second principle, "general relativity," "relativity," "the theory of relativity," "the principle of relativity," "the general principle of relativity," "the general theory of relativity," "the generalized postulate of relativity," "the extended theory," "the provisional formulation," (all used interchangeably) states that: "All bodies of reference... are equivalent for the description of natural phenomena (formation of the general laws of nature [mechanics and light's constancy]), whatever may be their state of motion."[118]

The subtle inconsistency in the use of terminology is a significant and unnecessary complication. It further obscures and confuses the already elusive meaning of these nebulous suppositions.

He basically does the same thing when referring to "reference frames." He also uses it interchangeably with "body of reference," "reference-body," "body," "reference system," "system of reference," "co-ordinate system," "system," "object," "conceivable system of reference," "framework," "ponderable body or mass or matter." A "reference-mollusc" is another one. He uses it to describe a non-rigid reference-body. A "material point" might be his best though. It's innately contradictory. There are probably others. But you get the idea.

You're never exactly sure what the different phrasing might mean or if it does have a different meaning or why he's changed it for that particular context. It would have been much easier, much clearer to just stay with the same original wording throughout, qualifying it if or when it's necessary. It's worth considering whether all this unjustified complexity, and relativity's in general, may have been intentional. I'm convinced it is.

His general principle integrates his special principle. So even if his special principle weren't invalidated by his later realization of light's variability,[119] it'd be unnecessary to retain it. But he insists that it's still viable, "where no gravitational fields exist,"[120] and he insists on keeping it despite that there are no places or conditions in the entire universe where gravity fields are not.

117. Einstein, *Relativity*, 67-68.
118. Einstein, *Relativity*, 69, 108.
119. Einstein, *Relativity*, 85.
120. Einstein, *Relativity*, 85, 104, 109, 171.

"This theory [the general principle] arose primarily from [his] endeavour to understand the equality of inertial and gravitational mass... [It gives us] good grounds for extending the [general] principle of relativity to include bodies of reference which are accelerated with respect to each other ["the law of the equality of inertia and gravitational mass"], and as a result we have gained a powerful argument for a generalized postulate [general principle] of relativity."[121]

His third principle is the, "more abstract... exact formulation of the general principle of relativity."[122] How can something be more abstract, that is to say generally more theoretical, while at the same time being exact, which is in essence precisely more practical? They're opposites. Classic Einstein.

His "exact formulation... of the general principle of relativity... replace[s] the provisional formulation."[123] It integrates the previous two versions. It states that, "*All Gaussian co-ordinate systems are essentially equivalent for the formulation of the general laws of nature.*"[124] Restating it another way for clarity: For all moving or nonmoving reference frames, which are innately non-Euclidean in nature, the laws of nature, including light's (presumed) constancy and sometimes gravity, hold true by use of the Lorentz transformations at each of the four spacetime variables of a curving Gaussian coordinate system.

Theoretically, all four spacetime variables of any reference frame would be bound to and dictated by light's (presumed) fixed velocity. This would cause dimension and time to change at every location throughout any moving reference frame. And all reference frames are experiencing some kind of motion. But a single reference frame with multiple, or really an infinite number of rates of time (due to light's presumed constancy), is conceptually absurd. It's unworkable in our real world. The notion is inherently conflicted. So is experiencing different rates of time when looking in different directions, which is in essence the same thing.

Besides, time doesn't actually exist. And neither does space. And spacetime's four-dimensional nature is itself an inconceivable concept. And its two-dimensional curving in our real, three-dimensional environment is also unfeasible, and contradictory. Any variation in a three-dimensional substance has to express as a fluctuation in its density. It cannot curve. Curvature is a property limited to one and two dimensions.

So what is it exactly that distinguishes each of these three superseding "principles" from one another? They all basically say the same thing. The laws of nature, really only light's constancy unless you're considering gravity, remain the same regardless of a reference frame's motion.

121. Einstein, *Relativity*, 77, 172.
122. Einstein, *Relativity*, 69, 108.
123. Einstein, *Relativity*, 108.
124. Einstein, *Relativity*, 108.

The same could be asked of "the law of the equality of inertial and gravitational mass" and "the principle of equivalence." They're the same as well. And they're actually included under "the general principle of relativity." Einstein indicates that.[125] But they're fictitious. Acceleration/braking and rotation do not in any way produce gravity. We easily established that earlier.

But his principles of relativity don't work either. They're all established on the incorrect premise of "the law of the propagation of light *in vacuo*" that maintains its constancy under all inferred conditions of motion, which is a demonstrably false. Conceptually, light is required to compound with motion as confirmed by all of the Michelson-Morley and Sagnac type experiments.

Its velocity also varies in gravity fields, just as Einstein contradictorily asserts.[126] And if motion and the rate of motion is a subject choice of each observer, as he also maintains,[127] how would this impact each Gaussian co-ordinate system's spacetime variables? Isn't that idea just as invalidating as light's compounding and variability? So his "principles" are not and cannot be laws of nature. They're nothing more than make-believe contrivances that only pertain to illusive hypothetical conditions that do not exist in reality.

125. Einstein, *Relativity*, 77.
126. Einstein, *Relativity*, 84-85, 104, 109, 171.
127. Einstein, *Relativity*, 16-18, 30, 67, 68...

Mass-Energy Relation

It's myth that Einstein originally conceived the exact equation for the mass-energy relation, $E=mc^2$. He maintains that the proper expression is, $E_{kinetic} = mc^2 / \sqrt{1 - v^2/c^2}$.[128] It utilizes the Lorentz transformation because of his belief in light's constancy. The only time he did publish the equation was apparently for a Time magazine article in 1946, "$E=mc^2$, the most urgent problem of our time." It was intended as a simplified analogy for the general reader.[129]

Depending on the definition and notation used for energy and mass and whether it was written in word or equation form, it may be inferred that it was actually Lorentz, or possibly Gilbert Lewis (American physical chemist, 1875-1946) and Richard Tolman (American mathematical physicist & physical chemist, 1881-1948), or even others who originally expressed the exact equation $E = mc^2$.

Einstein contends that classical mechanics is incorrect. It doesn't satisfy special relativity's requirements: "every motion has to be considered only as relative motion... [where the] general laws of nature (*e.g.* the laws of mechanics or the law of the propagation of light *in vacuo* [its constancy]) have exactly the same form in [all] cases [of relative motion].[130]

In accordance with the [special] theory of relativity the kinetic energy [energy derived from motion] of a material point of mass *m* is no longer given by the well-known expression, $mv^2/2$, but by the expression, $mc^2/\sqrt{1 - v^2/c^2}$. This expression [of kinetic energy] approaches infinity as the velocity *v* approaches the velocity of light *c*. The velocity must therefore always remain less than *c*, however great may be the energies used to produce the acceleration... In contrast to classical mechanics, the Lorentz transformation is the deciding factor..."[131]

The first question that immediately comes to mind is, how can the expression have any validity when light's velocity is variable? His assertion.[132] It's rooted in light's constancy, employing the Lorentz transformation to satisfy special relativity's requirement that the law of the conservation of energy should hold for all reference frames regardless of their relative motion.[133] If light's velocity is not fixed, it can't work.

But light's velocity is not only variable, it's also mechanically required to compound with the motion of its source and that of all other sources. It has no chance of ever remaining fixed.

128. Einstein, *Relativity*, 50.
129. "Mass-energy equivalence," Wikipedia, last modified Feb 20, 2023, https://en.wikipedia.org/wiki/Mass-energy_equivalence.
130. Einstein, *Relativity*, 16, 67-68.
131. Einstein, *Relativity*, 49-51.
132. Einstein, *Relativity*, 84-85, 104, 109, 171.
133. Einstein, *Relativity*, 51.

So his expression is fundamentally flawed. Without light's constancy, there can be no supernatural increase in mass. Nor can there be any metaphysical "slowing" in time's rate, or an object's length contraction, all impossibly and contradictorily becoming infinite at the speed of light.

But even if we assume the validity of light's constancy, his relation is still conceptually untenable. There's no physical mechanism or process by which the kinetic energy of moving objects can elicit an increase in mass with increasing velocity. The amount of material an object contains cannot just magically increase, appearing out of nowhere simply because it's inferred to have relative motion. That manifestation is limited to the theoretical realm, justified through math.

It seems that even after a hundred years, the debate over the validity of $E=mc^2$ is not yet over. Questions involving definitions and terminology persist that qualify mass as: relativistic or inertial or rest or fictitious, or a material point of mass, or point-mass.

"Point-mass" is a good one. It's another instance of meaningless technobabble. It's inherently contradictory. And it's impossible. A "point" is a location. It doesn't exist. It can't be material or have mass. Conversely, material or matter can't express as a point. It will always have height, width, and depth no matter how small. And if material exists, it will always have mass. That's what mass is.

"Rest mass" is another contradictory term. Nothing is at "rest." Everything is in motion whether it's the Earth's rotation, its rotation around the Sun, our solar system's motion through the galaxy, and our galaxy's motion through the universe. And if you're a big bang believer, there's the motion from its expansion. With the exception of the big bang, all of these motions are indisputable facts. So any "rest mass" is experiencing many different types of motion simultaneously. There is no such thing as "rest."

The same questions persist for the formula's energy: Is it relativistic, radiant, kinetic, inertial, rest, or beginning or after energy? Even an object's velocity or motion can be qualified differently.

The notion of equal relative motion is either impossible or theoretical. In an absolute sense, every location, or reference frame, is experiencing different rates of motion, like the rate of rotational velocity between our feet and our head. The notion has to apply at all scales.

But on top of that, Einstein remember, asserts that every motion is only relative motion and which object, or objects that are in motion, or the rate of their motion, is determined completely by the subjective choice of each observer.[134] So anyone can decide whether a particular mass is "rest mass" or "inertial mass" by simply choosing.

134. Einstein, *Relativity*, 16-18, 30, 67, 68...

And since mass supposedly increases with motion, this also means that each observer is subjectively determining how much material any particular object has. And even beyond that, your choice and my choice can be completely different at the same time. Meaning that the same mass can be both "rest" and "inertial" simultaneously, containing different amounts of material at the same time. All this is just nonsense.

You have to find it amusing that the "rest energy" of a motionless mass, that can never be motionless, is obtained by a constant, that can never be constant, that's presumed to be the fastest velocity in the universe, but isn't, that's also assumed to be unattainable by any amount of mass, which also isn't true.

But even more fundamental, remember Einstein has already invalidated his entire theory, along with his mass-energy relation, that's founded on light's constancy, with his realization that it's variable in gravity fields.[135] This makes it variable everywhere. Gravity fields are everywhere. They extend indefinitely. And every object, which always have mass, has self-gravity. So there's no way to disregard gravity fields as he says is the way to preserve not just special relativity but all of relativity.

The bottom line conclusion to be drawn from all this is that it's all subjective and theoretical. And it will always remain that way. What's not subjective or theoretical is that an object cannot acquire more material out of nothingness to increase its mass because it's been deemed to have relative motion. Nor is light's velocity fixed. So none of this has any practical value. It's just an ongoing abstract theoretical discussion with no resolution.

The relation, when properly qualified, does have real world applications when exploring the restricted conditions of subatomic particles subject to electromagnetic fields, as with accelerators. It's not the particles' kinetic energy from their motion that conforms to the equation. It's the increasing electromagnetic energy of the field that's accelerating particles. That's what corresponds to their increasing velocity and their increasing mass. The increasing strength of the field is inducing the particles with an increasing charge that increases their mass and size. This in turn slows their natural frequency, which is mistaken as a slowing of their rate of time.

It's not difficult to comprehend how an infusion of radiant energy could enlarge a particle. It's already congealed radiant energy. So more of it would naturally increase its mass, causing its natural frequency to decrease, not an incorrectly inferred rate of time.

135. Einstein, *Relativity*, 84-85, 104, 109, 171.

The reason the particle can only approach the speed of light but never attain or exceed it is because the electromagnetic field pushing it can't. Regardless of the amount of radiant energy it contains, it can never surpass the given speed of light set by the density of the field it inhabits, which on Earth is regulated by the compounded density of our gravitational or magnetic fields, the two aspects of the one universal field.

Without the speed-of-light limitation the electromagnetic field, a particle or any body of mass moved by chemical or mechanical forces or even gravitation has no limit on what its velocity could be. It's not being induced with charge that's in turn increasing its mass that would theoretically become infinite at the speed of light. Nor is its motion innately increasing its mass. Its mass remains constant while its inertia increases indefinitely.

So for the particular circumstances of accelerated particles subject to electromagnetic fields here on Earth, the relation is valid. But it only applies when the energy is electromagnetic in nature and the speed of light is qualified by the density of the universal field at the particle's location.

Conceptually, equating mass and energy should not at all be a problem. In a universe that's composed of only radiant electromagnetic energy that when congealed expresses as a subatomic particle, it should naturally be construed that radiant electromagnetic energy intrinsically has mass. But it's only perceptible by comparison. Why wouldn't it be more realistic to just say that E, defined as atomic energy, is equal to m, which is just mass, when multiplied by some constant that's established by observation that applies only to the universal field's density at our location here on Earth?

Finite Yet Unbounded

What most of us consider the big bang to be is Einstein's "finite and yet unbounded" non-Euclidean curving universe. He whimsically speculates that with the advent of non-Euclidean geometry we can now legitimately consider the possibility of a finite universe. He doesn't offer a reason why it should be finite. Which you'd think he might, especially when by definition it's infinite. He only suggests that, "The development of non-Euclidean geometry led to the recognition of the fact, that we can cast doubt on the infiniteness of our space without coming into conflict with the laws of thought or with experience (Riemann, Helmholtz)."[136] (Bernhard Riemann was a German mathematician, 1826-1866. Hermann von Helmholtz was a German physicist and physician, 1821-1894.)

This is the genesis justification for the big bang. But why would anyone want to contrive a solution for a conceptually impossible finite universe when by definition the universe has to be infinite? Even Einstein reasoned that ultimately the universe has to be considered infinite: "A larger box can always be introduced to enclose the smaller one. In this way space appears as something unbounded... As regards [to] its space [our quasi-Euclidean universe] would be infinite"[137] So why would he persist in promoting its finiteness? The answer may be found in the nature of his solution, which by any estimation has to be considered delusory.

Let's be generous and give him the benefit of the doubt and assume that he recognized the effect of self-gravity on a finite universe that was spherically three-dimensional. It would have everything coalescing toward its center. This would conflict with the observational fact of the universe's uniform, homogeneous, isotropic distribution of stars. (Galaxies and their uniform expression at the largest scale had not yet been discovered.) But rather than abandoning its impossible finiteness, he pursued a two-dimensional solution that amazingly has the three-dimensional geometry of a sphere somehow expressing two-dimensionally.

He employed "the three-dimensional spherical [two-dimensional] space which was discovered by Riemann."[138] It theoretically melds the three-dimensional properties of our real existence into the homogeneous isotropic curving qualities of a sphere's two-dimensional surface. In his mind, this would render the universe spherically limited but without three-dimensional borders. To him it could be both finite and without boundaries simultaneously.

A sphere's two-dimensional surface, unlike that of a (three-dimensional) sphere, doesn't even exist. Its two-dimensionality only defines the location of a plane.

136. Einstein, *Relativity*, 122.
137. Einstein, *Relativity*, 129, 158.
138. Einstein, *Relativity*, 125.

In some purely theoretical way, two-dimensional gravity would manifest isotropically, equally the same in all but now only two directions.

By changing the universe's geometry from three-dimensional to two-dimensional, gravity's force no longer dissipates in strength exponentially from its center out but would theoretically act uniformly via the isotropic two-dimensional surface of a sphere. This would theoretically cause its material to be uniformly distributed and allow him to invent a mathematical mechanism, a "cosmological term" or "constant," that equated to a universal repulsive force that could resist gravitational attraction and prevent his two-dimensional three-dimensional cosmos from collapsing in on itself.[139]

Deploying a uniform repulsive force, a constant, in three-dimensions would be unworkable. Gravity's condensing increases exponentially. So they'd be mismatched. It would still collapse the universe in on itself.

He later abandoned his constant in favor of Alexander Friedmann's (Russian mathematician, 1888-1925), "solution in which the 'world-radius' depends on time (expanding space)." He thought it, "was natural from a purely theoretical point of view." If "the 'world-radius' depends on time (expanding space)... [then] In that sense one can say, according to Friedman, that the theory demands an expansion of space."[140] And of course, we're all aware that he regarded his cosmological constant to be his biggest mistake.

For a self-involved individual with an overactive imagination who's intrigued by mathematical puzzlement, the allure of this convoluted resolution must have been irresistible. By contrast, an infinite, ageless, static universe evenly distributed with stars must have been intolerably mundane. It doesn't even present a problem to be solved, much less elicit any math. There are no issues other than the nature of infinity itself. So he manufactured one: how to reconcile the uniform distribution of the universe's material if the universe was finite.

If space were actually something, its uniform expansion in three dimensions would be just as impossible as gravity's uniform condensing. In our real world, they both have to obey the physical geometry of a sphere. Any expanding sphere with a fixed amount of material naturally dissipates in density exponentially per the inverse square law the same as the force of gravity. Neither can express uniformly, making their counterbalance by either a mathematical constant or its replacement, uniformly expanding space, an impossibility in our actual three-dimensional reality.

139. Einstein, *Relativity*, 122-127.
140. Einstein, *Relativity*, 153.

With the uniform distribution of the universe's galaxies an observational fact that directly conflicts with the physical law of the exponential diffusion of a sphere's material along with the coinciding exponential condensing from gravity, we have no choice but to conclude that our big bang universe cannot be expanding. Nor can it be finite. It's a physical impossibility, unless you delusively believe, like Einstein, that we can exist in two-dimensions. (See Diagram **7.0** Platonics 5a, 7.1 Inverse Sq Sphere 13a, 7.2 Inverse Sq Fields 8a & **7.5** Sphere Volume 4a)

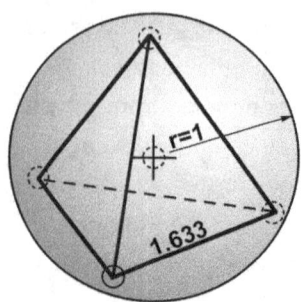 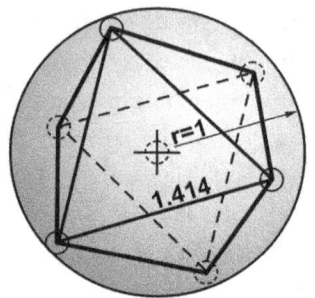 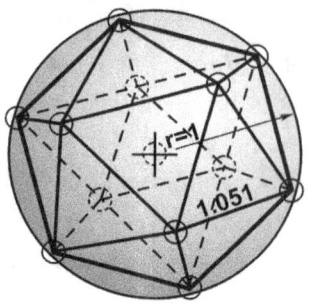

1. TETRAHEDRON   **2. OCTAHEDRON**   **3. ICOSAHEDRON**

PLATONIC SHAPES

The only three platonic shapes in nature that can be composed of equilateral triangles suggest that it's not possible for a spherical universe to be homogeneous. If we portion out any quantity of matter, be it subatomic particles or galaxies, into a spherical shape, we find that their uniform distribution is mathematically and physically impossible. Placing a galaxy at each of the triangles' vertices, or corners, and one at the sphere's center shows how their distribution would be equidistant around the circumspherical surface but always a lessor distance to the center galaxy. This principle of three-dimensional geometry simply reveals how the big bang universe, if it were actually three-dimensional, could not be homogeneous, but would always have to be denser towards its center, which would also establish an observable origin.

With a circumsphere radius of 1 of any unit of measure, the edge of each triangle, or the distance between each galaxy would be 1.633 for the tetrahedron, 1.414 for the octahedron, and 1.051 for the icosahedron.

(7.0 Platonics 5a)

150

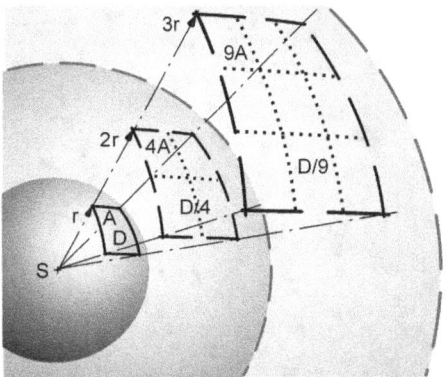

1. SPHERE'S SURFACE AREA & DENSITY
INTENSITY OR DENSITY OF AN EXPANDING SPHERE'S
SURFACE AREA, D α (IS PROPORTIONAL TO) $1/r^2$

INVERSE SQUARE LAW, SPHERE
The surface area of a sphere is $A = 4\pi r^2$. It increases
exponentially as it expands. When twice the radius
from its center, its area increases four times. Three
times the distance, nine times, and so on.

So the intensity or density (they're the same
thing) of any assumed flux (a quantity at the surface)
will dissipate exponentially following the inverse
square law where D (the density at a chosen radius)
= S (the original source density) / $4\pi r^2$ (the area of a
sphere). So its density twice the radius from its
center will be one-fourth its original. Three times the
radius, one-ninth, and so on, indicated in diagram **1**.

This same principle would have to apply to our
big bang universe that's assumed to be finite and
expanding. The material at each consecutive radius
throughout its interior would be simultaneously
diffusing exponentially per the inverse square law,
as represented graphically and qualified numerically
in diagram **2**.

But we observe a homogeneous and isotropic
universe. So it can't be finite, or expanding. This
simple indisputable fact of three-dimensional
spherical geometry by itself completely undermines
the big bang.

To preserve it, we've employed Einstein's belief
that it actually expresses two-dimensionally, like a
sphere's surface. Its homogeneousness is
maintained by confining galaxies to the surface of a
spherical universe that can then dissipate uniformly
with expansion. But there's no existence in two
dimensions. So the big bang can't be real, which
reduces it to nothing more than theoretical whimsy.

A more pragmatic direction would begin with an
infinite (non-expanding) universe that suggested
realistic alternatives for astronomical redshift.

(7.1 Inverse Sq Sphere 13a)

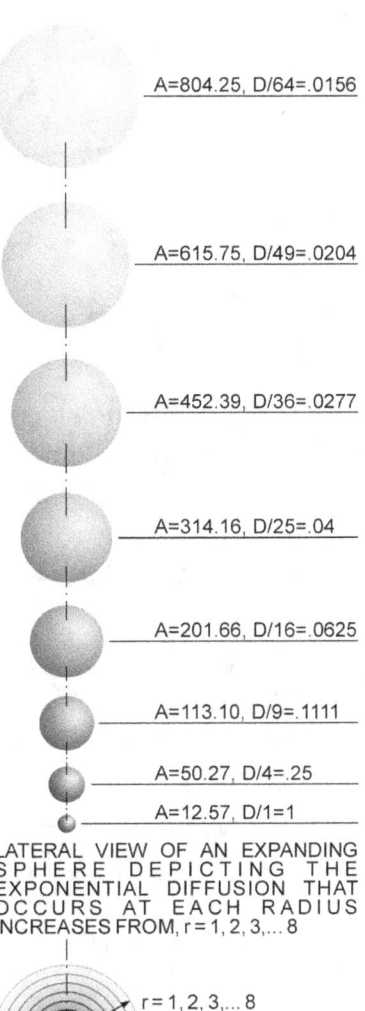

A=804.25, D/64=.0156

A=615.75, D/49=.0204

A=452.39, D/36=.0277

A=314.16, D/25=.04

A=201.66, D/16=.0625

A=113.10, D/9=.1111

A=50.27, D/4=.25

A=12.57, D/1=1

LATERAL VIEW OF AN EXPANDING
SPHERE DEPICTING THE
EXPONENTIAL DIFFUSION THAT
OCCURS AT EACH RADIUS
INCREASES FROM, r = 1, 2, 3,... 8

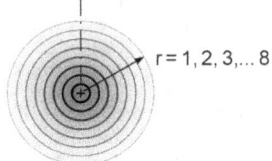

r = 1, 2, 3,... 8

SECTION VIEW THROUGH AN
EXPANDING SPHERICAL VOLUME
PORTRAYING THE EXPONENTIAL
DIFFUSION THAT OCCURS AT EACH
RADIUS SIMULTANEOUSLY AS IT
INCREASES FROM, r = 1, 2, 3,... 8

2. SPHERICAL EXPANSION
SURFACE AREA: $A = 4\pi r^2$
SURFACE DENSITY: $D \alpha 1/r^2$

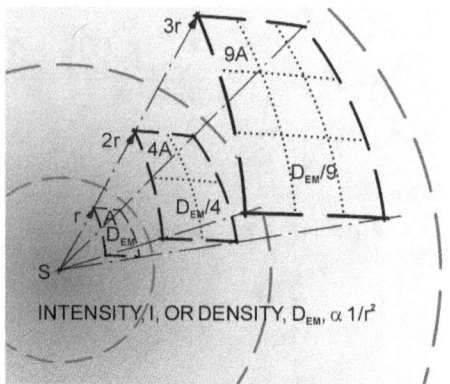

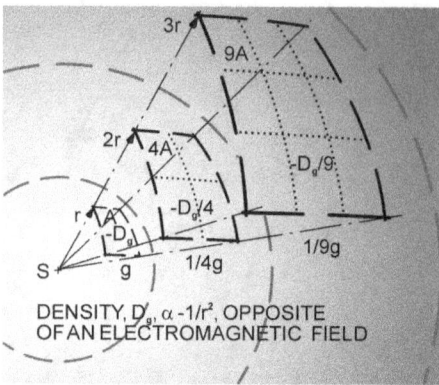

INTENSITY, I, OR DENSITY, D_{EM}, α $1/r^2$

DENSITY, D_g, α $-1/r^2$, OPPOSITE OF AN ELECTROMAGNETIC FIELD

1. ELECTROMAGNETIC FIELD

ELECTROMAGNETIC FIELD INTENSITY, WHICH IS DENSITY, DECREASES PROPORTIONAL TO THE INVERSE OF THE SQUARE OF THE RADIUS. THIS DIFFUSES ITS OUTWARD ACTING PRESSURE GRADIENT EXPONENTIALLY.

2. GRAVITY FIELD

GRAVITY FIELD INTENSITY, OR DENSITY, ALSO DECREASES PROPORTIONAL TO THE INVERSE OF THE RADIUS SQUARED. BUT ITS DIFFUSION IS INWARD. THIS INCREASES ITS OPPOSITE, NEGATIVE PRESSURE GRADIENT EXPONENTIALLY PER GRAVITY'S FORCE, g.

INVERSE SQUARE LAW, FIELD

An electromagnetic field (EM), depicted in section view as the diffusing background in diagram **1**, is subject to the inverse square law that's the product of the three-dimensional geometry of a sphere. So the field's intensity, which is the same as density that produces pressure which is force, twice the distance from its source is diluted by four times the area. This reduces its density to 1/4 the original. At three times the distance, it's spread over nine times the area, which reduces the density to 1/9 the original, and so on where D_{EM} (the density at a given radius) = S (the original density) / $4\pi r^2$ (the area of a sphere).

The big bang's radiant energy, correctly interpreted as a finite universal field, could never express uniformly as we presume. It's subject to the same physical geometry as every other field. It too would have to obey the inverse square law that'd have it diffusing exponentially with its expansion. The homogeneous isotropic expression that we observe is not physically possible for a finite, expanding universe. So in our actual three-dimensional, nontheoretical reality, the big bang is untenable.

The tangible, radiant, EM energy of our real universe that particles condense out of is all-pervasive, continuous, inseparable, and it varies in density. So the remaining ambient radiation that's not been drawn into a congealed particle has to thin inward, diffusing exponentially toward its center. This is what constitutes their, or collectively the bodies they compose, gravity field, portrayed in section view as the diffusing background in diagram **2**.

It's the opposite of an EM field. Its lowest density is reciprocal to an EM field's highest. Still bound to a sphere's inverse square law, its density, which is still intensity, which still equates to pressure and force, still has to dissipate exponentially. The gradient remains the same. It just expresses the opposite direction, diffusing inward instead of outward where D_g (the density at a given radius) = -S (the original point source strength or negative density established by a body's mass) / $4\pi r^2$ (the area of a sphere).

So at twice the distance from the center, its original negative density is diffused over four times the area, which is 1/4 the original that reduces the inward acting pressure by the same amount, decreasing gravity's force to 1/4g. At three times the distance, its negative density is spread over nine times the area, which is 1/9 as dense as its original that decreases the inward acting pressure the same, reducing gravity to 1/9g, and so on.

(7.2 Inverse Sq Field 8a)

152

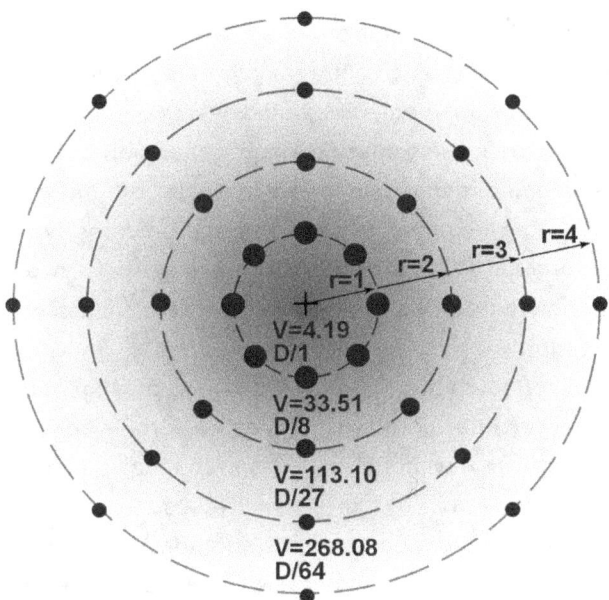

r=1 r=2 r=3 r=4

V=4.19
D/1

V=33.51
D/8

V=113.10
D/27

V=268.08
D/64

THE DENSITY OF THE VOLUME DIFFUSES EXPONENTIALLY

AN EXPANDING SPHERE
VOLUME OF A SPHERE, V = $4/3\pi r^3$
DECREASING DENSITY OF THE VOLUME, D α $1/r^3$

This conceptual depiction of a section view through a finite universe with a uniform distribution of galaxies at its expanding perimeter, represented by the dots, shows how its spherical two-dimensional surface, beginning at r=1, would expand uniformly with every galaxy moving away from every other while remaining an equal distance apart. But the density of its three-dimensional volume has to dissipate exponentially per the cube of the radius where D (volume density at a given radius) α $1/r^3$, as portrayed by the diffusing background and the disbursing dots.

This simply illustrates again how the big bang's uniform distribution of galaxies without some supernatural mechanism counterbalancing a sphere's innate diffusion like the constant exponential creation of new material out of nothing at its ever-expanding perimeter or a rate of expansion that somehow exponentially decreases from its center out is conceptually and physically impossible in a real three-dimensional nontheoretical world that obeys the natural laws of geometry.

(7.5 Sphere Volume 4a)

Consider this. If Edwin Hubble (American astronomer, 1889-1953, first to correlate the increasingly redshifted light of distant extragalactic objects with their decreasing brightness that were found to be evenly distributed over the entire sky) had introduced his misconstrued evidence for the universe's expansion before Einstein had advanced his two-dimensional solution for the universe, our natural inference from his observations would never have been that the universe was expanding. Without a premise that maintained its uniformity like a sphere's two-dimensional surface, there would have been no reason to draw that conclusion. It wouldn't be able to express uniformly.

It also would never have occurred to us that it was resisting the gravitational collapse of an entire finite universe. That couldn't express uniformly either. Both ideas would have been inconceivable without Einstein's preconceived finite yet unbounded universe that explained it away by expressing like a sphere's two-dimensional surface. We certainly would never have rationally jumped to his two-dimensional conclusion either. A solution that somehow allowed our two-dimensional three-dimensional existence is patently irrational.

We may have entertained the idea that the universe might be finite. But it would have been quickly dismissed. We would have naturally visualized a regular spherical universe in its normal three-dimensions expanding in a conventional manner, diffusing exponentially per the inverse square law as any three-dimensional expanding volume does. From our observation of the uniform distribution of galaxies and all magnitudes of their redshifts, we would have easily noted that galaxies were not dispersing from a center beginning point. So the universe couldn't be finite, which would leave us with an infinite and eternal universe while still searching for an adequate explanation of cosmological redshift.

But from here, still without a bias for a uniformly expanding two-dimensional universe, we might have reasoned that because of the runaway nature of gravitation, the natural coalescing of each galaxy's material would cause its continual condensing that produces an increasing infall velocity for that material that would be receding from every other galaxy. This would produce a commensurate redshift that would have to be higher for material closer to a galaxy's center. For smaller, fainter galaxies with less material obscuring their core, their redshift would make them only appear as more distant.

But rendered impotent by our extensive indoctrination to our big bang-relativity cosmology and groupthink's suppressive influence, uniformly expanding space has become a working premise that any new idea must conform to. We don't even see, much less question, the fundamental absurdities of a two-dimensional three-dimensional geometry, or of a three-dimensional space that somehow curves two-dimensionally, or that space somehow has an actual existence, or a three-dimensional, finite universe that somehow expands uniformly.

We can't recognize that these ideas only have validity theoretically as mathematical contrivances. So most of us continue to envision our big bang universe as an ordinary, three-dimensional spherical volume that expresses simple Euclidean properties where its expansion would dissipate normally unaware of these completely inane, contradictory requirements.

We still look out at the awe-inspiring universe with its homogeneous/isotropic distribution of galaxies and all magnitudes of their redshifts and not naturally conclude that it's ageless and infinitely vast, without a beginning or an end, where runaway gravitational coalescing proceeds unabated at each galaxy, perpetually collapsing, transmuting, and recycling all matter to radiant energy and back again. But instead, we insist that all galaxies must be receding from us and every other at a uniform rate in a curving non-Euclidean universe that somehow manifests two-dimensionally like a sphere's surface, despite all of the undeniable incongruities, including that in two dimensions there is no actual existence. We then justify that conclusion by incorrectly reasoning through our impenetrable groupthink indoctrination that a spiral galaxy's material must be orbiting around its center in a near circular path without any inward migration, while also ignoring Einstein's presumed gravitational redshift from both mass and rotational velocity so that galactic redshifts can remain an indication of each entire galaxy's recessional velocity, and certainly not an indication of their material's ever-increasing infall velocity.

Again, it cannot be overstated that with either uniformly expanding space, or the cosmological constant, the fact remains that our real universe expressing the two-dimensional, isotropic properties of a sphere's curving surface is an inconceivable reality, a consequence of only the theoretical curving, non-Euclidean geometry of a sphere's surface that has no practical application. A three-dimensional existence can never physically express two-dimensionally. Fusing the two is complete nonsense. It cannot happen.

His fanciful reasoning is no different from his misperception of gravitational effect as a consequence of space's non-Euclidean curvature. Even if it was actually something, it could never curve two-dimensionally. It's also physically impossible to curve any three-dimensional volume of something. It can only vary in density.

It's interesting how his manipulation of his contrived, non-Euclidean geometry has allowed him, in his imagination, at the largest scale to manufacture a constant two-dimensional curvature for space's solution that makes gravitational attraction uniform to satisfy his cosmological constant, or expanding/stretching space, so that the coalescing and the eventual collapse of the entire universe can be averted. But at smaller scales, the same curving non-Euclidean geometry also allows him to concoct a two-dimensional nonuniform space that is not of constant curvature to satisfy the same requirement of gravitational coalescing for matter. It seems that any desired end that can be imagined can be achieved through the illusion of two-dimensional non-Euclidean geometry. (See Diagram 17 Curvature 6a)

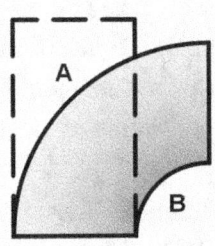
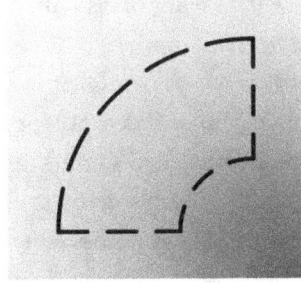
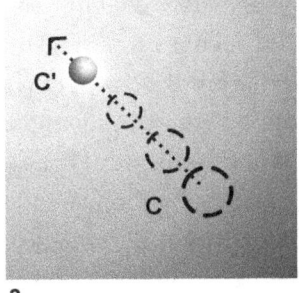

1   **2**   **3**

CURVATURE & DENSITY

If we were to envision the shape in diagram **1** as three-dimensional, giving it the same depth as its width, we can see how if it were bent over where its two-dimensional surfaces, at **A** & **B**, now curved, its interior would have to vary in density, as represented in section view by the diffused shading that dissipates toward the outer curved side with a greater radius, **A**. Note how the density along both the outer and inner curved surfaces would remain constant as **A** diffused and **B** condensed when bent.

If we were to imagine that shape as part of a larger volume of the same material, as depicted in diagram **2**, we can more easily see how any three-dimensional volume of something can never curve but can only vary in density. Curvature is a property of only one and two dimensions, which don't actually exist in reality. Without the third dimension, they only define a location of a line or a plane.

These principles of curvature and density would also have to hold true for the three-dimensional volume that's commonly refer to as space. It could never curve. It could only vary in density. But space is the nothingness between objects. It's nonexistent by definition, and it would also be nonexistent if it were a curving two-dimensional plane. So there's nothing there to vary or curve. But there is something that does correspond exactly to all space: radiant electromagnetic energy. Everywhere that space is, so is the all-pervasive universal field of radiation that comprises the electromagnetic spectrum. This field is continuous, its continuity uninterruptible, and it's average volume is fixed per any given quantity. And its density varies.

Because of its innate continuity, the density of the radiation that's been drawn together to form a particle has to diffuse. But it diffuses inward toward the particle's center, still dissipating exponentially per the inverse square law because of the three-dimensional geometry of a sphere. It's the compounding of those inward diffusing fields of the particles themselves, and the objects they compose, that's responsible for gravitation.

Referring to diagram **3**, if we were to place a small object into the decreasing density of a gravity field of a much larger object, it would begin moving with acceleration in the direction of least resistance directly toward the larger object as it mechanically sought to find equilibrium in the field's decreasing density. It'd naturally move from the highest density toward the lowest. But it would also be contracting in size. Its atoms also naturally coalesce and condense within the field's decreasing density, as indicated by the sequence of dashed circles representing the object's motion from **C** to **C'**. In reality, each object would move toward the other in proportion to the size of their gravity fields toward a common center of mass because of the compounding effect of their gradient.

(17 Curvature 6a)

Almost everyone fails to see the fallacy of his concocted, non-Euclidean illusions. But because he's held in such high esteem and because of the apparent success of his predictions of gravitation, some will certainly argue that his assessment has to be correct. His unrealizable explanations just present us with simple analogies that those of us of a lower intelligence can understand.

Well, that's not necessarily true. People visualize space, meaning see in three dimensions, differently and their ability to perceive it varies greatly. It's a common difficulty, even beyond any of Einstein's "learning disabilities." Our perception of gravitation as the product of space's two-dimensional curvature underneath three-dimensional bodies as opposed to the product of a spherical, radial, three-dimensional decrease in density in the universal field's radiant energy around every body elicits an entirely different response to the process.

Just try visualizing the different solutions for a supermassive black hole like the ones thought to reside at the center of most spiral galaxies. Do you see "space" as curving two-dimensionally like an immense funnel created by an enormous but unseeable mass equivalent to billions of stars that's located somewhere in the direction of its never-ending bottom where objects are sucked down into it, stretching like spaghetti as they pass over an imagined event horizon, endlessly falling, accelerating toward the infinite (mathematical) condition of a singularity, never to be seen again?

Or do you perceive "space" as actually being the three-dimensional nothingness between bodies that corresponds to the all pervasive field of radiant energy that comprises the universe that innately decreases in density exponentially around each body that naturally causes in their mechanical pursuit of equilibrium in its ever-decreasing density, which produces their ever-increasing coalescing/condensing that eventually manifests as a mature spiral galaxy's whirlpooling vortex of infalling material that's constantly migrating inward toward its common center of mass, not a supermassive black hole, that increases nuclear reactions that eventually transform all matter back into the radiant or plasma energy it originated from, which is then radiated back out and /or expelled as huge bipolar jets?

It does seem that our ability to perceive three-dimensionally is directly related to our level of consciousness. It wasn't until the renaissance that peoples' awareness had even evolved enough to accurately visualize in three dimensions and correctly draw in perspective with its converging parallel lines and foreshortened surfaces. That was only five hundred years ago. But as our insight continues to slowly increase, our spatial skills improve and our ability to accurately perceive in three dimensions becomes easier and more acute.

Fortunately for us, elevating our consciousness is not all that difficult. It begins by simply being willing to give up our preconceived notions. The one that we're faced with right now is the genius behind Einstein's solutions. If he had correctly perceived space's three-dimensionality, not just have known that it was three-dimensional, he would have presented it in terms of varying density, not the curving, two-dimensional non-Euclidean nonsense associated with his finite yet unbounded universe or his gravity's Gaussian coordinate system. And he would never have endorsed Riemann's "discovery" of space's two-dimensionality. It's not geometrically or physically or conceptually possible.

Einstein's unimaginable notions are indicative of the current problem plaguing cosmology, and other science, where common sense is often abandoned in favor of mathematical pursuits. Just because an equation can be fashioned to yield a desired result does not mean that it is, or can be, representative of our actual physical reality. Any mathematical portrayal of a fantasy is still no less a fantasy. But we insist on endorsing altogether impractical theories only because they've been "evidenced" with the apparent legitimacy of an equation.

In a similar way, computer simulations also give a false sense of legitimacy. Many times their high degree of accuracy, a consequence of their digital nature, fools us into accepting erroneous results. They look valid when often they're anything but. Our extensive computer modeling, in any field of study, also illustrates how math can be utilized to simulate almost any desired condition that may or may not represent our actual substantive environment. That's not to say that all of the conditions of our real universe cannot be accurately modeled mathematically. But any workable equation does not necessarily correspond to physical fact and may not even be conceptually fathomable.

So we can't automatically assume that mathematical solutions reflect our actual material existence. But we can accept that math is only a language, just another way to express a concept that requires real world verification, but nothing more. Whatever can be described mathematically should also be easily describable in common, nonmathematical terms to anyone with a freshman high school education. If it can't be easily described, the person doesn't understand it themselves or it's nonsense or they're being deceptive.

This is an excellent way of proofing the legitimacy of any idea. If the theory of light's constancy only works in one dimension then it's not valid in our real world. If we can't actually exist in two-dimensions then our universe can't be expanding and the big bang theory is not valid. If two-dimensional space can't actually exist, because it's two-dimensional and it's nothingness, then its curvature beneath bodies can't be facilitating their attraction.

It's not difficult to understand our insistence on an instant mathematical response to every question before a sound solution has even been conceptualized. Rather than calmly perceiving nature's simple truths, we're obsessively driven to complication, preferring the challenge of intricate puzzles and complex equations. The compulsiveness innate to our mind's programming has us squelching any inspired revelation, always rejecting it as implausible, and diving headlong into as much complexity as we can fabricate. This quickly becomes apparent when working at our computers. They exacerbate our compulsive natures, which is really just our unconscious attempt to avoid reality.

This excessive cognitive urge, which we all suffer from, only further obscures any real answer that's usually right in front of us, readily apparent, waiting to be discovered. It's always recognized later as common sense. Surely, it'd be wise for us to step back for a moment, get our heads out of the math, collect ourselves, and take a candid objective look at what relativity is actually proposing. It truly is ludicrous.

Look again at Einstein's finite yet unbounded universe. How can it be both finite and infinite at the same time? His supposition is once again at odds with itself. And when "explaining" the problem of its space, he contends that, "space (space-time) has an existence independent of matter or [gravity] field." But then he says the opposite, "space... has no separate existence." He adds, "If we imagine the gravitational field... removed, there does not remain a space... but absolutely *nothing*..." Then he says, "There can be no such thing as empty space, *i.e.* a space without a [gravity] field."[141] Do you follow any of that? All it is, is contradictory nonsense. And it's all in the same paragraph.

Space by definition is the nothingness outside of or between objects. The very fact that his universe is finite defines it as an object. It's something with measurable dimensions. This implies that it has to be in whatever is external to it. And since it's physically impossible for something to exist in nothing, or without space, then whatever is outside of his finite universe would also have to be space, but a space without a gravitational field, which he contends is not possible.

If space without a gravity field isn't possible then the effects of his special relativity aren't possible either. He insists they can only exist outside of gravitational fields. The contradictions seem endless.

He also argues that any finite space can always be enclosed by a larger finite space, causing it to be seen as unbounded or infinitely vast. "A larger box can always be introduced to enclose the smaller one. In this way space appears as something unbounded."[142] We might agree with that, maybe.

141. Einstein, *Relativity*, 176.
142. Einstein, *Relativity*, 158.

But again, he's in conflict with his previous assertion that there exists no space outside of the space of his universe. Beyond that, the very notion of a finite universe is at conflict with itself. The universe by definition means the entire whole. This would naturally have to include the infinite space outside of the finite space.

Most everyone, at least at some deeper subconscious level, intuitively recognizes this. So why not clearly address the issue of infinity at its most fundamental level, and simply accept from the outset that the universe has to be infinitely vast. All of these gyrations are nothing more than contradictory semantical banter that has no practical value.

Einstein "explains" that in his universe, someone can, "draw lines or stretch strings in all directions [meaning spherically, radially, three-dimensionally] from a point... At first, the straight lines which radiate from the starting point diverge farther and farther from one another, but later they approach each other, and finally they run together again at a 'counter-point' to the starting point. Under such conditions they have traversed the whole spherical [two-dimensional] space."[143] This is a physically impossible scenario. It's also inherently conflicted. The strings are first diverging three-dimensionally, which can only continue indefinitely. But then he has them somehow magically start to converge two-dimensionally.

He again condescendingly admonishes us that visualizing the two-dimensionality of his universe's, "... space means nothing else than that we imagine an epitome of our 'space' experience, *i.e.* of experience that we can have in the movement of 'rigid' bodies. In this sense we *can* imagine a spherical space."[144]

Of course, all this is ludicrous, even delusional. It makes absolutely no sense. And no matter what he believes is attainable through contrived two-dimensional three-dimensional space and curving non-Euclidean geometry, we can't seriously be expected to accept this. Any objective individual with minimal visualization skills can quickly see that a three-dimensional universe that expresses two-dimensionally is conceptually absurd. It's an inconceivable reality. You really have to wonder if he isn't secretly testing the limits of human gullibility. Yet it's this very concept, even though plainly illusive, that forms the foundation of big bang orthodoxy.

With the elimination of the cosmological constant, replaced by Friedmann's idea that space is expanding with an increasing radius that corresponds to time, it's important to note yet another elemental contradiction. Not only does this notion of time, which is supposedly fixed by the overall expanding motion of the universe, fundamentally conflict with the variability and subjectivity of time for objects in motion, as Einstein contends, but if the expanding universe underwent a brief inflationary

143. Einstein, *Relativity*, 125-126.
144. Einstein, *Relativity*, 125.

period then slowed and has now started accelerating again, as is currently believed, then its rate of time would have to be fluctuating. And if our rate of time is fluctuating then the speed of light cannot be fixed but also has to be fluctuating. It seems that no one ever seriously considered this basic conflict.

What's more, space's expansion during its presumed early inflationary period is theorized to have proceeded faster than light, which, of course, is not supposed to be possible. How would this have affected time's rate? With light's constancy rooted in relativity's math, it doesn't matter whether space is stretching or not, which is the fabricated mechanism to justify faster-than-light velocities. It's only the relative motion between objects that counts, which according to the equations would cause time to become infinitely slow at the speed of light.

Still, none of this is even pertinent. Time itself is nonexistent aside from the natural frequency of the object's cyclic motion that we've selected to establish its rate. It's not an independent aspect or constituent or property of the universe. Nor can it arbitrarily change with our subjective choices of motion. Nor can its rate change as a variable in an equation. Nor do the past and future somehow coexist concurrently with the present moment. Nor is it proceeding as a function of the presumed accelerating expansion of the universe. Nor can it run in reverse with the universe's inferred future collapse. Time's rate cannot in any way be divorced from the physical reality that established it, which we choose.

The whole point of the big bang's two-dimensionality is so that his cosmological constant that was later replaced by uniformly expanding space could resist the universe's inward collapse from its self-gravity while maintaining the universe's uniformity. But he's again having problems correctly perceiving dimensions.

He inconceivably imagines the entire universe as two-dimensional, confined to a sphere's surface. But he envisions its overall inward collapse from its self-gravity as three-dimensional. How would that work? If the universe by definition is everything and it's only two-dimensional, how could it ever include or be affected by a third dimension?

Its counteracting expansion has the same problem. It's also three-dimensional. It would coincide with an ever-increasing radius of a spherical volume that also couldn't exist if the universe were actually two-dimensional.

But again, in two dimensions there is no actual physical existence. Two dimensions only define the location of a plane. So there's no mass to create the self-gravity that would cause the universe's exponential condensing and ultimately its collapse. The same would have to be true for its expansion. If it's impossible for anything to exist two-dimensionally then there wouldn't be anything there, including his space that he defines as something, to expand to counter its self-gravity from its nonexistent mass.

But let's say our finite big bang universe was actually three-dimensional and somehow expressed uniformly, which it can't. That's physically impossible. But this is how almost everyone visualizes it. Most have never even thought about it being two-dimensional or considered how ludicrous the notion actually is.

So, theoretically, if our universe was a three-dimensional finite sphere that somehow became uniformly populated with galaxies, then to maintain that uniformity its expansion could never be uniform. The universe's self-gravity would still condense exponentially, which wouldn't match the universe's uniform expansion, no matter its rate. So its galaxies would still begin dispersing.

The only way their distribution could remain uniform is if expansion precisely compensated for self-gravity's increasing condensing. It would have to impossibly dissipate exponentially, just like gravity, proceeding the fastest at its origin where gravity would be strongest while exponentially decreasing to its perimeter, exactly matching gravity's decreasing force. Try visualizing that from its inception.

Regardless of how you conceive our universe's dimensionality and whether you believe it to be uniformly distributed with galaxies or not, the question of how its expansion could ever prevent its overall collapse still remains. The force of its expansion would always have to match or exceed gravity's. And if that's true, how would the universe's primeval radiant energy have ever condensed into mass in the first place? It could never have allowed its initial coalescing.

For matter to exist, wouldn't expansion's force always have to be less than gravity's? And if that's true, wouldn't the universe collapse in on itself immediately after its inception? Or does it suggest that our universe would had to have somehow come into being at some size with its matter already intact? That's how Einstein envisions it.

It would seem that the only way for expansion to prevent the collapse of our assumed finite yet unbounded universe is if shortly after its big bang inception it would have had to slow for a while to allow for radiant energy's initial condensing into matter. But after that, it would have to start accelerating again. This is generally what's believed, excluding the contradictory faster-than-light inflationary period. It's also imagined that its expansion is still accelerating and will continue indefinitely, eventually overcoming gravity's force, diffusing all matter back into radiant energy.

But what is it that's responsible for these metaphysical fluctuations in universal expansion? We have no idea, maybe some form of exotic dark energy. But stuck with our big bang-relativity dogma, we can't even entertain the possibility of alternatives. Only a crackpot would do that. So we continue doing all these backflips, contriving all these ad hoc workarounds, trying everything we can think of to justify it. You'd think that any rational, objective individual, free of groupthink and indoctrination, would have long ago given up and abandoned their unworkable finite yet unbounded, expanding curving two-dimensional universe and had started looking for other viable explanations for astronomical redshift.

Just try visualizing how a real three-dimensional universe might appear if all of its matter was endlessly coalescing and collapsing unabated because of gravity's runaway nature. Would it express as a single large sphere of space evenly populated with galaxies like what Einstein imagined for his finite yet unbounded, two-dimensional three-dimensional, delusion that'd collapse in on itself if it were not for his cosmological constant or its uniform expansion? Or might gravity's unrestrained coalescing in an infinitely vast environment naturally manifest as an endless number of evenly distributed galaxies that are naturally coalescing into whirlpools of condensing infalling material that's perpetually recycled into radiant energy and back? This is exactly what we observe as variations of elliptical and irregular galaxies that eventually evolve into spirals.

Unlike Einstein's fanciful speculation, it can be easily envisioned how the huge jets of radiant and plasma energy that spew from the center of mature spirals might tend to initially organize galaxy clusters while fostering the great voids in between them. Their expanding higher density radiation could be increasing the field density between galaxies and their clusters, creating vast intervening regions that would inhibit their coalescing in some areas while promoting it in others. Eventually, as their radiation slowed and cooled, it would revert back to ordinary matter and begin coalescing again, perpetuating an endless recycling process.

With the fact that there appears to be an uncountable number of galaxies with each demonstrating gravity's runaway exponential coalescing, not just one immense sphere of exponentially coalescing material, you'd think that would easily evidence an infinite, ageless universe. One that's in a non-expanding, steady-state condition that's dynamically recycling a ceaseless inflow of coalescing-collapsing material at each galaxy that seemingly remains static and balanced over billions of years while maintaining the homogeneous distribution that we see at the largest scale.

And with our own relentless infall toward our own galactic center, the product of our own galaxy's runaway coalescing-collapsing, together with the recessional velocity from the unrelenting infall of all the coalescing-collapsing material at every other galaxy, we now have a rational explanation for cosmological redshift that doesn't rely on universal expansion that preserves its innate unlimited eternal nonfinite nature.

But instead of squarely facing the obvious and mundane, we'd much rather indulge in delusive illusions, imagining that galactic redshifts originate from the recessional velocity of each entire galaxy being whisked away from every other, implausibly receding because of the incomprehensible uniform expansion of a spherically shaped finite non-Euclidean curving universe that somehow manifests two-dimensionally like a sphere's surface where, incredibly, if we had a powerful enough telescope we could in some metaphysical way look out, in any direction, radially, and see our backside.

Supposedly, for some unknown reason, this unfathomable universe began smaller than an atom. Then, for some unknown reason, it burst out somehow faster-than-light in a big bang. Then, for some unknown reason, its expansion slowed for some unknown period of time. But now, for some unknown reason, its repulsive dark energy, that's conveniently undetectable, is overcoming its dark matter, that's also conveniently undetectable, which is presumed to comprise the vast majority of the universe's mass, causing it to start accelerating again. All the while, gravity's runaway exponential coalescing is still allowed to collapse material three-dimensionally at every other scale despite the universe's presumed two-dimensionality.

Unification

It's widely held that after general relativity Einstein spent the balance of his life pursuing a grand unified theory or a theory of everything where the three other presumed forces of nature, the electromagnetic, the strong nuclear, and the weak nuclear forces, would be unified with gravity all under one comprehensive understanding. But with the realization that gravitation occurs not because of the impossible two-dimensional curvature of a nonexistent, inconceivable, four-dimensional spacetime but because of an object's mechanical reactive search for equilibrium in the ever-decreasing density of the all-pervasive electromagnetic field that comprises the universe, it becomes clear that gravitation is simply the product of the one electromagnetic force.

There's only one thing in the universe, radiant electromagnetic energy, the radiation that comprises the electromagnetic spectrum. The entire universe is nothing but a single field of radiant energy. It could be qualified in two ways, when it's been condensed into a particle that we define as matter or when it remains radiant, expressing as some frequency of the electromagnetic spectrum. (See Diagram 8 EM Spectrum 9a)

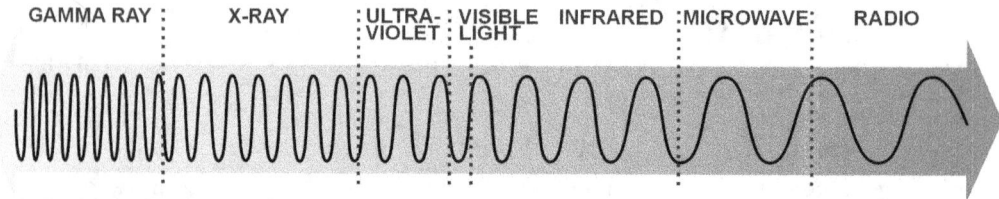

GAMMA RAY | X-RAY | ULTRA-VIOLET | VISIBLE LIGHT | INFRARED | MICROWAVE | RADIO

1. ELECTROMAGNETIC SPECTRUM

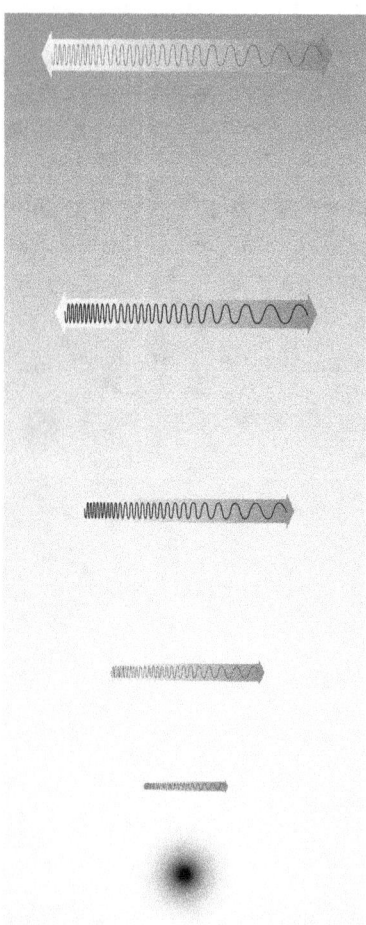

2. WAVELENGTH & AMPLITUDE CONDENSE EXPONENTIALLY IN A GRAVITY FIELD

ELECTROMAGNETIC SPECTRUM

The universe is essentially comprised of only one thing, radiation, which is the radiant energy of the universal field of the electromagnetic spectrum. Wavelength or frequency increases or decreases without limit at either end of the spectrum, as portrayed in diagram **1**.

When this radiation "spontaneously" condenses toward a single point, it forms a subatomic particle. The remaining ambient radiation not drawn into the particle dissipates inward, not outward, but still diffuses exponentially per the inverse square law because of the simple geometry of a sphere, as depicted in section view by the diffusing background in diagram **2**. This inward diffusion defines every particle's or object's gravity field that's literally but not practically felt over an infinite distance. The inward exponential decrease in field density, which is always less in between particles or any object, causes them to naturally move toward one another as they mechanically seek equilibrium, constantly pushed by the higher density toward the lower.

The condensed radiation that's been drawn together to form the particle does not exhibit a surface but increases in density exponentially to a maximum at its center also because of the simple geometry of a sphere. This defines its charged field that lies at the center of its much larger gravity field. Its increasing density naturally repulses, pushing away or holding other particles at bay.

Like everything else, the wavelength and amplitude of the universal electromagnetic spectrum expand or contract depending on its density where the measurement is recorded. Stronger gravity fields that are less dense naturally result in shorter wavelengths and smaller amplitudes at any given location. The problem is, without a baseline you can never compare a measurement from one location to another. Only someone at third location receiving the other two at the same time could perceive a difference.

(8 EM Spectrum 9a)

If matter is characterized as the substance that composes all objects, and radiant energy is everything that's not matter, then space, which is everywhere objects are not, and radiant energy coincide exactly. But since space doesn't actually exist, one might conclude that the gravitational effects it's presumed to cause must actually be originating from the universal field.

Because an electrical charge is defined in terms of intensity, which is the same as density, a charge can also be assigned to the universal field. Since it varies in density, its denser regions could be considered positively charged as compared to those regions of lesser density. They could be regarded as negatively charged. Those variations in charge, or density, would produce a current. This reasoning may seem elemental to some. But there's an important nuance here that exposes the flaw in mainstream interpretations.

Matter also manifests in only two ways, either as an electron or as a proton. Both of which being stable can be considered the only fundamental particles. For our purposes, we can ignore the myriad of ever-increasing unstable particles whose real existence is still only conjecture. A neutron can be construed as composed of both an electron and a proton.

Since these particles exhibit mass and since all mass inherently manifests a gravity field (the innate decrease in the universal field's density around the mass) then a particle's mechanical reactive movement as it naturally seeks equilibrium in its varying density, or intensity, could be understood as the product of charge and current. It dissipates in density, or intensity, which is strength consistent with the inverse square law, as all electromagnetic and gravity fields do. Being spherical, they're bound to the inherent geometry of a sphere.

Since density and charge are relative and because the higher density of any individual particle (congealed electromagnetic radiation) is greater than the decreasing density of its ambient electromagnetic radiation (its gravity field) it would have to be considered as positively charged by comparison while its gravity field would have to be considered negatively. Even though the electrical charges of an electron and proton have been interpreted as being equal, the electron's charge has been arbitrarily, and incorrectly, assigned as negative to explain their apparent attraction, opposites attract.

Despite that the assumed forces between subatomic particles have been measured to a very high degree and that their interactions are theorized to be controlled by exchange forces carried by numerous "virtual particles," their apparent attraction and repulsion that determine their binding and relative distances is actually the product of how their fields interact.

The increasing field density of every particle's charge is positive by comparison and repulsive in nature, while the decreasing density of their gravity fields is negative by comparison and "attractive" in nature. It still repulses but inward. This elemental electromagnetic relationship governs all their interactions with forces that obey the inverse square law that's the product of the physical geometry of a sphere.

Because a proton's gravity field is considerably greater than its charged field, it's governed more by its gravity field. Its lesser density mechanically pushes it toward other protons and electrons and they toward it as their fields combine.

For an electron, which has about 1,837 times less mass than a proton, its gravity field is proportionally smaller and its charged field is relatively larger. So the higher density of its charged field naturally repulses other electrons. They're mechanically pushed away from each other as their fields compound.

But the opposite is true when they interact with protons. The decreasing density of their compounded gravity fields is greater than the increasing density of their compounded electromagnetic fields. So they're mechanically pushed toward one another as they persistently seek equilibrium until they find their "ground state." This occurs when the outward repulsive effect of their compounded charge fields is equal to the inward repulsive effect of their compounded gravity fields.

Neutrons are created by the initial pairing of two protons. The decreased density of their combined gravity fields is greater than the increasing density of their combined charged fields. So they're naturally pushed together. The decreased density of their compounded gravity fields then draws in an electron. Their compounded fields tightly pressing it inward, holding it to either one or both of the paired protons. This is what we define as a neutron. But the three together still have a negative or a decreasing field density. They have just the right amount to push in and capture a second electron to create a balanced state.

The "attraction" between any two or more quantities of matter, be it a subatomic particle or massive body, is only a consequence of the compounding of their inward diffusing gravity fields. This creates a region of lesser density directly between them, or toward a common center of mass for multiple bodies, that mechanically pushes them toward one another as they naturally seek equilibrium. And the repelling effect between particles and even massive bodies is also the natural repulsive effect of their interacting charged fields that are of increasing density. It's just the compounding of these varying field densities that determines how all particles, and the bodies they compose, interact and their distances from one another.

Ask yourself, at what point do the ongoing chemical reactions of atoms and molecules that combine to create compounds transition from the reactions of the presumed nuclear forces to the coalescing process of gravitation? When put in this context, we can more readily see how the processes are and have to be the same. All chemical reactions can only be the product of the compounding conditions of a particle's negatively charged gravity field of decreasing density and positively charged electromagnetic field of increasing density, which can be manipulated by either adding or subtracting radiant energy.

The apparent attraction of the protons and neutrons that are densely clustered in an atom's nucleus is mistakenly thought to be caused by what is called the strong nuclear force. It's also incorrectly regarded as simultaneously repulsive in nature. The reason why this "attraction" appears to be facilitated by a separate force other than gravity and why it seems so much stronger than gravity is because the decreased density of a single proton's gravity field is much greater without the increased density of the electromagnetic field of a captured electron. All objects are composed of nearly an equal number of protons and electrons, and neutrons. So the strength of a single proton's gravitational field will always be proportionally greater than the strength of the gravity field of any size object.

The radiant energy of the electromagnetic spectrum comprises a universal field of radiation that is the universe. What we consider matter is just the condensing of this field into a "particle." Ultimately, there's no such thing as matter. There's only radiation and condensed radiation.

Because it's continuous, unbreakable, inseparable, and indivisible the radiation that's congealed into a particle leaves a deficit in the universal field that surrounds it. That deficit has to decrease in density exponentially because of the immutable geometry of a sphere that's bound to the inverse square law. But its decreasing density diffuses inward, not outward. This defines every particle's, and the objects they compose, gravity field, which pushes them together as the gravity fields combine.

The particle's innate outward diffusion defines its electromagnetic field, which pushes them and the objects they comprise apart. But it's much smaller than gravity's field. The compounding of both fields is responsible for the simultaneous inward and outward repulsion of particles, and the objects they compose, through their mechanical reactive search for equilibrium. This simple understanding easily explains the contradictory aspect of the attractive and repulsive nature of the inferred strong and weak forces while revealing how they're actually the natural result of the interacting fields of electromagnetism. (See Diagram **18.1**, **18.2**, **18.3**, **18.4** Atom 8a)

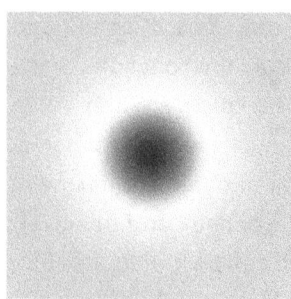

1. PROTON

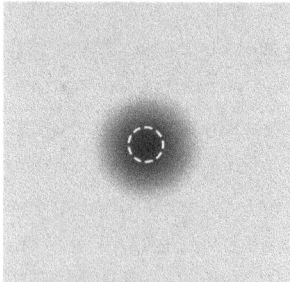

2. ELECTRON

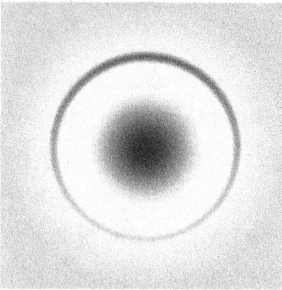

3. HYDROGEN ATOM

ATOMS

Protons should not be considered tiny physical particles within an electromagnetic field but as the field itself. There's no surface where the field stops and matter begins. The field becomes progressively more dense until it peaks at its center, represented in section view by the darker diffused circle in **1** & **3**. But because that proton field has condensed out of the universal field of electromagnetic radiation, the ambient radiation not drawn into the proton has to decrease in density around it diffusing exponentially like any field because of the geometry of a sphere.

But its diffusion disperses inward not outward, which defines its gravity field, depicted in section by the diffusing background in **1** & **3**. Because the decreasing density of a proton's gravity field is larger than the increasing density of its electromagnetic field, the gravity field's compounding with the fields of other particles tends to push them together as they naturally pursue equilibrium, mechanically seeking the lowest density that always lies directly between them. So protons should actually be considered negatively charged.

Convention has protons positive charged and electrons negative. Apparently, this has been mostly an arbitrary designation. But it doesn't correspond to physical reality. It's one of the reasons why gravitation and electromagnetism are not recognized as being the same effect.

Electrons should also be considered as having condensed out of the universal field. Its charge is considered equal to that of a proton. But its mass is 1/1837th as much. So it yields a much smaller gravity field, indicated by the small white dashed circle. For graphic clarity, it's shown proportionally much larger than it would actually be.

Being that the decreasing density of its gravity field is smaller than the increasing density of its electromagnetic field, it has a repulsive effect that when compounded with the fields of other electrons tends to push them away. So in reality it's positively charged. With the electromagnetic field of the electron still smaller than the gravity field of the proton, the compounding of their fields still pushes them toward one another.

An atom's electrons should not be envisioned as small objects that rapidly orbit the nucleus as always portrayed. They're more accurately conceived as having been pressed down and smeared out all over and around the entire nucleus, spherically, three-dimensionally, by the decreasing density of the universal field enveloping it, the atom's gravity field. It's compressed to a level where the repulsive effects of all the fields balance out and find equilibrium, as is implied in the section view through a hydrogen atom that has only one electron and one proton.

(18.1 Atoms 8a)

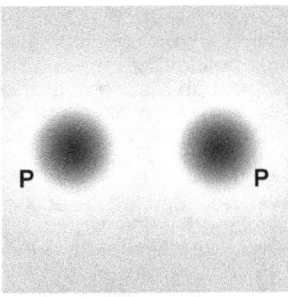

1. TWO PROTONS

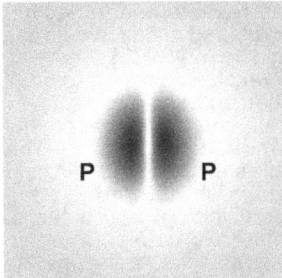

2. PAIRED PROTONS

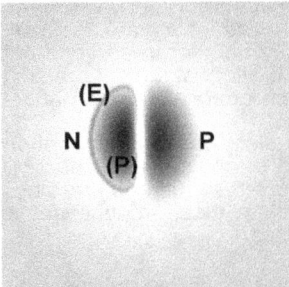

3. NEUTRON & PROTON

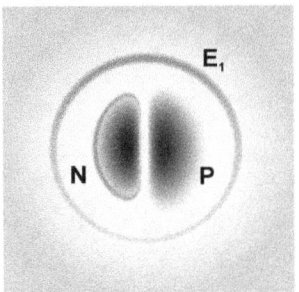

4. HYDROGEN ISOTOPE

(18.2 Atoms 8a)

NEUTRONS & ISOTOPES

A neutron can be considered a merging of a proton and an electron. The compounding of their two electromagnetic fields and their two gravity fields are theoretically balanced to yield no charge, that is if it could stand alone. Its neutral charge suggests that the electromagnetic fields of protons and electrons are half as strong as their combined gravitational fields.

For numerical convenience, if we assume the strength of a proton's gravity field is (-1), negative because of the field's decreasing density, and we know that an electron's is 1/1837th of that (-.00054) then their electromagnetic fields would have to be half of (-1) + (-.00054) or (+.50027), positive because of its increasing density. So a proton's relative charge would be its gravity field (-1) plus its electromagnetic field (+.50027) or (-.49973). And an electron's relative charge would be its gravity field (-.00054) plus its electromagnetic field (+.50027) or (+.49973).

Neutrons usually only exist, though, through the initial pairing of two protons, located at **P**. The compounding of the decreasing density of their fields, (-.49973) + (-.49973) or (-.99946), first draws them together mechanically as they naturally seek equilibrium. Then the even higher decrease in density of their combined fields draws in and tightly holds an electron, which is positively charged (+.49973), located at **E**, to create, or define, a neutron, located at **N**.

It's likely that the electron may move back and forth between protons or at times envelop both at once. But the three together still have a negative charge, or a field of decreasing density of (-.49973), that can draw in another electron (+.49973), located at E_1, to achieve a balanced state, in this case deuterium an isotope of hydrogen.

The actual distance to the electron would be over 60,000 times the radius of the nucleus. At the scale depicted that would put it more than 100yds away. The important principle that's trying to be conveyed here is that it's the sequence in which the particles assemble, which is facilitated by the relative densities or actual charge of their fields, that is responsible for the creation of a neutron. Otherwise, you'd just end up with a hydrogen atom.

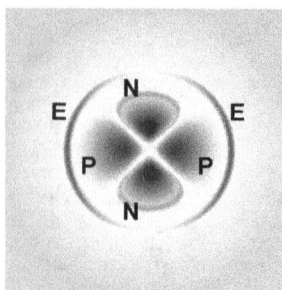

1. HELIUM ATOM

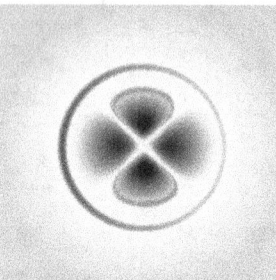

2. NEGATIVE ION (-E)

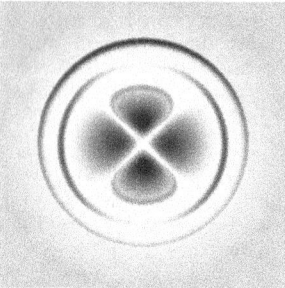

3. POSITIVE ION (+E)

IONS

The actual charge of an ion is also opposite of convention. If we begin with a ground state helium atom, shown theoretically in a section view through its center, the electromagnetic and gravity fields of its two protons, located at **P**, and two electrons, at **E**, balance to neutralize its charge. Its neutrons, at **N**, already a combination of an electron and proton, remain neutral.

If one of the electrons is removed, as depicted in section view in **2**, the density of its combined fields would be decreased where its relative charge, as calculated in the previous diagram, would be (-.49973), where its gravity fields dominate, which would tend to draw in other particles, making its charge negative.

If an electron were added, as represented in section view in **3**, the density of its combined fields would be increased. Its relative charge would be (+.49973), where its electromagnetic fields dominate, which would tend to push away other particles, making its charge positive.

A decreasing density in the universal field, a gravity field, is a negative charge that tends to push inward. The increasing field density of a particle's electromagnetic field is a positive charge that tends to push outward. It's the inherent repulsive nature of a particle's, or any object's, electromagnetic field that mechanically causes them to seek equilibrium in the universal field that innately decreases in density around every particle, or object.

Their reactive search for the lowest density in their combined fields that always lies directly in between them, or toward a common center of mass for multiple objects, causes them to move toward one another in an apparent attraction. It's the same repulsive effect of their interacting fields that pushes or holds them apart when they attain equilibrium.

Protons and neutrons and electrons are not bound together or repelled by imaginary strong and weak nuclear forces that are magically transmitted by unseen massless particles. Gravitation resulting from electromagnetism is simply governing all their interactions.

(18.3 Atoms 8a)

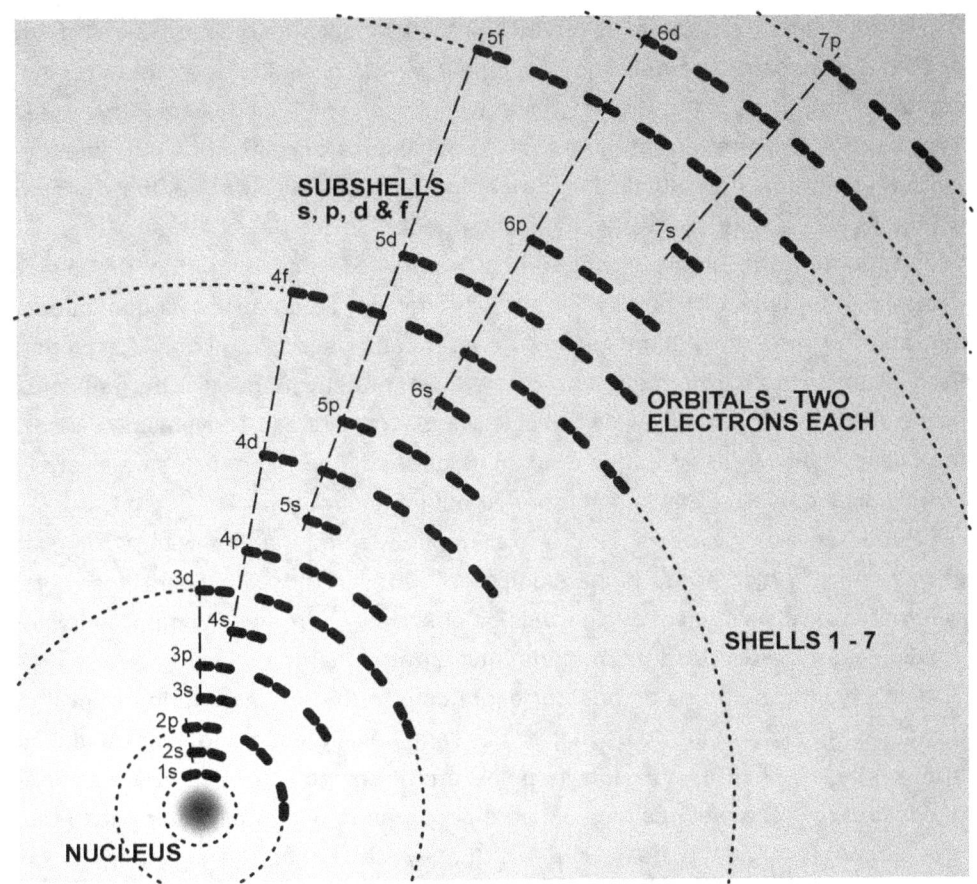

Labels in figure:
5f, 6d, 7p
SUBSHELLS s, p, d & f
5d, 6p, 7s
4f
6s
5p
4d
5s
4p
3d
4s
3p
3s
2p
2s
1s
NUCLEUS
ORBITALS - TWO ELECTRONS EACH
SHELLS 1 - 7

AUFBAU OR BUILD-UP PRINCIPLE OF ELECTRONS

This diagram shows how the electrons of all the known elements theoretically distribute themselves around an atom's nucleus according to the build-up principle. It depicts seven shells (1 -7) and their four subshells (s, p, d & f) that contain either 1, 3, 5 or 7 orbitals. Each orbital consists of two electrons.

Electrons tend to fill lower "energy" levels, or gravitate toward the nucleus, pushed inward and smeared out all around it by the increasing force of the decreasing density of the atom's gravity field, portrayed in section by the inward diffusing background, while the repulsive nature of their electromagnetic fields holds them apart. It always keeps them, along with the protons and neutrons at the nucleus, at their maximum distance from one another, which increases from the center out as gravity's force decreases exponentially.

It's important to note the electrons' outward diffusion. They disperse exponentially because of the geometry of a sphere. It's also important to know that they're not paired up side by side in each orbital as the diagram implies. Their repulsive nature ensures their even distribution over the entire nucleus. The same is true in the outward direction. Gravity's ever-decreasing field density pushes them inward causing them to nestle in between one another, naturally pursuing their most balanced and stable distribution that ends up forming shells and subshells that express symmetrically at each consecutive level.

(18.4 Atoms 8a)

The notion of a charge associated with field density, or intensity, can be applied at the scale of celestial bodies. A galaxy's gravity field is the product of the compounding of all of the individual gravity fields of all of its coalescing material. This sets up an energy potential that establishes a current from the galaxy's perimeter where its gravity field is denser, more positively charged, to its center where it's less dense, more negatively charged.

The material's reactive mechanical search for equilibrium in the ever-decreasing density, or intensity, of a galaxy's gravity field causes its continuous migration inward toward the galaxy's center, always moving from the higher density, positive charged field, toward the lower density, negative charged field, as a current. The decreasing field density toward a galaxy's center eventually causes the collapse and transformation of all of its material into radiant/plasma energy. For elliptical galaxies, this is first radiated out in all directions.

But later as it evolves into a well-defined spiral galaxy with a bipolar expression, it's ejected as high-velocity jets. Considered as a field, these jets being densest at its point of origin would be positively charged as compared to the gravity field that collapsed the material that created them.

We could imagine field lines corresponding to the jets emanating from the galactic center where they'd be at their densest, having just made their transition from matter to field, up through its poles and then arching back down as the radiant energy reconstitutes back into ordinary matter, eventually returning to the less dense field at the galactic plane that would be negatively charged by comparison. (See Diagram **19.1**, **19.2** G Recycling 5a)

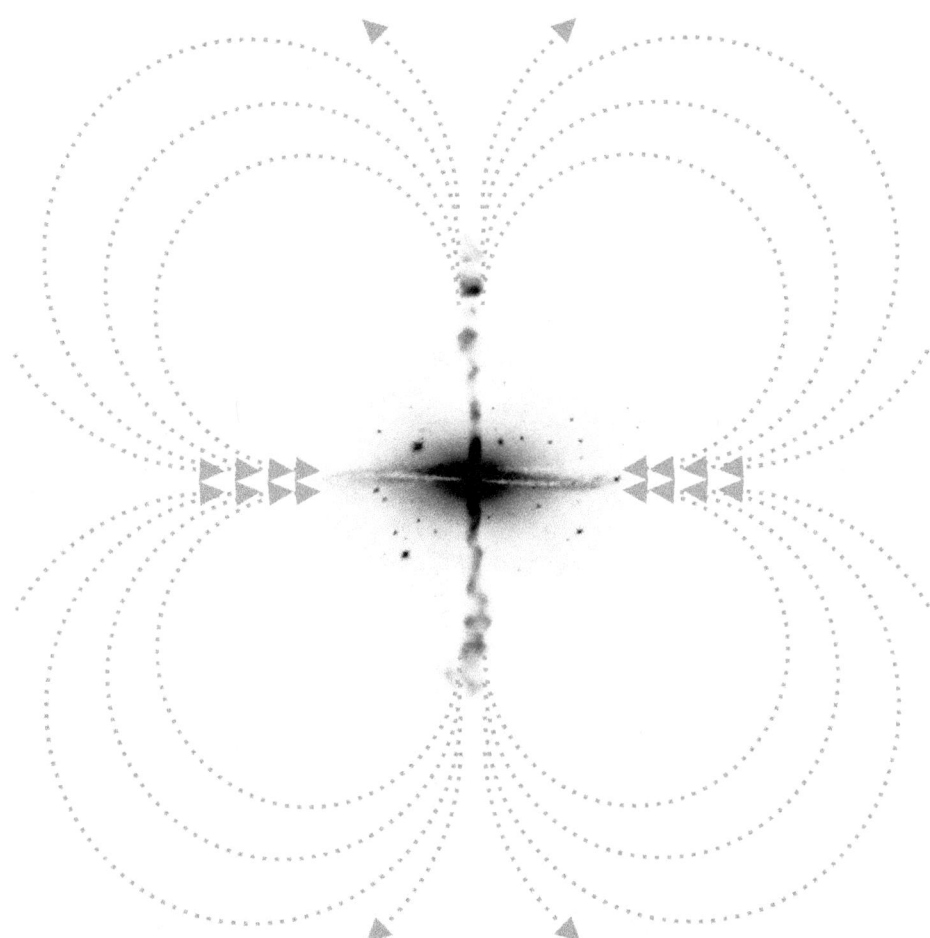

GRAVITY'S PERPETUAL RECYCLING

At the largest scale, protogalactic material begins coalescing spherically, three-dimensionally. Any asymmetry causes it to start rotating, which begins to form an axis. Because the resulting centrifugal force is less at the ensuing poles, it coalesces more readily down the axis toward its center while the more perpendicular material tends to flatten out into the whirlpooling disks that define a fully developed spiral galaxy, shown in side view.

But its coalescing doesn't stop here. Gravitation is a runaway process. It continues unabated facilitated by its exponential condensing that naturally allows its vortex to flatten out. As the material spirals in toward the galactic center, the compression becomes so great that fusion reactions are triggered that begin converting it back into the radiant energy it originated from.

It's then ejected out in huge, high-velocity jets in a bipolar fashion, shown as visible. As the radiation slows and cools, it eventually congeals back into ordinary matter and begins gravitating back toward its or another nearby galaxy, the dotted arrows. The material of every galaxy is subject to this never-ending process of perpetual recycling, which is the real source of cosmological redshift: the Doppler effect from the material's ever-increasing recessional velocity from every other galaxy coupled with our own infall velocity at our own galaxy.

The image is a highly modified black and white negative of galaxy ESO 510-G13 taken by NASA and the Hubble Heritage Team with the HST.

(19.1 G Recycling 5a)

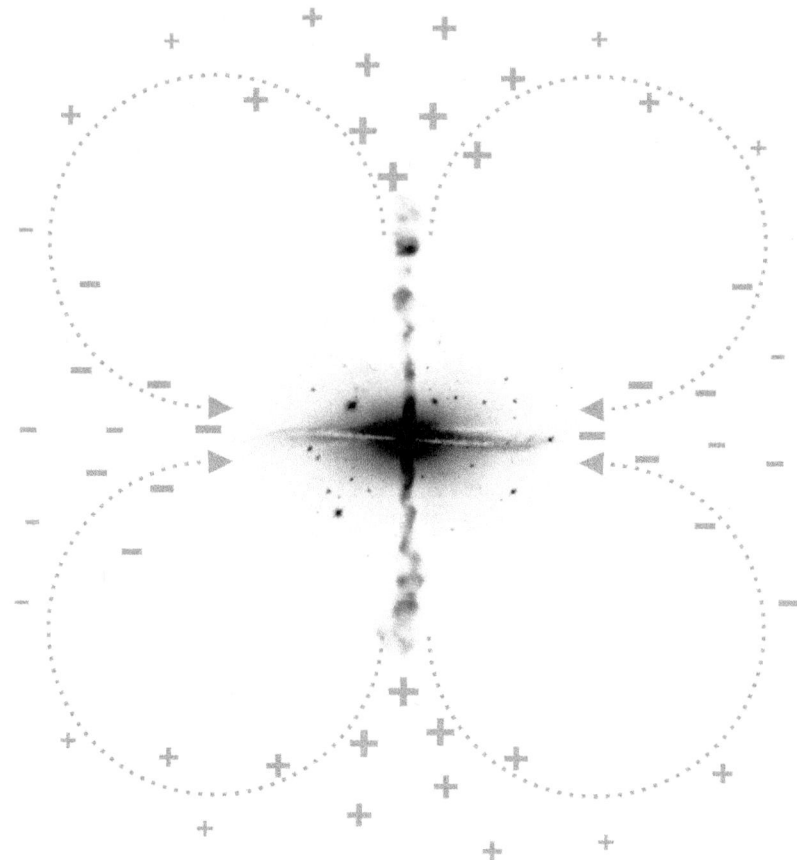

GRAVITY'S BIPOLAR CHARGE

The universal field's exponentially decreasing density that naturally occurs at every object, subatomic particle or galaxy, which is their (electromagnetic) gravity field, manifests a charge and current. This is explicit in the ambient field of mature spiral galaxies.

Their material constantly spirals inward toward their center along their plane in a flattened whirlpooling vortex that's facilitated by their gravity field's ever-decreasing density that's bound to the inverse square law of spherical geometry. That inward, ever-decreasing density corresponds to a negative charge.

The ever-increasing pressure from their gravity field's ever-decreasing density begins to set off fusion reactions near the galaxy's core that eventually convert all matter back into the radiant electromagnetic energy it spawned from. It's then spewed back out perpendicularly in enormous bipolar jets. These jets begin with the highest field density. So they're positively charged by comparison. But they continuously diffuse.

As the radiation slows and cools, it reconstitutes back into ordinary matter. It then naturally starts to gravitate back to its or an adjacent galaxy to repeat the perpetual recycling process. Opposing charges are at their greatest at the galaxy's center as the transformations occur.

The image is a highly modified black and white negative of galaxy ESO 510-G13 taken by NASA and the Hubble Heritage Team with the HST.

(19.2 G Recycling 5a)

Ultimately, there's only one force in nature, electromagnetism. Whether it's at the subatomic or galactic scale, simply accepting gravitational coalescing as the natural consequence of a constant mechanical reactive search for equilibrium in the varying density of the one universal field, gravity is fully reconciled with electromagnetism.

When the universe's radiant electromagnetic energy condenses into "matter," it becomes a compact spherical field that repulses and diffuses outward. The remaining ambient radiant energy that's left outside it that it originated from is decreased commensurately. Its loss naturally compels it to become "matter's" reciprocal. It acts oppositely. It's a field that repulses and diffuses inward. This is a gravitational field. It's no more complicated than that.

Unification only requires that we recognize that "space" is not an actual property of the universe. It doesn't exist. This precludes it from being an attribute of Einstein's cryptic, inconceivable, four-dimensional, non-Euclidean spacetime continuum that impossibly expresses as a nonexistent two-dimensional plane that curves underneath three-dimensional massive bodies to somehow facilitate their attraction. In reality, space is the nothingness between objects. In three dimensions, it corresponds to the universal field. It exists everywhere objects do not.

The universal field's radiant energy, like everything else that's three-dimensional, has an actual existence. So it cannot curve two-dimensionally as a nonexistent plane. Because a field's physical continuity is impossible to disrupt, it has to vary in density. It's the universal field's inherent exponential gradient around objects, the natural result of spherical geometry, that causes their apparent attraction, and repulsion, through their mechanical reactive search for equilibrium. With this fundamental realization, all else falls effortlessly into place.

To be sure, this is far too simple and rational for those academic elites who are bound to established orthodoxy. Perverted by groupthink and blinded by indoctrination, they can't see it. And they don't want to see it. Anything that threatens their entrenched ideology has to be stamped out, immediately. That's how something so easy, eminently practical, and so overtly obvious can remain unrecognized for so long.

There's no genius here, just ordinary common sense that anyone can understand. Elites hate that too. The human mind without fail rejects the realistic and pragmatic in favor of delusion and complication. If we aren't conquering difficult problems, that are always of our own making, we could never feel superior. This is always the subconscious imperative of those who secretly feel otherwise.

It may very well take generations before electromagnetism is fully accepted as the only force in nature. But those who are willing to concede the obvious and confront their conditioned beliefs now can take solace in confidently knowing the truth.

Gist

It may not be true after all that you can't fool all the people all the time. Everyone believes relativity to be the factual preeminent work of a certified genius when in reality it's little more than self-deluded fantasy. It must also be true that the taller the tale, or the bigger the lie, the more people believe it. It's just the way the human psyche works.

For science, this psychological principle might be stated: the more absurd the theory the more it's revered. Relativity's postulates are so absurd they're laughable. But it persists anyway. It even forms the foundation of our entire cosmology despite that its untenability is blatantly obvious, right in front of us for everyone to see. Just a cursory, but objective, investigation easily reveals the false premises, fundamentally impaired logic, and self-conflicting concepts that undermine virtually all of it, including the big bang.

Those reputed experts preaching relativity could never admit, not even to themselves, that they don't comprehend it. Their indoctrination may have had them study and learn it, committing it to memory. But they do not and cannot understand it because it's not rationally comprehensible.

For those who delusively believe that they really do grasp it, not realizing that memorization is not comprehension, the very best they could ever hope to achieve is a parroting of its ludicrous tenets. They may be able to regurgitate it as if it were fact. But it's just not intelligible or in any way rationally interpretable. They might agree that it's difficult. But they don't realize the obvious. It's so hard because it's nonsense. It does not work. I'm convinced Einstein was aware of its nullifying failures. But he persisted in huckstering his "new physics" anyway.

When applying a little objective reason to his convoluted logic, it quickly becomes apparent that his emphasis is not on conveying his presumed insights but on rationalizing his outrageous ideology at any cost. With a patronizing narrative that's wrought with incoherent complication, he intentionally misdirects and obscures rather than reveals and explains while implicitly employing an "Emperor's New Clothes" ruse that effectively dupes any avid elitist.

So instead of calling its bluff, openly challenging its overt deficiencies, and labeling them as such, we're easily swayed by its appearance and snake-oil presentation over fact, substance, and reason. Impressed with its facade, especially its excessive mathematical gyrations, we're quickly taken by Einstein's high-sounding technical rhetoric as he condescendingly assures us in an overconfident, serious manner with tortuous, deceitful explanations while implying only the wise and intelligent are capable of comprehending his "new physics."

Embarrassed we can't make sense of his irrational but legitimate-sounding discourse, we never even consider disputing his outlandish assertions. Instead, we stand there nodding our heads in agreement, anxiously endorsing his foolish ideas, secretly pretending that they're logical and that we too understand his nonsense, fearful that someone may realize we really have no idea.

Our elevation of Einstein to genius, possibly even the most intelligent person to have ever lived, has created a shameless snowball effect. Throngs are now coming forward clamoring that they too have conquered relativity. Their insecurity has them ridiculing anyone who might suggest that it's impractical, referring to them as a crackpot or a simpleton or even an idiot mentally unequipped to grasp it. Threatened by positions they can't refute, they resort to insults. That's always the strategy of lower consciousness.

So it's completely understandable why we rarely see anyone seriously question its validity. Only someone outside the profession with nothing to lose is in a position to step forward and expose its flaws. Why risk appearing ignorant and a simpleton to those who can't see or incur the wrath of others who secretly doubt their own understanding when everyone knows only the brilliant and sophisticated, the educated elite, are capable of comprehending its meaning. It's much easier to just stay in line and continue touting groupthink dogma, hoping to not be discovered.

The truth is, his deliberate effort to deceive and obfuscate to rationalize his incoherent theories at any cost while being fully aware of their untenability goes beyond an ordinary narcissistic obsession. Some may be inclined to ascribe a sociopathic aspect to his behavior.

What's just as troubling is his wanting spatial awareness, his inability to perceive spatially, to visualize in three dimensions as most of us do naturally. It'd be a mistake to attribute this to his dyslexia (a debilitating learning disorder that it's widely known he struggled with). This is much different.

Regardless of how his personality flaws are characterized, they get at the source of what's responsible for the confused mess that's relativity. No one active in the field can now consciously allow themselves to ever arrive at such a controversial yet reasonable and obvious conclusion, much less dare to state it publicly. It'd be tantamount to committing professional suicide.

Still, it can easily be shown that in our real nontheoretical world of three actual dimensions relativity fails in all areas. Whether it's his incorrect underlying premise of light's constancy that only works theoretically in the one abstract dimension of linear motion, or his mass-energy equation that's based on light's impossible fixed velocity, or his incorrect inference of a "principle of equivalence" that has acceleration/braking and rotation creating gravity fields despite that they aren't the same at all, or his inference of an inconceivable four-dimensional "spacetime" that's an amalgamation of nonexistent space with nonexistent time that impossibly expresses two-dimensionally as a nonexistent plane that impossibly curves as it dents underneath three-dimensional massive bodies to cause their attraction, or his fanciful speculation of a finite yet somehow unbounded non-Euclidean universe that also impossibly expresses two-dimensionally like the surface of a sphere so it can remain uniform, none of these elemental discrepancies have even a remote possibility of ever being workable. And there's much more.

To suggest Einstein was spatially challenged would be a huge understatement. Yet, incredibility, we rush to preemptively excuse all of his dimensional incongruities, insisting that they're esoteric genius. Some have even elevated him to such a height that for them he's revered as something of a Messiah.

When confronted with relativity's discrepancies, disciples will usually argue that they can be reconciled but only through its complicated math, which they'll often quickly concede to not fully comprehend themselves. But they don't comprehend it, or relativity in general, because it's not comprehendible. It's representative of an inconceivable reality. There's no amount of even the most sophisticated math that can ever conceptually validate:

- how the speed of light can be constant when it departs a moving source at 186,000mi/s in every direction (three-dimensionally) at the same time,
- how the speed of light can simultaneously remain constant both in the direction of motion and perpendicular to or opposite the direction of motion when time's slower rate has to apply equally over the entire reference frame,
- how the speed of light can remain fixed when objects contract only in the direction of motion while light radiates in all directions three-dimensionally,
- how the speed of light can be variable in gravity fields but fixed outside of them when gravity fields extend indefinitely and every object from particles to galaxies or even the entire presumed finite universe has its own self-gravity,
- how the speed of light can be variable in a gravity field of curved space of constant density when its velocity can only vary in a medium that varies in density,
- how Einstein can assert that relativity would completely unravel if it were found that the speed of light was not constant while still insisting its velocity is variable in gravity fields,
- how time can be an independent constituent of the universe set by the universe's expansion but can be altered by an observer's subjective choices of motion,

- how time can change independent of the physical reality that established it when we're solely responsible for determining its rate through our selection of an object with periodic motion,
- how someone in motion, experiencing a slower rate of time, would not see everything outside of their reference frame as contracting in size to maintain light's constancy or record a speed of light that exceeds 186,000mi/s,
- how time's rate would slow, becoming larger with motion, when an object's corresponding contraction in the direction of motion and light's slower velocity would require a smaller or faster running time,
- how someone in motion halfway between two simultaneous events would see the one in the direction of travel first when motion is suppose to cause time to run slower not faster,
- how that same person would see the event in the direction of travel first when any change in time's rate would have them observing both events simultaneously but only at a different time than if stationary,
- how it's impossible for an occupant of a moving reference frame to detect their contraction when they could easily observe any angle between the vertical and the horizontal parallel to the direction of motion increase with increasing velocity,
- how the relative motion of any body can be a subjective choice of each observer when any arbitrary choice for celestial bodies would violate the laws of gravitation and conflict with the choices of other observers,
- how a particle's mass increases not from the charge that's induced by subjecting it to an electromagnetic field but only from its increasing velocity,
- how mass and energy can be interchangeable when the energy type is anything other than electromagnetic,
- how an accelerating object can become infinitely large at the speed of light while simultaneously becoming infinitely small because of its contraction in the direction of motion,
- how the relativistic effects of special relativity, time's slowing, length contraction, and the increasing mass of accelerating objects are only valid outside of gravity fields where light's velocity is constant when gravity fields extend indefinitely and every object from particles to our entire presumed finite universe has its own self-gravity,
- how gravitational time dilation can be valid when it's derived from special relativity's relativistic effects that are only valid outside of gravity fields,
- how Einstein's mass-energy equation, $E=mc^2$, that's dependent on light's fixed velocity can have any validity outside of an accelerator,

- how gravitation can be mechanically propagated instantaneously by spacetime's non-Euclidean curvature while simultaneously propagated at the speed of light by a force that expresses similar to electromagnetic energy,
- how gravitation can be propagated by curved space when by definition space is the nothingness between objects,
- how three-dimensional space can curve when curvature is a property of only one and two dimensions,
- how space can express as a two-dimensional plane when a plane by definition is only a location that can have no existence,
- how Einstein can assert that light's velocity remains constant as it curves following spacetime's two-dimensional geodesic while also contending its refracted slowing that has its velocity varying with position in gravity fields,
- how Euclidean properties can only apply at smaller everyday scales while curving non-Euclidean properties govern at larger scales when gravity is supposed to curve space at all scales over an infinite distance,
- how a single object, or reference frame, can express an infinite number of rates of time at all of its spacetime coordinates simultaneously,
- how space can stretch under gravity while the objects in it contract.
- how space can stretch while its density remains constant,
- how space can stretch in the direction of massive objects but contract in the perpendicular direction causing infalling objects to spaghettify,
- how an object can endlessly fall toward the infinite condition of a singularity when in three dimensions an infalling object would arrive almost instantly at a common center of mass,
- how acceleration can be inferred as gravity when its reaction is uniform, acts one-dimensionally, is mechanical and instantaneous, and it doesn't coalesce and gravity's reaction is nonuniform, acts three-dimensionally, presumably acts at the speed of light, and it does coalesce,
- how acceleration can be inferred as gravity when gravity doesn't require any motion,
- how acceleration-created gravity can be real when an object's motion and the rate of that motion is subjectively determined by each observer,
- how any of the effects of special relativity can manifest for accelerating objects when those effects can only occur outside of gravity fields, which would have to include acceleration-created gravity fields,

- how a perpendicular ray of light projected from someone experiencing acceleration can travel a curved path, evidencing an acceleration-created gravitational field, when each quantum of the projected light would always travel a straight path from the location of its origin,
- how the centrifugal force of a rotating object can be inferred as gravity when it acts outward two-dimensionally and gravity's acts inward three-dimensionally,
- how the centrifugal force of a rotating object can be inferred as gravity when it becomes stronger with distance while gravity's becomes weaker,
- how the centrifugal force of a rotating object can be inferred as gravity when it disperses objects outward while gravity coalesces objects inward,
- how the centrifugal force of a rotating object can be inferred as gravity when it requires the rotation of an object while gravity's force requires no rotation,
- how the centrifugal force of a rotating object can be inferred as gravity when its reaction is mechanical, which is essentially instantaneous, while gravity's is presumed to act via waves and/or particles at the speed of light,
- how any of the effects of special relativity can manifest for rotating objects when those effects can only occur outside of gravity fields, which would have to include centrifugal-created gravity fields,
- how an object can have three conflicting types of gravity fields, one originating innately from its mass, another from its acceleration, and a third from its rotation,
- how the perimeter of a rotating disk can contract while its radius remains constant,
- how the surface of a rotating sphere can contract while its radius remains constant,
- how it is that time's rate can vary from the center to the edge of a rotating disk, or from the poles to the equator of a rotating sphere, due to the varying rate of their rotation from the center out when they're of a single reference frame,
- how redshift, established by an atom's reduced frequency due to its rapid rotating motion around a massive body, can indicate gravitational potential when time's rate would vary from zero at the body's poles to its slowest rate at its equator,
- how redshift, established by an atom's reduced frequency due to its rapid rotating motion around a massive body, can indicate gravitational potential when the phenomena of special relativity can only occur outside of gravity fields,
- how Einstein can claim that if redshift does not indicate gravitational potential then relativity would be untenable, but still accept that space is expanding based on galactic redshifts as an indication of recessional velocity,
- how it can be a subjective choice for each observer that the Earth is not spinning about its axis but that the entire solar system along with the rest of the universe is revolving around it,

- how Einstein can define three principles of relativity, special, general, and the exact general, when each succeeding principle supersedes the previous and a principle is supposed to be fundamental unchanging truth that serves as a foundation,
- how inertial mass and gravitation mass can be legitimate when there's only mass, a measure of the amount of material an object contains,
- how Einstein can first identify inertial mass and gravitation mass as different so that he can then later determine them to be the same through his "principle of equivalence,"
- how the universe can ever be considered finite when by definition it would include the infinite space beyond its visible space,
- how the universe can be finite yet unbounded when finite and unbounded are inherently conflicting qualities,
- how the universe can express two-dimensionally like a sphere's surface when existence in two dimensions isn't even conceptually possible,
- how strings can be stretched straight from a single point, radiating out in all directions three-dimensionally while also diverging two-dimensionally, confined to the universe's curving spherical surface, until they reach their maximum separation where they begin to converge, finally running together at a counterpoint to their starting point,
- how a two-dimensional universe that expresses like the surface of a sphere can have a radius when conceptually it would be outside all that we can know,
- how someone could look out in any direction (three dimensionally) with a powerful enough telescope and see the backside of our galaxy,
- how a real three-dimensional finite spherical universe can expand uniformly when the volume of a sphere can never express uniformly,
- how a real three-dimensional finite spherical universe can expand uniformly when the inherent geometry of a sphere's expanding volume and its self-gravity would cause it to dissipate and condense exponentially per the inverse square law.

But an unsentimental objective consideration of each of these elemental and unresolvable contradictions will reveal their folly and prove relativity to be totally unworkable.

We can summarize relativity's absurdities another way by reviewing what changes Einstein would have to implement to correct his understanding. First and most important, he'd have to let go of his incorrect premise, light's constancy, and accept that it's conceptually and physically impossible in our real three-dimensional world. Light's velocity has to compound motion.

Then in addition to its compounding, he'd need to fully embrace its variability under all conditions. It refracts through the varying density of any electromagnetic/gravity field. It appears that he discerns this but can't bring himself to let go of its fixed velocity. So he tries to qualify it after the fact as existing outside of gravity fields, which is impossible. Gravity fields are everywhere.

Had he embraced light's compounding and variability, he would have never concluded any of his principles of relativity. They were all founded and are dependent on the assumption that light's constancy is a natural law.

He'd also need to realize that there's no such thing as time. That it's not a property of the universe. It doesn't exist. We create it ourselves. We establish its rate through our selection of objects with periodic motion to use as reference. So it cannot change with motion because of light's (assumed) fixed velocity.

In concert with this, he'd need to also recognize that motion cannot be arbitrarily decided on a whim. All motion, but especially gravitational motion, is not necessarily relative. So time's dilating rate (or length's contraction or mass's increase) cannot be changed by our subjective choices of motion.

Had he also correctly perceived light and time's three-dimensionality, he would have realized that time dilation, length contraction, and the increased mass accelerating objects only work theoretically, in the one abstract dimension of linear motion. In our real three-dimensional environment their rates inherently conflict with those in every other direction than directly forward.

With this understanding he would have realized that subatomic particles subjected to electromagnetic fields acquire a charge that increases their size and mass, which slows their natural frequency, not their rate of (nonexistent) time. He would have then seen that the mass of accelerated particles, and more generally any object, does not increase as its kinetic energy from motion increases due to light's constancy. It's really just the increasing electromagnetic field that's accelerating the particles that's inducing them with a charge, which is increasing their mass that's causing their natural frequency to decrease proportionately.

He also could have seen that the same concept would apply to all those airborne clock experiments that appear to demonstrate time's changing rate. It's really the charge induced to the atomic clocks' cesium atoms from the motion through the Earth's magnetic field that's increasing their mass that in turn is slowing their rate of operation, not time's rate.

For those experiments with clocks situated at different elevations where time appears to run faster with altitude, it's only a reduction in the size of their cesium atoms in the decreasing density of the Earth's magnetic field that's increasing their natural frequency that in turn causes the clocks' rate the run faster at higher elevations.

Another concept he'd need to embrace is space's nonexistence. It's actually the nothingness between objects. So it can't have any affect on them. But along with that, he'd need to recognize that there is something that does exist that corresponds

188

to all space, the universal field of radiant electromagnetic energy. It's the fundamental constituent of the universe. But it's not two-dimensional. It's three-dimensional. So it cannot curve. Any reduction in it causes its density to vary because its continuity is uninterruptible, inseparable, indivisible.

From here, he would have had the necessary insight to perceive how the universal field has to diffuse inward exponentially when it congeals into a particle because of three-dimensional spherical geometry. This is what defines a gravity field. He could have then realized that it's the compounding of every particle's, and the objects they compose, gravity field that causes their runaway gravitational coalescing because of their natural mechanical pursuit of equilibrium. They're constantly pushed by the highest density toward the lowest that always lies directly in between them toward their common center of mass.

With the basic understanding that the universal field is the elemental constituent of the universe, that it's uninterruptible, that particles condense out of it, and that the exponential gradient of the inverse square law's spherical diffusion has to apply to both outward and inward dispersion, he could have then recognized that there's only one force in nature, electromagnetism. In the most rudimentary but rational way, this resolves all unification problems that he's said to have desperately pursued for the balance of his life.

Had he also correctly perceived that acceleration/braking and rotation do not create gravity fields, he would have never proposed that they're equal to real mass-created gravity fields. He would have then also recognized that there is no such thing as either inertial or gravitational mass, that there's only mass, which has inertia when in motion and weight when resting on the surface of a more massive body. So he would never have first erroneously concluded that mass can be either inertial or gravitational and then mistakenly conclude that they're actually the same thing and then contrive a "principle of equivalence" to enshrine that errant reasoning.

Without a "principle of equivalence" and without light's constancy he would never have reasoned his version of gravitational redshift that's dependent on both. And without his gravitational redshift, he never would have asserted that relativity would be untenable if it were found that his source of redshift was incorrect.

And finally, had he been able to correctly perceive that our three-dimensional universe could never express two-dimensionally like a sphere's surface, he would never have proposed a finite and yet unbounded, non-Euclidean universe. He would have correctly perceived it as infinite.

Without a finite universe, he'd then have no need for his cosmological term, or its (unworkable) substitute Friedmann's expanding space, to prevent it from collapsing in on itself from self-gravity. And without space's expansion, he could then reason that Edwin Hubble's redshift observations could not be the product of each entire galaxy's recessional velocity. But that they're more likely the product of the recessional velocity of each galaxy's constant infall of ever-coalescing material that's continuously condensing and collapsing due to gravitation's innate runaway nature.

This could have led him to correctly perceive the universe as endlessly vast, eternal, unchanging, and static, but in a very dynamic state of equilibrium that's perpetually recycling matter to energy and back through each galaxy's relentless coalescing, the resultant collapse and ejection, and the eventual reconstitution of its material.

I'm confident that most of you who still disagree with this assessment would reverse your position if you would undertake your own truly independent, factual study of relativity, examining not what others have written trying to justify it, but only Einstein's writings. I can assure you, it's not beyond your grasp, as is usually condescendingly implied by those who claim to have conquered it. I would guess that most of those actively campaigning for relativity have read very little of Einstein's own material, but have just bought into what our presumed authorities profess about it, assuming they must be better qualified, more capable, and that they've correctly considered it.

Those reputed experts, many of whom are self-proclaimed, are not freethinkers with unbiased perception and original insight. For years or even decades, they've been subjected to intense academic conditioning and groupthink's coercing peer pressure. This has severely obstructed their analytic abilities. They're left only with their own confused programming, not insight and understanding. Secretly hoping to advance their position and authority, they try to demonstrate their superior understanding by parroting relativity's indefensible doctrine.

For anyone who's unduly influenced by preprogrammed beliefs, as we all are to some degree, but who still has an authentic desire for the truth, in any field of study, it only requires a willingness to free ourselves from our preconceived notions by first admitting that we really don't know. That's all it takes. We then only need to allow the result of our investigation to be what it wants to be, not what we'd like it to be. Accept it as it is no matter the ramifications.

But instead, our controlling confirmation bias unconsciously has us twisting the evidence into palatable rationalizations that we believe will be politically acceptable, personally profitable, or egotistically gratifying. Obviously, this should be the objective of any empirical research. It's simple and obvious, but it's more difficult than it sounds.

I strongly encourage you to accept the challenge of independent research and decide for yourself. But read carefully. Einstein has quite a talent for obfuscation and misdirection, which is the real source of confusion, not the subject matter. Still, with even a small measure of objectivity, I'm certain that those of you with enough courage to undertake such a task will be just as surprised as I was to discover how groundless and irrational our entire relativity-based cosmology actually is, but more so, how delusive and credulous humanity can be.

Academic elites will never make such a recommendation. Advocating for their sacred scripture, they'll argue that you'll just be wasting your time. Relativity's been proven beyond any doubt. You'll be much better off remaining a follower like the rest of us where there's comradeship and safety in numbers. This is what we all believe. It must be correct. If it wasn't, we'd all be wrong. And that can't be. God forbid groupthink's entrenched institutional "consensus" from being tossed aside in favor of just a little common sense.

Coda

Even from just a cursory analysis of his book *Relativity: The Special and the General Theory*, it's clear that Einstein had a steadfast, unyielding, even neurotic obsession to promote his "new physics," presumably to secure a name for himself. This had him interpreting reality through a self-serving bias that led to fantastic speculations and completely irrational conclusions that he justified with misleading and intentionally ambiguous arguments that are further masked with condescension, unnecessary complication, and elusive math.

It's also evident that he didn't arrange his book in the order that the ideas came to him, as he claims. This would have revealed the step-by-step progression of his "reasoning." It appears instead that he backed into much of it with torturous and ludicrous logic as he struggled to rationalize and obscure his outlandish ideology.

It's apparent that he first adopted the premise of light's constancy. Then he constructed his "new physics," from that premise. He did not discover how the universe works. He fabricated a mythology about how it works.

He never confirmed or established light's fixed velocity, or proved its constancy, as many believe. His proof is that, "Every child at school knows, or believes he knows, that this propagation takes place in straight lines with a velocity c = 300,000 km./sec... In short, let us assume that [this] simple law... is justifiably believed by the child at school."[145] And that's essentially the extent of the evidence all of relativity is founded on.

When he later undertook the challenge of gravitation, it appears that he predetermined that the relativistic effects of special relativity would be compatible with and incorporated into his solution despite the inherent contradictions. Rather than allowing gravity's solution to naturally emerge, it seems that he first reached a mathematical resolution through years of trial and error that some still contend doesn't actually work. Then, despite the unresolvable conflicts, he manufactured explanations that attempt to integrate and justify special relativity. His obsession with achieving that end led to convoluted, illogical reasoning in support of impossible conclusions. But it remains unrecognized.

His apparent mathematical success with gravitation and his mass-energy relation, but that he reasoned and described the effects incorrectly accounts for much of the persistent controversy and bewilderment over relativity and our ongoing cosmological confusion in general. In spite of his many conspicuous errors, we're easily duped into buying his narrative because the math appears correct. But also our innate obsession with fantasy, and even more so, our longing for acceptance by peers and those elites claiming understanding, drives an

145. Einstein, *Relativity*, 21.

eagerness to subjectively interpret experimental results through a consuming confirmation bias in support of his completely impractical and unfathomable doctrine.

Our quandary is not dissimilar to Ptolemy's (a second century Greek astronomer and mathematician) geocentric solution to heavenly motion. It had the entire universe revolving around the Earth. The retrograde motion (seemingly backward movement) of some of the planets' orbits was explained away with smaller circular paths superimposed on larger circular orbits. All of which he could validate almost perfectly through mathematics. Of course, the solution was entirely incorrect. But it remained the accepted model for nearly fourteen hundred years.

And here we find ourselves with Einstein's presumably irrefutable mathematical solutions to gravitation and his mass-energy relation but explanations that are absolutely unworkable. We gullibly take it for granted that if the math seems correct then the associated rationale behind it must be also, even though it's clearly physically impossible.

Because we have little appreciation for actual content and results, we're often easily deceived by appearance and performance over substance, fact, and reason. Swayed by his patronizing, presumptuous, overconfident discourse, along with his relentless promotion, but now also lacking an ability for factual objectivity because of groupthink's corrupting influence and indoctrination, we're unable to distinguish between truth and fantasy. Relativity sounds right, having all of the correct buzzwords and portentous technical rhetoric, and looks good because of the deluge of complicated equations. But any factual investigation quickly reveals that it's completely hollow.

After stripped away its obscuring math and exposing its deceitful narration, an objective dissection of his "explanations" reveals relativity to be completely fabricated and totally irrational, the product of a restless individual's invented imaginings, unsound logic, an inability to perceive spatially, patently incorrect premises, and just as significant, an unrelenting narcissistic desire to justify his concocted theories at any cost.

Despite all this, we still absolutely refuse to acknowledge its shortcomings. But instead, we rush headlong to further promote his already highly overinflated status, believing him to be the consummate scientist, maybe the smartest person to have ever lived, while secretly judging ourselves as somehow deficient because of our inability to comprehend his untenable ideas.

Part Two
The Big Bang Bamboozle_____

It's been said that the understanding of what something is first begins with the understanding of what it is not. But many in the sciences, especially cosmology, consider it improper and ill-mannered to criticize another's ideas or work or demonstrate its shortcomings. It's kind of like the unspoken agreement in a country club. No one is allowed to challenge another's delusions.

Fear of criticism often originates from a general sense of insecurity. For cosmology, that insecurity would not be unfounded. With the implausibility of the big bang, and relativity, we can easily understand the concern. Many would consider it catastrophic if their irrationality and invalidating flaws were exposed for all to see.

Just imagine if you will the profound embarrassment and humiliation, the long list of defunct projects, the years of wasted research, the huge wealth of lost funding, and especially the number of squandered careers if it were suddenly recognized, understood, and undeniably established that the big bang, along with relativity, was conceptually and physically impossible. After dedicating so much of their professional lives to the study of what turned out to be a completely imaginary universe, do you believe that more than just a handful would have a constitution that would permit such a realization? That's not very likely.

What's more, proposing or even just endorsing the exploration of such an oppositional and scandalous notion would be tantamount to professional suicide. You'd be summarily excommunicated for threatening the status quo.

Inhibiting the free exchange of ideas in any way of any scientific endeavor is ludicrous. Never is it justified, even if it's to preserve someone's reputation, even after they've passed. Still, we have without doubt succumbed to political tactics that routinely trade truth for personal gain and peer acceptance. Politics, of course, has no place in real science. Besides, the truth always prevails in the end. And those who resist it inevitably just end up looking foolish.

To omit any negative criticism, an explanation of why something cannot be, is not rational inductive reasoning. This is crucial because it only requires one observation of conflicting factual evidence to disprove any entire theory. If any part of an idea is false then the entire idea must also be false, which includes the reasoning behind it.

Empirical investigation in any scientific venue requires ruthless objectivity and a steadfast ability to confront hard facts regardless of our personal desires, bias, or especially the consequences of our findings. Also, employing rigorous unyielding logic can itself provide irrefutable proof even when the interpretation of evidence appears to be subjective. Those proceeding in any other way are engaged in pseudoscience. They're just pretending to be researchers.

And really now, who'd want to continue any research that's based on invalid premises, ill-conceived theories, or faulty reasoning? That'd be silly. And you'd think any sensible individual would be grateful to have their false pursuits identified so they could make the necessary corrections and enjoy a meaningful career. But helping another to realize their errors, even though it may be purely constructive criticism, is risky. Rarely is it welcome, much less the suggestion of reasonable alternatives. Usually, it's met with fierce resentment. Any threat to conditioned beliefs is a threat to the ego.

And even though we'd all generally agree with this assessment, when personally confronted with our own foolishness, we all still react the same. It's just human nature. The human personality would rather continue pursuing a lifetime of the counterfeit, striving for artificial prestige, false power, and acceptance from others, especially peers, rather than experiencing the unparalleled satisfaction of true discovery.

Discovering truth is simple, albeit difficult. It only requires that we eliminate our belief in that which is untrue. But first we have to summon enough courage to admit to ourselves that it's possible that we may actually harbor false beliefs. That's not easy.

Most people "believe" in the big bang. They've not factually considered it, so have little reason to doubt it. They're also unaware of Einstein's pivotal influence. It's based entirely on his whimsical speculation that the universe might be finite.

His adoption of Riemann's "discovery" of two-dimensional space allowed him to theorize a universe that expressed two-dimensionally like a sphere's surface so that gravity could be resisted uniformly by his cosmological constant, a mathematical term that would coincide with the homogeneous distribution of the universe's material. In three dimensions, the gravity of a finite universe would condense its material exponentially inward toward its center.[146]

After Edwin Hubble correlated the increasingly redshifted light of distant (appearing) extragalactic objects with their decreasing brightness in the late 1920's, we've interpreted these redshifts as originating from the recessional velocity of each entire galaxy being swept away by expanding space, or more recently space's stretching. So Einstein accepted Friedmann's idea of expanding space as a substitute for his cosmological constant.[147]

---

146. Einstein, *Relativity*, 122-127.
147. Einstein, *Relativity*, 152-154.

But a three-dimensional, finite universe that was expanding would be diffusing exponentially. So here we are a century later still stuck with an utterly inconceivable, contradictory, finite yet unbounded, curving non-Euclidean, uniformly expanding big bang universe that impossibly expresses two-dimensionally like the surface of a sphere. And we absolutely refuse to see the absurdity of this nonsense, holding fast instead to our false beliefs despite their obviosity.

Wonder

A fundamental axiom for success, no matter the endeavor or how you define it, is that 95% of all people are incorrect in their assessment of reality. Consequently, they're unsuccessful at whatever it is they're attempting. To be successful then, we might naturally choose to not model our pursuits after anyone else's. Doing exactly the opposite would be well advised. The 5% of successful people understand this.

And even though they continue to flounder, unsuccessful people, not surprisingly, justify their unprofitable pursuits by citing their overwhelming numbers. So they continue traveling their unproductive paths deluded by the belief that they can't all be wrong even when it can be clearly shown they are. Successful individuals are also aware of this paradox.

Well, our currently accepted big bang-relativity dogma, which not unexpectedly is endorsed by virtually all of academia, is not exempt from this axiom either. What began as Einstein's fanciful musings of a finite yet unbounded spherical universe, which would naturally collapse in on itself because of gravitation but was instead held in check, mathematically, by his cosmological term that equated to an omnipresent uniformly counteracting force that would preserve the universe's static homogeneous condition if it exhibited two-dimensional properties like a sphere's surface,[148] has evolved into today's widely accepted but hopelessly convoluted and totally unworkable big bang logic.

Supposedly, our entire finite universe somehow condensed itself down to a size smaller than an atom, or began from nothing. Then for some unknown reason, it exploded into nothingness, not even empty space. Its expanding material somehow spread outward uniformly, impossibly expressing two-dimensionally like a sphere's surface. But in an attempt to explain early galaxy formation in its otherwise homogeneous condition, most have adopted the view that just after its inception it somehow underwent a brief period of very rapid, faster-than-light inflation. But then it slowed for some period of time while still continuing its uniform expansion. But now it's imagined to have started accelerating again. It's thought that an undetectable repulsive dark energy that comprises about 68% of the universe is expanding space at an ever-increasing rate as it begins to overcome the gravitational attraction of an unseeable dark matter that comprises about 27%. The remaining 5% being ordinary matter.[149]

To the nonprofessional even this quick generalization sounds suspect and highly improbable. Less affected by indoctrination and free of groupthink peer pressure,

---

148. Einstein, *Relativity*, 122-127.
149. "Big Bang," Wikipedia, last modified Mar 19, 2023, https://en.wikipedia.org/wiki/Big_Bang.

it's much easier for them to recognize the gimmickry and lack of common sense. But as we continue to patch together evermore surreal rationalizations, a noose of irrationality will be tightening. Eventually, it will compel us to confront the impossibility of the big bang along with virtually all of relativity.

Let's consider some of the main arguments for and against the big bang.

Basic Arguments For:
- Cosmological redshift - Each galaxy's redshifted light is theorized to be the product of the big bang's uniform accelerating expansion that was initially thought to originate from a Doppler effect from each galaxy's recessional velocity. But now it's believed that the big bang's expansion is stretching space that stretches a galaxy's light toward the red end of the electromagnetic spectrum.
- The existence of cosmic microwave background radiation - Predicted by theory to be the byproduct of one of the earliest phases of the big bang.

That's it, only two. The first is conceptually impossible, and the second is inferred from that conceptual impossibility.

Basic Arguments Against:
- The universe by definition is infinite. It cannot be finite.
- There's no explanation for what initially triggered the big bang.
- There's no explanation for what came before the big bang.
- The entire universe was created from nothing.
- The universe originated from an infinitely small singularity.
- Or, all the matter and radiant energy of the entire universe was unimaginably condensed into a tiny volume smaller than a single atom just prior to its eruption.
- There's no explanation for what caused the universe to condense into such a compact state, or singularity.
- There's no explanation for what's outside of the big bang universe.
- The big bang has space inconceivably expanding into nothingness, not even empty space.
- Space itself, being nothingness, cannot expand or stretch.
- It's conceptually impossible for our three-dimensional universe to express two-dimensionally in the shape of a sphere's surface.
- It's conceptually impossible to exist in two dimensions.

- It's conceptually impossible for a three-dimensional spherical universe to express two-dimensionally like a sphere's surface so that its rate of expansion could slow, or two-dimensionally like a flat plane so that its rate of expansion could remain constant, or two-dimensionally like a saddle shape so that its rate of expansion could accelerate.
- These three potential two-dimensional "analogies" of the big bang universe have no real world counterpart in our actual three-dimensional existence.
- It's conceptually impossible for any three-dimensional volume of something to curve. Curvature is a property of only one or two dimensions.
- The notion that the big bang's space is uniformly curved into the shape of a sphere's surface at the largest scale to counter gravitation is conceptually at odds with the notion that space's nonuniform curvature at massive bodies causes gravitation.
- It's conceptually impossible for a spherical universe to not have a center.
- It's conceptually impossible for an expanding universe to not have an origin or a center beginning location.
- It's conceptually impossible for an expanding universe to not have a center beginning location but still have a radius.
- It's conceptually impossible for an expanding universe's origin or center beginning point to be everywhere, or halfway between every other galaxy.
- It's conceptually impossible for a two-dimensional universe to be confined to the surface of a sphere but still have a radius.
- Being able to look just another billion years past the farthest galaxy to observe the universe's primeval beginnings is conceptually impossible.
- Being able to look out in any direction and see the universe's primeval beginning is conceptually impossible.
- Being able to look out in any direction and see the backside of our own galaxy is conceptually impossible.
- Being able to see the backside of our galaxy in any direction is conceptually at odds with being able to see the universe's primeval beginnings in any direction.
- Being able to set out straight in any direction and arrive back where you started is conceptually impossible.
- A homogeneous, finite, three-dimensional big bang is not possible because of the inherent nature of a sphere.
- A homogeneous, finite, three-dimensional big bang is not possible because of self-gravity's exponential condensing.
- A homogeneous, finite, three-dimensional big bang is not possible because of expansion's exponential diffusing.

- If a homogeneous, finite, three-dimensional big bang was possible and did exist, we'd see a condensing two-dimensional array of galaxies across the entire sky that was least dense in the direction of our outward-bound direction of travel where the universe's perimeter would be its closest. It'd be most dense exactly opposite our outward-bound direction of travel in the direction of the universe's origin where the greatest number of galaxies would be. We'd see this same pattern whether our finite, universe was expanding or not, or uniform or not.
- For an expanding spherical universe to remain homogeneous it'd have to be constantly creating new matter exponentially out of nothing at its expanding perimeter.
- An expanding spherical universe that constantly created new matter at its expanding perimeter would exhibit a spectrum of galaxies whose ages ranged from oldest at its center to newest at its expanding perimeter.
- Einstein's cosmological constant would not be able to counteract the exponential condensing of the self-gravity of a real three-dimensional finite universe. They're mismatched.
- Space's uniform expansion would not be able to counteract the exponential condensing of the self-gravity of a real three-dimensional finite universe. They're also mismatched.
- The self-gravity of a three-dimensional finite universe would cause the strength of gravity to be different at every body.
- The expansion of a three-dimensional finite universe would cause the strength of gravity to be different at every body while varying over time.
- The universe's rate of uniform expansion could never fluctuate to allow matter's initial coalescing.
- The universe's uniform expansion would never have allowed matter's initial coalescing.
- The universe's uniform expansion would inherently conflict with its simultaneous condensing into bodies from gravitation.
- Any density increase in the universe's primeval radiation would not have led to coalescing.
- Cosmological redshifting from a galaxy's recessional velocity/space's stretching would place us at the exact center of a three-dimensional universe.
- If space doesn't exist then galactic redshifts can't originate from its stretching.
- If space actually did exist and its stretching did produce redshifting then any massive body that wasn't receding that would still be surrounded by their gravity's stretched space should exhibit a redshift from that stretching, but they don't.
- If space actually was stretching, it'd conflict with an object's, or a galaxy's relativistic condensing with motion.

- If galactic redshifts did actually originate from space's stretching then the pattern of an object's spectral lines should appear as progressively stretching in a progressively stretching spectrum, not just displaced toward the red end of a static spectrum.
- If cosmological redshifting originates from space's stretching then blueshifted galaxies must originate from space's contraction, which is inexplicable.
- If redshifting from even one other potential source, like tired light's energy loss over distance, atoms with an induced charge, relativistic time dilation, light's slower velocity in gravity fields, gravitational redshifting from light's loss of energy, Einstein's gravitational redshifting from relativistic time dilation, redshifting from the Doppler effect from the recession velocity of galaxies, and redshifting from the Doppler effect from the recessional velocity of each galaxy's constant infall of coalescing material, proved to be viable then it would conflict with our current interpretation of redshifting from space's stretching and we could never know how much redshift to attribute to each. This would prevent us from ever determining the universe's rate of rate expansion, or even inferring that it was expanding, which would completely invalidate the big bang.
- Astronomers only use the redshift from elliptical galaxies when presuming to measure the universe's rate of expansion. The redshift from all the other types is too inconsistent, which suggests that redshift does not originate from universal expansion but from some other source most likely a galaxy's rotation and/or its self-gravity's constant infall of ever-coalescing material.
- The big bang's faster-than-light inflation period violates relativity's cardinal tenet, which is not spatially dependent.
- The compounding effect of space's assumed uniform expansion would instantly cause the matter at opposite sides of the universe to exceed the speed of light, violating relativity's cardinal tenet.
- The notion of a symmetrical, 16 billion light-year cosmic horizon is premised on faster-than-light velocities and a two-dimensional non-Euclidean existence.
- Being able to see almost all the way back to the universe's inception 14 billion years ago but that it has a radius of about 46 billion years suggests a compounding of light's velocity with space's motion in violation of relativity's cardinal tenet.
- Being able to look back and observe the universe's inception is conceptually impossible.
- Being able to look back in time more than 13 billion years almost to the universe's inception would have to make the minimum age of the universe just over 26 billion years old.

- The cosmic microwave background radiation could originate from any source, not necessarily the primeval beginnings of the big bang, and would have to if the big bang is a fallacy.
- The correlation of the cosmic microwave background radiation with galaxies suggests that it originates from galaxies.
- The correlation of the cosmic microwave background radiation with galaxies conflicts with the belief that it originates from the primeval beginnings of the big bang.
- The cosmic microwave background radiation that we can observe is a portrait of how it looks only from our vantage point on Earth. It would look different from everywhere else.
- The cosmic microwave background radiation that we observe is a portrait of how it looks today. Its distribution would appear completely different just after a big bang.

Given this huge disparity, you have to wonder why anyone in their right mind would ever choose to believe in a conceptually impossible, curving non-Euclidean, finite yet somehow unbounded universe that somehow expands uniformly by somehow expressing two-dimensionally like a sphere's surface. It's far more rational to simply accept that cosmological redshift can't be indicative of an entire galaxy's recessional velocity/space's stretching and that we inhabit an infinitely vast and ageless universe.

Deflating the Balloon

Just try to imagine the infinite. You can't. It's a concept that's beyond the grasp of the human mind. Look at an image of the Hubble Deep Field or Ultra Deep Field and try to imagine the universe extending on indefinitely forever without limits. It's just not possible. And then try imagining the infinite number of galaxies associated with infinite size. Same thing. The two trillion that's now being claimed for the big bang is almost as impossible to grasp. But in addition to size and number, we also have to contend with time. Try imagining an eternal universe with no beginning or end, no inception, no beginning of time, no origin, a universe that's always been and always will be. That too is unfathomable. (See Diagram **21** HUDF 2a)

HUBBLE ULTRA DEEP FIELD

Peering into this photo trying to imagine the universe extending on forever gives us some idea of how unfathomable the concept of infinity actually is. With our inherent inability to grasp it, we can see how we're naturally inclined to accept ideas we can comprehend like a finite, spherically-shaped, expanding big bang universe despite that the homogeneous distribution of its matter would be absolutely impossible in our real three-dimensional world.
Credit: NASA, ESA, and S. Beckwith (STScI) and the HUDF Team.

(21 HUDF 2a)

With the idea of foreverness outside our ability to comprehend this sets up an inherent bias for the conceivable, the tangible, the realizable - the finite. Something we can perceive and work with. Accepting a notion as innately incomprehensible is unappealing. So naturally we find ourselves preferring bounded, ponderable, determinable cosmological solutions that we can manage and solve, like a finite universe, despite that it may make no sense whatsoever and have absolutely nothing to do with reality.

But opting for a finite universe presents a myriad of other problems like what's outside the big bang? Its space is inconceivably expanding into nothingness, not even empty space. Just try to imagine that, a place or location that doesn't even have empty space. Einstein and others argue that space is something.[150] It would have to be if it could actually expand, or stretch, or curve. But in reality by definition space is the nothingness between objects. So there's nothing there to be expanding or stretching, or curving.

Even though we're forced to conclude a beginning to the big bang, there's still no rational explanation for what actually came before it or what triggered it. It's all pure speculation. And no matter what we come up with, we're still left back at infinity. And we have to accept that the universe either spontaneously arose from nothing, a concept that isn't physically possible in today's universe, or all its matter and energy was somehow compressed into a tiny volume smaller than an atom, leaving us with the question: What caused its condensing before it burst forth?

Given its immense unfathomable size, it's extremely difficult to imagine how the big bang could have compressed down into such a compact state. It's been calculated that each of our bodies contain about 200 billion atoms. We can't envision how just those could be compressed to the size of a single atom. There are about seven billion of us on the planet now. Our mass comprises an insignificant fraction of the total mass of our planet. Who knows how many atoms it might contain. And it's a small fraction of the entire mass of our solar system. And our solar system is only one of about 200 billion in our galaxy. And our galaxy is only one of about two trillion in the visible universe. That's a whole lot of atoms to have to squish down to the size of just one.

So given our innate bias for the finite, we can see how Riemann's "discovery" of two-dimensional, three-dimensional space could have fueled Einstein's whimsical speculation of a finite yet unbounded universe that has our three-dimensional existence confined to the two-dimensional isotropic properties of a sphere's surface so that its gravity could be uniformly resisted, mathematically, by a constant that prevents its collapse.

150. Einstein, *Relativity*, 155-159.

It's interesting how the big bang's space at the largest scale can be uniform, curved in the shape of a sphere's surface to resist its self-gravity. But at smaller scales, the same space is nonuniform, curved two-dimensionally around every body to facilitate gravity. Isn't that a fundamental conflict?

Apparently, a sphere's inherent inability to express uniformly was not a part of his reasoning that led to a two-dimensional three-dimensional universe, just gravity. Also a sphere's inherent inability to expand uniformly must not have influenced his later acceptance of Friedmann's proposal to substitute uniformly expanding space for his cosmological constant.

Despite that the big bang absurdly expresses two-dimensionally while remaining three-dimensional, it's just conceded without question that this impossibility is somehow reconciled through the magic of non-Euclidean curving geometry. The commonly used model that attempts to illustrate it is that of a balloon that's uniformly covered with dots - the same one we're all familiar with. Each dot represents a galaxy. So as the balloon expands with air each dot moves away equally from every other dot.

The dots' relative motion might at first appear consistent with our observation of the uniform dispersion of galaxies and all magnitudes of their redshifts. We choose to believe that their redshift indicates equal recessional velocity from one another over radial recessional velocity from us that would improbably put us at the exact center of the universe.

But any objective consideration quickly reveals that there can be no practical relationship between the balloon's expanding two-dimensional surface and our actual three-dimensional existence. Employing the isotropic quality of a sphere's surface as the universe was a concocted device Einstein exploited to make gravity uniform so that it could be resisted by a mathematical constant, or later uniformly expanding space.[151]

An isotropic, homogeneous, finite universe can only be a theoretical consequence of and is restricted to the curving two-dimensional surface of a sphere. It can in no way physically manifest in a real three-dimensional environment. A two-dimensional existence is nothing more than theoretical at best. Two dimensions only define a plane that has no thickness or height. It's a location. (See Diagram **22** Balloon 8a, 7.0 Platonics 5a, 7.1 Inverse Sq Sphere 13a, 7.2 Inverse Sq Fields 8a & 7.5 Sphere Volume 4a)

151. Einstein, *Relativity*, 122-127.

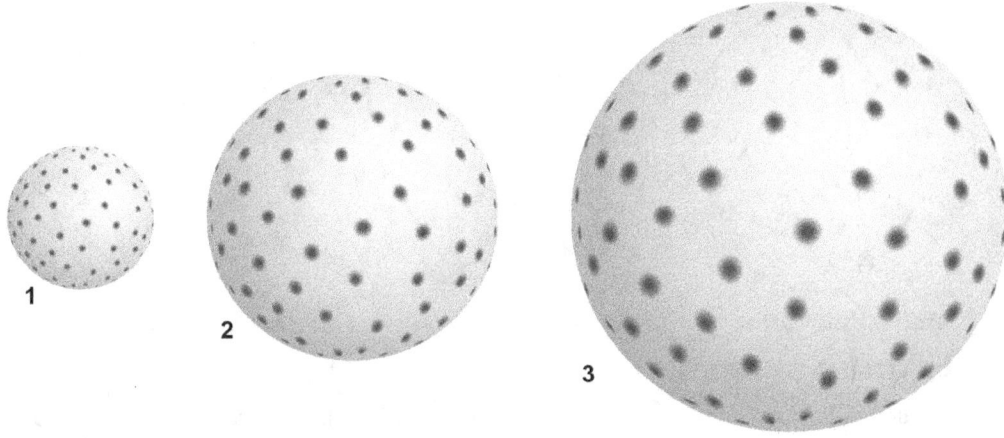

1

2

3

THE BIG BANG'S BALLOON ANALOGY

A balloon with dots on its surface portraying galaxies is representative of Einstein's whimsical musings of a two-dimensional finite yet somehow unbounded non-Euclidean curving universe. With expansion substituted for his cosmological constant that countered gravity's condensing, it has become the big bang. As the balloon fills with air, the dots move away from one another uniformly with the balloon's expansion. The problem is, there's no counterpart for this analogy in three dimensions. Uniform expansion can only take place theoretically on the two-dimensional surface of an expanding sphere.

It's accepted that the universe is homogeneous and isotropic, the same in all directions, just like a sphere's surface. But for a three-dimensional sphere, no conditions exist in our real world where its volume can even express uniformly. Its innate geometry can never be homogeneous. And any expansion would cause the density of its material to diffuse from its center out exponentially per the inverse square law. And it's the same for its self-gravity. It would also cause the exponential condensing of its material.

One way an expanding spherical universe might remain uniform is if it were constantly creating new material exponentially at its ever-growing perimeter that exactly countered its exponential outward diffusion. But even then, we'd still see a difference. With the new material coalescing later, the universe's outer region would be populated with younger and fewer galaxies and with fewer spiral galaxies than toward its center.

Another potential way a finite three-dimensional expanding universe might remain uniform is if its expansion decreased exponentially from its center out, exactly compensating for a sphere's inherent geometry. Theoretically, that might work. But it'd require two types of coexisting expansion, the one that decreased exponentially compensating for a sphere's geometry and another that was uniform that maintained its expansion. Not very likely.

(22 Balloon 8a)

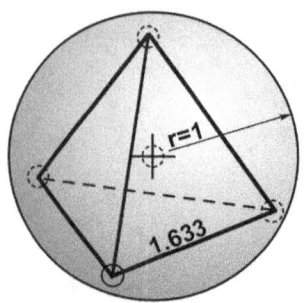 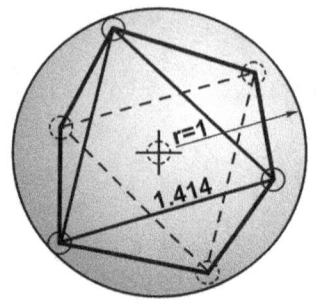 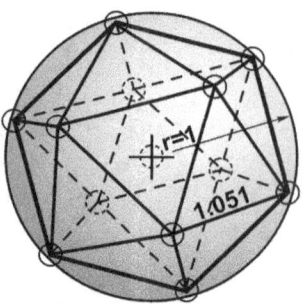

1. TETRAHEDRON   **2. OCTAHEDRON**   **3. ICOSAHEDRON**

PLATONIC SHAPES

The only three platonic shapes in nature that can be composed of equilateral triangles suggest that it's not possible for a spherical universe to be homogeneous. If we portion out any quantity of matter, be it subatomic particles or galaxies, into a spherical shape, we find that their uniform distribution is mathematically and physically impossible. Placing a galaxy at each of the triangles' vertices, or corners, and one at the sphere's center shows how their distribution would be equidistant around the circumspherical surface but always a lessor distance to the center galaxy. This principle of three-dimensional geometry simply reveals how the big bang universe, if it were actually three-dimensional, could not be homogeneous, but would always have to be denser towards its center, which would also establish an observable origin.

With a circumsphere radius of 1 of any unit of measure, the edge of each triangle, or the distance between each galaxy would be 1.633 for the tetrahedron, 1.414 for the octahedron, and 1.051 for the icosahedron.

(7.0 Platonics 5a)

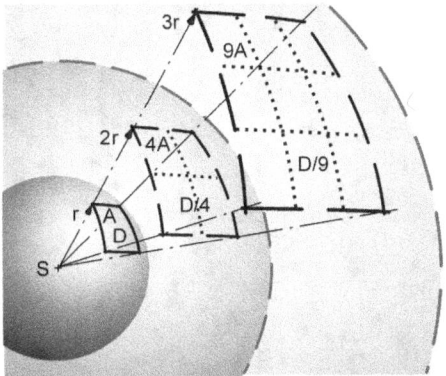

1. SPHERE'S SURFACE AREA & DENSITY
INTENSITY OR DENSITY OF AN EXPANDING SPHERE'S
SURFACE AREA, D α (IS PROPORTIONAL TO) $1/r^2$

INVERSE SQUARE LAW, SPHERE
The surface area of a sphere is $A = 4\pi r^2$. It increases
exponentially as it expands. When twice the radius
from its center, its area increases four times. Three
times the distance, nine times, and so on.

So the intensity or density (they're the same
thing) of any assumed flux (a quantity at the surface)
will dissipate exponentially following the inverse
square law where D (the density at a chosen radius)
= S (the original source density) / $4\pi r^2$ (the area of a
sphere). So its density twice the radius from its
center will be one-fourth its original. Three times the
radius, one-ninth, and so on, indicated in diagram **1**.

This same principle would have to apply to our
big bang universe that's assumed to be finite and
expanding. The material at each consecutive radius
throughout its interior would be simultaneously
diffusing exponentially per the inverse square law,
as represented graphically and qualified numerically
in diagram **2**.

But we observe a homogeneous and isotropic
universe. So it can't be finite, or expanding. This
simple indisputable fact of three-dimensional
spherical geometry by itself completely undermines
the big bang.

To preserve it, we've employed Einstein's belief
that it actually expresses two-dimensionally, like a
sphere's surface. Its homogeneousness is
maintained by confining galaxies to the surface of a
spherical universe that can then dissipate uniformly
with expansion. But there's no existence in two
dimensions. So the big bang can't be real, which
reduces it to nothing more than theoretical whimsy.

A more pragmatic direction would begin with an
infinite (non-expanding) universe that suggested
realistic alternatives for astronomical redshift.

(7.1 Inverse Sq Sphere 13a)

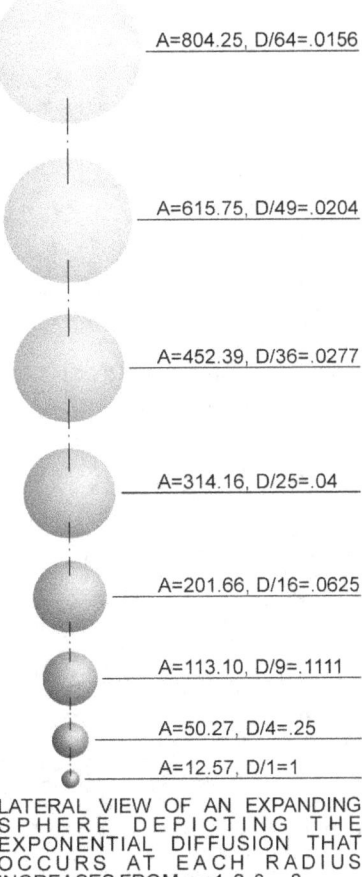

A=804.25, D/64=.0156

A=615.75, D/49=.0204

A=452.39, D/36=.0277

A=314.16, D/25=.04

A=201.66, D/16=.0625

A=113.10, D/9=.1111

A=50.27, D/4=.25

A=12.57, D/1=1

LATERAL VIEW OF AN EXPANDING
SPHERE DEPICTING THE
EXPONENTIAL DIFFUSION THAT
OCCURS AT EACH RADIUS
INCREASES FROM, r = 1, 2, 3,... 8

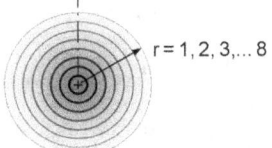

r = 1, 2, 3,... 8

SECTION VIEW THROUGH AN
EXPANDING SPHERICAL VOLUME
PORTRAYING THE EXPONENTIAL
DIFFUSION THAT OCCURS AT EACH
RADIUS SIMULTANEOUSLY AS IT
INCREASES FROM, r = 1, 2, 3,... 8

2. SPHERICAL EXPANSION
SURFACE AREA: $A = 4\pi r^2$
SURFACE DENSITY: $D \alpha 1/r^2$

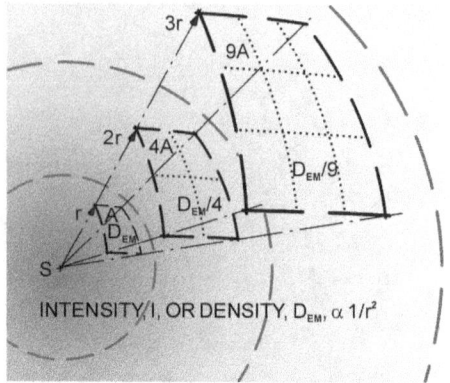

INTENSITY, I, OR DENSITY, $D_{EM}, \alpha\, 1/r^2$

DENSITY, $D_g, \alpha\, -1/r^2$, OPPOSITE
OF AN ELECTROMAGNETIC FIELD

1. ELECTROMAGNETIC FIELD
ELECTROMAGNETIC FIELD INTENSITY, WHICH IS DENSITY, DECREASES PROPORTIONAL TO THE INVERSE OF THE SQUARE OF THE RADIUS. THIS DIFFUSES ITS OUTWARD ACTING PRESSURE GRADIENT EXPONENTIALLY.

2. GRAVITY FIELD
GRAVITY FIELD INTENSITY, OR DENSITY, ALSO DECREASES PROPORTIONAL TO THE INVERSE OF THE RADIUS SQUARED. BUT ITS DIFFUSION IS INWARD. THIS INCREASES ITS OPPOSITE, NEGATIVE PRESSURE GRADIENT EXPONENTIALLY PER GRAVITY'S FORCE, g.

INVERSE SQUARE LAW, FIELD
An electromagnetic field (EM), depicted in section view as the diffusing background in diagram **1**, is subject to the inverse square law that's the product of the three-dimensional geometry of a sphere. So the field's intensity, which is the same as density that produces pressure which is force, twice the distance from its source is diluted by four times the area. This reduces its density to 1/4 the original. At three times the distance, it's spread over nine times the area, which reduces the density to 1/9 the original, and so on where D_{EM} (the density at a given radius) = S (the original density) / $4\pi r^2$ (the area of a sphere).

The big bang's radiant energy, correctly interpreted as a finite universal field, could never express uniformly as we presume. It's subject to the same physical geometry as every other field. It too would have to obey the inverse square law that'd have it diffusing exponentially with its expansion. The homogeneous isotropic expression that we observe is not physically possible for a finite, expanding universe. So in our actual three-dimensional, nontheoretical reality, the big bang is untenable.

The tangible, radiant, EM energy of our real universe that particles condense out of is all-pervasive, continuous, inseparable, and it varies in density. So the remaining ambient radiation that's not been drawn into a congealed particle has to thin inward, diffusing exponentially toward its center. This is what constitutes their, or collectively the bodies they compose, gravity field, portrayed in section view as the diffusing background in diagram **2**.

It's the opposite of an EM field. Its lowest density is reciprocal to an EM field's highest. Still bound to a sphere's inverse square law, its density, which is still intensity, which still equates to pressure and force, still has to dissipate exponentially. The gradient remains the same. It just expresses the opposite direction, diffusing inward instead of outward where D_g (the density at a given radius) = -S (the original point source strength or negative density established by a body's mass) / $4\pi r^2$ (the area of a sphere).

So at twice the distance from the center, its original negative density is diffused over four times the area, which is 1/4 the original that reduces the inward acting pressure by the same amount, decreasing gravity's force to 1/4g. At three times the distance, its negative density is spread over nine times the area, which is 1/9 as dense as its original that decreases the inward acting pressure the same, reducing gravity to 1/9g, and so on.

(7.2 Inverse Sq Field 8a)

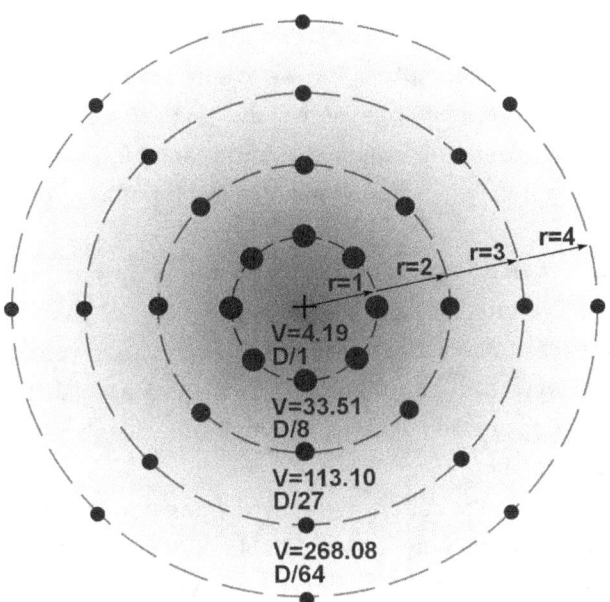

V=4.19
D/1

V=33.51
D/8

V=113.10
D/27

V=268.08
D/64

THE DENSITY OF THE VOLUME DIFFUSES EXPONENTIALLY

AN EXPANDING SPHERE
VOLUME OF A SPHERE, $V = 4/3\pi r^3$
DECREASING DENSITY OF THE VOLUME, $D \propto 1/r^3$

This conceptual depiction of a section view through a finite universe with a uniform distribution of galaxies at its expanding perimeter, represented by the dots, shows how its spherical two-dimensional surface, beginning at r=1, would expand uniformly with every galaxy moving away from every other while remaining an equal distance apart. But the density of its three-dimensional volume has to dissipate exponentially per the cube of the radius where D (volume density at a given radius) $\propto 1/r^3$, as portrayed by the diffusing background and the disbursing dots.

This simply illustrates again how the big bang's uniform distribution of galaxies without some supernatural mechanism counterbalancing a sphere's innate diffusion like the constant exponential creation of new material out of nothing at its ever-expanding perimeter or a rate of expansion that somehow exponentially decreases from its center out is conceptually and physically impossible in a real three-dimensional nontheoretical world that obeys the natural laws of geometry.

(7.5 Sphere Volume 4a)

A plane's inherent nonexistence is a rudimentary fact of basic geometry, yet it seems to remain unrecognized. The assumption that we live in a curving non-Euclidean, three-dimensional environment that somehow expresses two-dimensionally is now just taken for granted. Evidently, we must actually believe that we all exist in an undetectable, two-dimensional version of our three-dimensional universe.

We might reason that it's largely because our indoctrination has us unsuspectingly following Einstein's lead. But also, we allow ourselves to be fooled by the math, trusting that because it's been put to an equation it must have some factual basis in reality. This, along with the sophistication, or more precisely, the complication of the concept suggests legitimacy, which we unwittingly also accept as evidence of its actual existence.

The truth is, Einstein's, and Riemann's, supernatural two-dimensional three-dimensional, non-Euclidean geometry can only remain a fantasy. It could never express in our real physical world. I suspect that some may now be discreetly awakening to this realization.

But rather than fully embracing the absolute impossibility of existing in a non-Euclidean two-dimensional universe that expresses as the surface of a sphere, which would completely invalidate the big bang, the consequences of such a recognition have us blindly accepting the equally impossible notion that our normal, three-dimensional, Euclidean universe can expand uniformly. We can almost envision how that might work. But it doesn't.

Objections to the two-dimensional balloon modeling are dodged by insisting that it's just a metaphor or an analogy to our actual physical condition. It's not to be taken literally. But analogous to what? The question can't be answered because there's no actual physical counterpart in our real world. We simply cannot envision three-dimensions in a two-dimensional environment. It's conceptually impossible, an unfathomable reality.

Einstein skillfully evades the issue with a condescending, intentionally obscuring explanation, admonishing us that visualizing his two-dimensional three-dimensional, "...space means nothing else than that we imagine an epitome of our 'space' experience, *i.e.* of experience that we can have in the movement of 'rigid' bodies. In this sense we *can* imagine a spherical space."[152] Does that have any rational meaning?

He continues to obfuscate by asserting that we can, "draw lines or stretch strings in all directions [spherically in three dimensions] from a single point... At first, the straight lines which radiate from the starting point diverge farther and

152. Einstein, *Relativity*, 125.

farther from one another, but later they approach each other, and finally they run together again at a 'counter-point' to the starting point. Under such conditions they have traversed the whole spherical space."[153]

This incomprehensible, self-conflicted rationalization is nonsense to say the least. He has the strings diverging three-dimensionally but somehow converging two-dimensionally. How does that work? It's nothing more than a desperate attempt to justify a concocted, purely abstract speculation that cannot have any practical feasibility in our physically real, three-dimensional cosmos. And we're supposed to accept that this prerequisite dogma is genius, coming from quite possibly the smartest person ever.

To insist that this account must somehow physically manifest because it can be formulated mathematically is not a rational argument. Mathematics, non-Euclidean or otherwise, being in essence only a language, can be manipulated to simulate almost any desired condition whether it's conceptually sound or exists in physical reality or not. And following the math, which has evolved into a prescript for discovery, is no better than chasing your tail.

Proof of an idea lies in its rationality based on empirical evidence, which is always testable in our real world, not in its ability to be expressed mathematically. If it's not physically possible to alter a three-dimensional volume of something so that it can acquire the properties of a curving two-dimensional surface then it simply cannot exist in our actual physical universe no matter how much we might wish otherwise.

So, if you believe in the big bang then you have to adhere to its tenet that our three-dimensional universe can manifest two-dimensionally. Your only other choice is to abandon your discipleship, retreat back to a position of a normal three-dimensional finite spherical universe, and try to convince yourself that it can express and expand uniformly in violation of the physical geometry of a sphere, the principle of entropy, and the laws of gravitation.

We can be sure that most people do not realize that the entire big bang is based on and, in essence, originates from Einstein's wholly untenable, non-Euclidean, finite yet unbounded two-dimensional three-dimensional spatial delusions. But we can be just as sure that many professionals have long ago lost sight of this as well. If we were now not so committed to this concocted fantasy, having built layer upon layer of ancillary theories on it, this single issue would no doubt cause its instant rejection whether an acceptable alternative existed or not.

153. Einstein, *Relativity*, 125-126.

We also like to refer to other impossible two-dimensional shapes beside a sphere's surface to explain the big bang's potential rates of expansion. For its slowing (closed), we use a sphere's surface. For a constant rate of expansion (flat), we use a flat plane. And for its continued acceleration (open), we like to employ a hyperbolic paraboloid, a saddle shape.[154] None of these shapes, analogies or not, have any validity in our actual three-dimensional reality. They're two-dimensional. A real three-dimensional environment cannot curve. It can only vary in density. (See Diagram **23** U Curvature 5a & 17 Curvature 6a)

154. "Shape of the universe," Wikipedia, last modified Feb 26, 2023, https://en.wikipedia.org/wiki/Shape_of_the_universe.

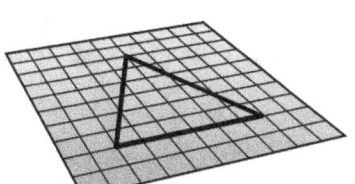 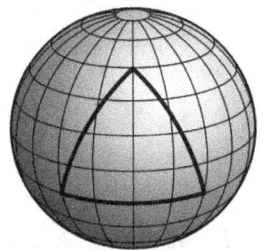 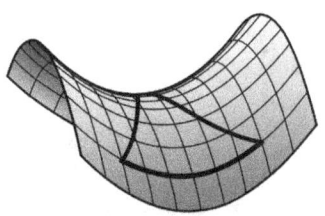

1. FLAT
NO CURVATURE
EUCLIDEAN
ANGLES = 180°
CONSTANT EXPANSION

2. CLOSED
POSITIVE CURVATURE
NON-EUCLIDEAN
ANGLES > 180°
SLOWING EXPANSION

3. OPEN
NEGATIVE CURVATURE
NON-EUCLIDEAN
ANGLES < 180°
ACCELERATING EXPANSION

THE UNIVERSE'S CURVATURE

Einstein suggests that the total mass of the universe, which includes its energy, determines its curvature. The above diagrams that are similar to the University of Oregon's illustrate again how he confuses density, a property of three dimensions, with curvature, a property of two dimensions.

When the universe's total mass is the same as its critical mass (the amount of mass necessary for its expansion to remain constant), the universe is said to have a density parameter (Ω_o) = 1, which indicates a constant rate of expansion. It's considered to be flat like a plane without curvature. It would be Euclidean in nature where parallel lines remain parallel, diagram **1**.

When the universe's total mass is greater than its critical mass, its density parameter would be > 1, which indicates a slowing rate of expansion. It's regarded as closed with a positive curvature like a sphere's surface. It'd be non-Euclidean in nature where parallel lines would not remain parallel but would converge, diagram **2**.

When its total mass is less than its critical mass, its density parameter would be < 1, which indicates an accelerating rate of expansion. The universe would be open with a negative curvature like a hyperbolic paraboloid. It'd also be non-Euclidean but where parallel lines would diverge, diagram **3**.

So the universe must have begun not as an extremely hot dense small homogeneous sphere of mass/energy, but actually as a sphere's two-dimensional surface that somehow embodies mass. A short time after it erupted, it would have begun to slow. But then it's believed to have started accelerating again through a brief period of faster-than-light inflation. After that, it would have again slowed for a while, but now it's believed to have once again started accelerating. So it must have been oscillating back and forth through each of these different shapes as its rate of expansion fluctuated over time. How does that work?

The problem is that none of these two-dimensional shapes has anything to do with the universe's actual three-dimensional structure. It cannot manifest two-dimensionally, despite Einstein and Riemann's delusions of a two-dimensional three-dimensional reality. If you want to argue that these two-dimensional shapes are just analogies then you have to show how they can have counterparts in our real three-dimensional world, not just assert that they must exist because of the math.

We don't have a problem at all with all this two-dimensional three-dimensional curving nonsense, not because it's logical but because we've been indoctrinated to it. It's our accepted doctrine. And it will never be questioned or challenged. It's critical to our big bang-relativity belief system. Without it, it completely collapses.

If you decide that our universe is actually three-dimensional then you're eventually going to have to accept that a sphere, much less an expanding one, can never express uniformly. So our universe can't be finite or have a beginning. It must be infinite and ageless and galactic redshifts have to be indicative of something other than universal expansion like the recessional velocity of each galaxy's perpetually recycling, infalling material.

(23 U Curvature 5a)

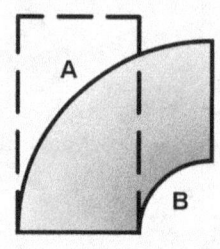

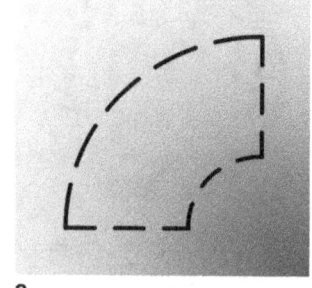

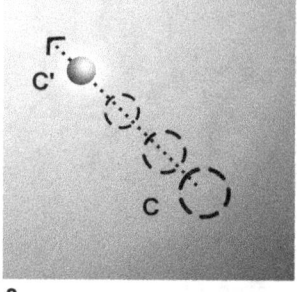

1   2   3

CURVATURE & DENSITY

If we were to envision the shape in diagram **1** as three-dimensional, giving it the same depth as its width, we can see how if it were bent over where its two-dimensional surfaces, at **A** & **B**, now curved, its interior would have to vary in density, as represented in section view by the diffused shading that dissipates toward the outer curved side with a greater radius, **A**. Note how the density along both the outer and inner curved surfaces would remain constant as **A** diffused and **B** condensed when bent.

If we were to imagine that shape as part of a larger volume of the same material, as depicted in diagram **2**, we can more easily see how any three-dimensional volume of something can never curve but can only vary in density. Curvature is a property of only one and two dimensions, which don't actually exist in reality. Without the third dimension, they only define a location of a line or a plane.

These principles of curvature and density would also have to hold true for the three-dimensional volume that's commonly refer to as space. It could never curve. It could only vary in density. But space is the nothingness between objects. It's nonexistent by definition, and it would also be nonexistent if it were a curving two-dimensional plane. So there's nothing there to vary or curve. But there is something that does correspond exactly to all space: radiant electromagnetic energy. Everywhere that space is, so is the all-pervasive universal field of radiation that comprises the electromagnetic spectrum. This field is continuous, its continuity uninterruptible, and it's average volume is fixed per any given quantity. And its density varies.

Because of its innate continuity, the density of the radiation that's been drawn together to form a particle has to diffuse. But it diffuses inward toward the particle's center, still dissipating exponentially per the inverse square law because of the three-dimensional geometry of a sphere. It's the compounding of those inward diffusing fields of the particles themselves, and the objects they compose, that's responsible for gravitation.

Referring to diagram **3**, if we were to place a small object into the decreasing density of a gravity field of a much larger object, it would begin moving with acceleration in the direction of least resistence directly toward the larger object as it mechanically sought to find equilibrium in the field's decreasing density. It'd naturally move from the highest density toward the lowest. But it would also be contracting in size. Its atoms also naturally coalesce and condense within the field's decreasing density, as indicated by the sequence of dashed circles representing the object's motion from **C** to **C'**. In reality, each object would move toward the other in proportion to the size of their gravity fields toward a common center of mass because of the compounding effect of their gradient.

(17 Curvature 6a)

218

Beyond matter's inherent inability to even exist two-dimensionally, there's a host of other contradictory problems associated with a universe that manifests in the shape of a sphere's surface. It's said that the universe has no beginning location, or no origin, no center, or that it's halfway between every galaxy, or that its center beginning point is everywhere, but that it does have a radius.

And every galaxy is supposedly located at that radius. Well how can it have a radius without having a center? Or how can its center be everywhere? And if its center is everywhere, how can it have a radius? Or if the center is in between every galaxy, how can every galaxy be located at its radius? There's no tangible answer to these questions.

Einstein argues that its radius is equal to time, but still maintains there's no origin.[155] If there's no origin then there can be no beginning to time, or a radius, and time can't be equal to the universe's radius if it doesn't exist. Theoretically, or mathematically, a two-dimensional shell can continue to shrink indefinitely, never arriving at a central beginning point.

If there was a beginning to time then it would had to have occurred at the origin of a universe that would had to have been three-dimensional and spherical. At some point after its eruption, it would had to have transitioned over to the two-dimensional big bang shell that's now being asserted. How is that explained?

We believe that if we could look back just another half billion years or so past the farthest galaxy, in any direction, we'd be able to see the universe's primeval beginnings. How can that be true if the universe's origin began at a single point smaller than an atom? How can that be true if the universe's center beginning location remains outside its two-dimensional realm? How can that be true if the universe doesn't have a center, or its center is halfway between every galaxy, or it's everywhere? How can that be true if we can look out in all directions and see the backside of our own galaxy? With faster-than-light travel impossible and the big bang occurring 14 billion years ago, wouldn't the universe's inception be lost forever?

It's just astonishing that with all its inherent contradictions, unfathomable propositions, and torturous logic that any rational individual dedicated to science can even entertain, much less propose, that we actually inhabit a curving non-Euclidean, uniformly expanding, finite but somehow unbounded, homogeneous and isotropic, two-dimensional three-dimensional universe that manifests like a sphere's surface, and then declare the idea brilliant.

155. Einstein, *Relativity*, 152-153.

Breaking Bread

Another comparison that's commonly used to convey the big bang's uniform expansion is a rising loaf of raisin bread. This may sound somewhat comical, but it has more validity than the balloon model. As the bread expands, the raisins that represent galaxies all move away from one another.

This effect, free of the Earth's gravity, might at first appear consistent with theory. But let's be clear, any expanding spherical volume of something, be it a loaf of bread or a finite universe, cannot express uniformly in a real three-dimensional environment. Its matter has to diffuse from expansion or condense from gravity exponentially because of a sphere's inherent geometry, which would also yield an identifiable center.

In a limited way, the raisin's motion approximates an expanding universe. But it wouldn't correlate at all with our universe's homogeneous distribution of galaxies and their redshifts and no perceptible origin. The raisins also wouldn't correspond at all with the dots on the expanding balloon analogy. (See Diagram **24** U Expansion 10a & **25.1**, **25.2** U 3-D Expansion 7a)

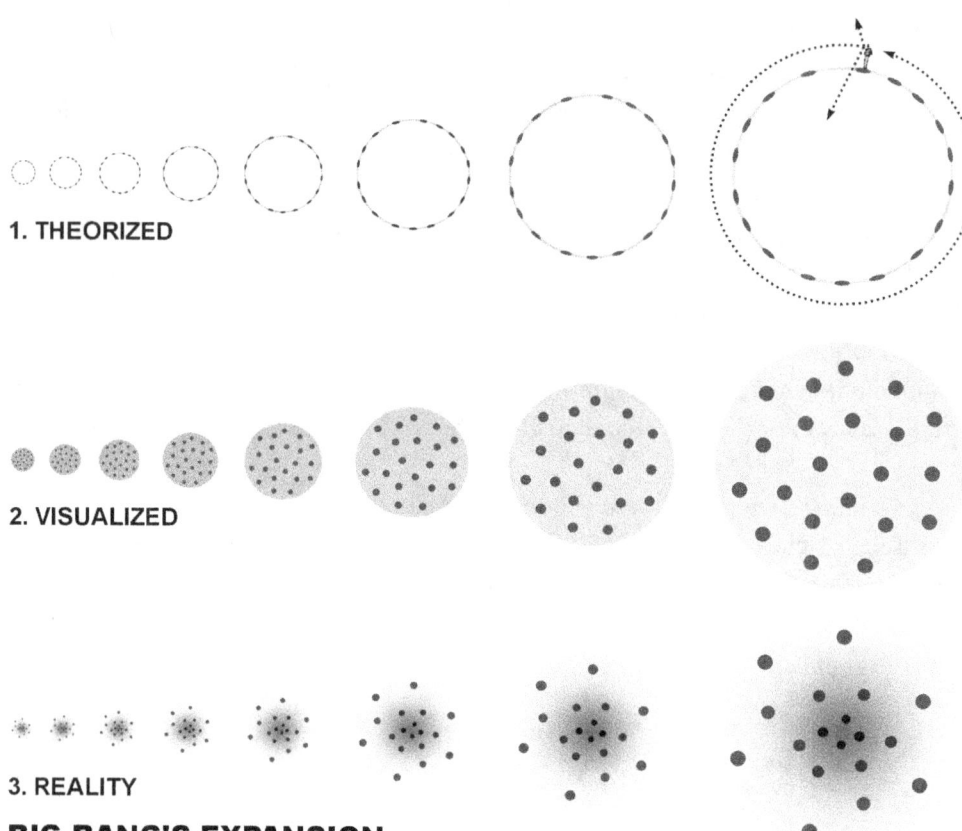

1. THEORIZED

2. VISUALIZED

3. REALITY

BIG BANG'S EXPANSION

To counter the effect of its self-gravity, Einstein has his finite yet unbounded, big bang universe expressing two-dimensionally with non-Euclidean curvature like the surface of an expanding balloon, conceptually illustrated in section view in diagram **1**. The flattened dots represent galaxies. But two dimensions has no height, not even the thickness of a thin shell like the line depicting it. So how can anything even exist? He contends that the big bang has no center but still has a radius that's equal to time, and that strings stretched straight, in any direction, will ultimately return to their beginning point. How does any of that actually work?

Some maintain that it's not possible to look "down" across the universe's center and see the other side or look "up" into the nothingness above. But they still insist that they can look out, in any direction, and almost see the universe's inception while also asserting, consistent with Einstein, if their telescope was powerful enough they'd be able to see the backside of our own galaxy. How can any of this be considered rational, much less genius?

Because the two-dimensionality of the big bang is unfathomable, most people actually visualize it like a rising loaf of raisin bread where the raisins represent the uniform distribution and expansion of galaxies, as portrayed also in a section view in diagram **2**. The problem is, it's impossible for a spherical volume to ever express homogeneously as our real universe does. Plus, its self-gravity and expansion would cause the exponential condensing and exponential diffusion of its material, also depicted in section in diagram **3**.

If the big bang can't express two-dimensionally and an expanding three-dimensional spherical universe can't express homogeneously then it can't be finite or expanding. It must be infinite. And galactic redshifts most likely originate from each galaxy's own ceaseless infall of gravitating material that's perpetually coalescing, condensing, collapsing, ejecting, and reconstituting because of the inherent runaway nature of gravity.

(24 U Expansion 10a)

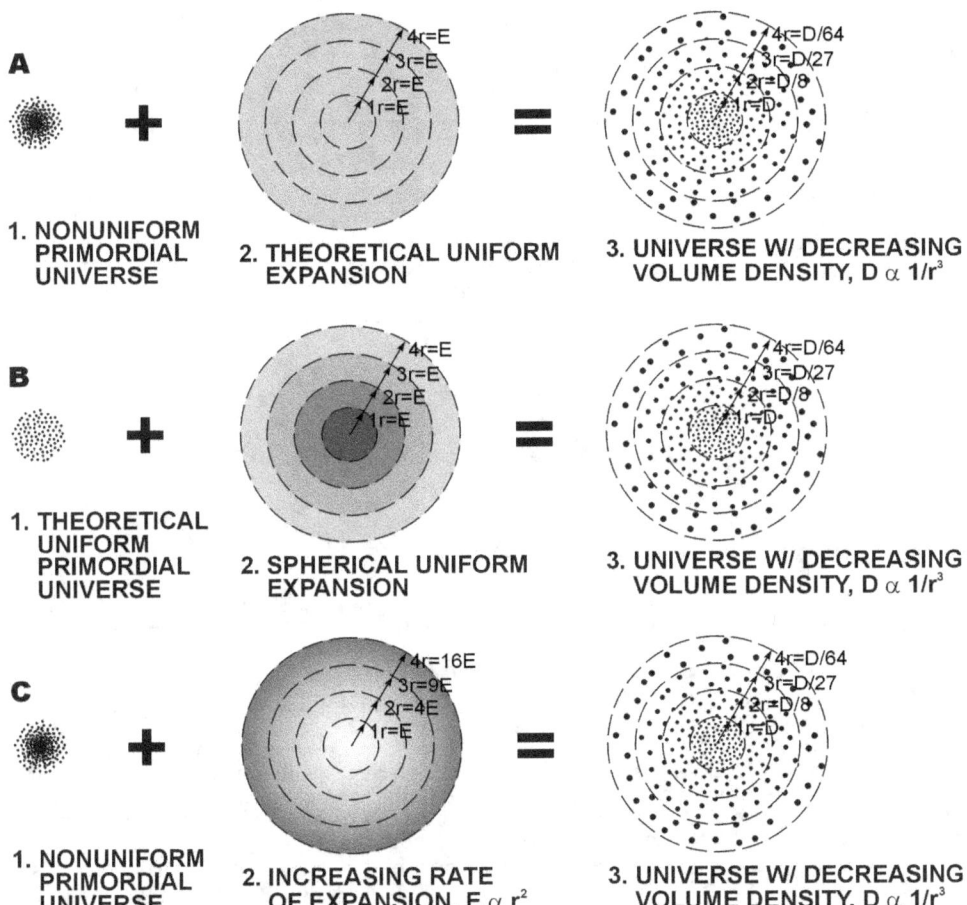

A

1. NONUNIFORM PRIMORDIAL UNIVERSE

2. THEORETICAL UNIFORM EXPANSION

3. UNIVERSE W/ DECREASING VOLUME DENSITY, $D \propto 1/r^3$

B

1. THEORETICAL UNIFORM PRIMORDIAL UNIVERSE

2. SPHERICAL UNIFORM EXPANSION

3. UNIVERSE W/ DECREASING VOLUME DENSITY, $D \propto 1/r^3$

C

1. NONUNIFORM PRIMORDIAL UNIVERSE

2. INCREASING RATE OF EXPANSION, $E \propto r^2$

3. UNIVERSE W/ DECREASING VOLUME DENSITY, $D \propto 1/r^3$

BIG BANG'S 3-D EXPANSION

The material of finite big bang universe that was three-dimensional and spherical, whether its expansion was theoretically uniform or not, would have to dissipate exponentially. Row **A** begins with its primaeval material assumed as nonuniform, gravitationally condensed exponentially, shown in section view at diagram **1**. The dots could represent galaxies or any quanta of matter including subatomic particles. If space were actually something and it could actually expand uniformly, as depicted in section view at **2**, then the result would have to be a diffusing universe as suggested in section at **3**. A nonuniform distribution subject to uniform expansion would produce a nonuniform result. Row **B** theoretically assumes a uniform primordial big bang that ignores gravity but is subject to uniform expansion in three dimensions. The result is still nonuniform because the inherent geometry of an expanding sphere has its perimeter diffusing exponentially. Row **C** shows how the volume of a nonuniform primordial big bang subject to the real effects of three-dimensional expansion that increases exponentially would produce exponential diffusion. Darker grey represents space's exponentially increasing expansion.

(25.1 U 3-D Expansion 7a)

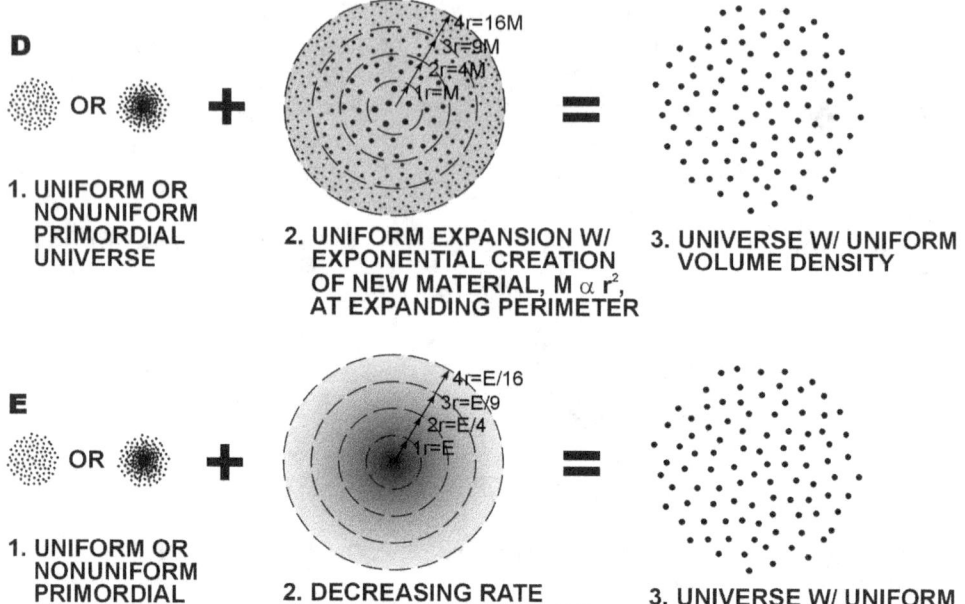

D

OR ⬤ **+**

1. UNIFORM OR NONUNIFORM PRIMORDIAL UNIVERSE

4r=16M
3r=9M
2r=4M
1r=M

2. UNIFORM EXPANSION W/ EXPONENTIAL CREATION OF NEW MATERIAL, M α r², AT EXPANDING PERIMETER

=

3. UNIVERSE W/ UNIFORM VOLUME DENSITY

E

OR ⬤ **+**

1. UNIFORM OR NONUNIFORM PRIMORDIAL UNIVERSE

4r=E/16
3r=E/9
2r=E/4
1r=E

2. DECREASING RATE OF EXPANSION, E α 1/r²

=

3. UNIVERSE W/ UNIFORM VOLUME DENSITY

BIG BANG'S 3-D EXPANSION - CONTINUED

One way a uniformly expanding, three-dimensional, spherical universe that began either uniform or nonuniform, still depicted in section view at **D-1**, could express uniformly, as portrayed in section at **3**, is if it were constantly creating new material at its expanding perimeter at an exponentially increasing rate, implied by the section view at **2**, where the dots could again represent either galaxies or subatomic particles. But this would leave us with an exponential increase of young galaxies and exponentially fewer spiral galaxies at the universe's expanding perimeter, which we don't observe.

The only other potential way a three-dimensional big bang could express uniformly is if its expansion decreased exponentially exactly compensating for a sphere's inherent geometry, as suggested in section by the diffusing background at **E-2**. The darker the shading the faster the expansion. But it'd still have to be commingled with some degree of uniform expansion. If I were a big bang follower, this is the argument I'd use to justify my belief. It's simple, almost rational, and comprehendible. Unfortunately, it's still completely unworkable. But at least it's not nonsense.

Some won't necessarily disagree with any of this reasoning. They just won't believe it's applicable, other than how it may have influenced Einstein to conceive his finite yet unbounded, two-dimensional solution to circumvent the problem of self-gravity's exponential condensing. They don't realize they've been indoctrinated to his fantasy, have uncritically memorized it, and are now perfectly comfortable existing three-dimensionally in the unfathomable nonexistent two-dimensional realm of a sphere's expanding surface.

But those who've not been indoctrinated to all of our big bang-relativity foolishness who realize that our three-dimensional universe could never express two-dimensionally, they can easily perceive these fundamental conflicts and quickly realize that there's no way our real universe can be expanding. Nor can it be finite. But that it must be infinite, and it must be eternal. And galactic redshifts have to originate from something other than an entire galaxy's recessional velocity being swept away by stretching space like the ceaseless coalescing and condensing of each galaxy's material due to the runaway nature of gravity.

(25.2 U 3-D Expansion 7a)

There are three reasons why an expanding universe can never express uniformly. First, Its expansion suggests it was smaller at an earlier time and had a beginning. So in a real three-dimensional environment subject to the natural laws of geometry, it'd have to be a spherical volume. It can't be flat, or somehow be curving two-dimensionally. And it can't be infinite.

A sphere's inherent three-dimensional properties make the homogeneous distribution of its material physically impossible. Its radius is always smaller than the distance between a uniform distribution of points on its surface. So for a real three-dimensional spherical universe, it would have to be arrayed with galaxies that were uniformly distributed around its surface but became increasingly more dense towards its center. This increasing density would be observable and generate an easily discernible center location. (See Diagram 7.0 Platonics 5a)

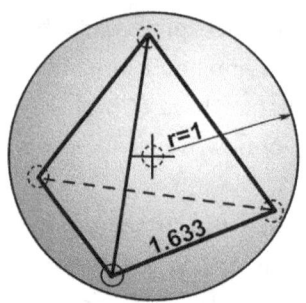 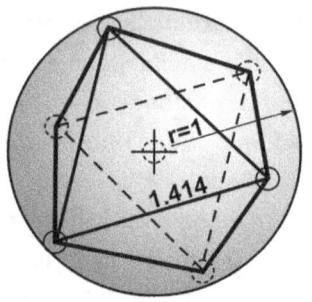 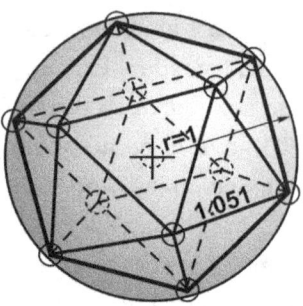

1. TETRAHEDRON   **2. OCTAHEDRON**   **3. ICOSAHEDRON**

PLATONIC SHAPES

The only three platonic shapes in nature that can be composed of equilateral triangles suggest that it's not possible for a spherical universe to be homogeneous. If we portion out any quantity of matter, be it subatomic particles or galaxies, into a spherical shape, we find that their uniform distribution is mathematically and physically impossible. Placing a galaxy at each of the triangles' vertices, or corners, and one at the sphere's center shows how their distribution would be equidistant around the circumspherical surface but always a lessor distance to the center galaxy. This principle of three-dimensional geometry simply reveals how the big bang universe, if it were actually three-dimensional, could not be homogeneous, but would always have to be denser towards its center, which would also establish an observable origin.

With a circumsphere radius of 1 of any unit of measure, the edge of each triangle, or the distance between each galaxy would be 1.633 for the tetrahedron, 1.414 for the octahedron, and 1.051 for the icosahedron.

(7.0 Platonics 5a)

226

A sphere's self-gravity would also make the homogenous distribution of its matter physically impossible. Every object generates its own self-gravity that innately condenses its material exponentially toward its center of mass per the inverse square law because of a sphere's inherent geometry.

In the same manner, the galaxies of a normal three-dimensional finite spherical universe would naturally coalesce towards its center, condensing exponentially. This would also produce a distinct center. (See Diagram **26.1** U Center 5a)

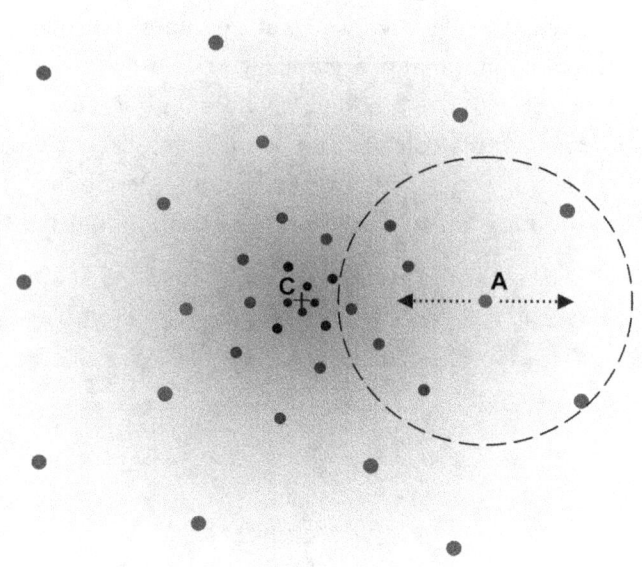

LOCATING A FINITE UNIVERSE'S CENTER
PROPERTIES OF A SPHERE & SELF-GRAVITY

If the big bang does not express metaphysically like the two-dimensional surface of a sphere but manifest normally with three tangible dimensions then it'd have to be subject to the inherent properties of a sphere and the laws of self-gravity, which would cause its material to be dispersing and condensing exponentially, portrayed in section view by the diffusing grey background and the dispersing dots that represent galaxies.

If our galaxy was located at **A**, we'd be able to quickly determine the location of the universe's center origin if it actually had one, which would confirm its finiteness and expansion. It'd be at **C**, identified easily by a pattern of increasing density of galaxies across the entire sky. The location where the density culminates to its highest would correspond to its origin.

Suggesting that the light from the galaxies near its origin is just too faint or beyond our ability to see for other reasons, the horizon implied by the dashed circle, wouldn't be a viable argument. We already insist that we're viewing back to within a few hundred thousand years of the universe's inception when we observe the cosmic microwave background radiation. But even if the origin itself was not visible, its precise direction would still be identifiable. It'd be in line with and beyond the same location where the density of galaxies peaked, exactly opposite the lowest density.

All we see, though, is an endless homogeneous uniform expanse of galaxies. So our universe can't be finite or expanding. It has to be infinitely vast and ageless, unless you believe that it can express two-dimensionally or have rashly decided that the big bang is no longer finite, but now must be infinite. An infinite universe that's expanding makes no sense whatsoever. You're faced with the impossibility of finding explanations for all of its primordial conditions that are dependent on it having a beginning and size, density, and pressure, which all become meaningless in an infinite reality.

(26.1 U Center 5a)

228

A sphere's expansion would make the homogenous distribution of its matter physically impossible as well. A fundamental property of a sphere in our actual three-dimensional environment is that its surface area increases exponentially as its radius increases incrementally.

If an area of that surface were filled with a fixed number of points evenly arrayed then their density would diffuse exponentially as it expands. This is the essence of the inverse square law that would have to apply to an expanding spherical big bang universe of fixed mass that existed in three real dimensions.

The density of a uniform distribution of galaxies at any defined radius would decrease exponentially with incremental expansion according to the principles of geometry of a sphere. Their dispersion would naturally generate an origin. Both would be easily identifiable. (See Diagram 7.1 Inverse Sq Sphere 13a, 7.2 Inverse Sq Fields 8a, 7.5 Sphere Volume 4a & 26.1 U Center 5a)

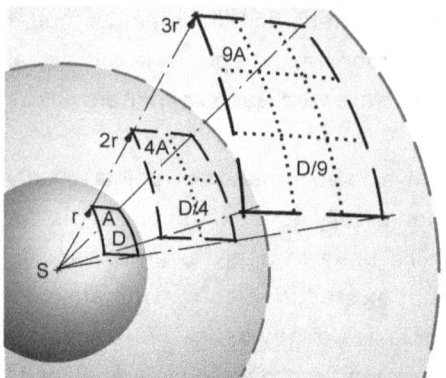

1. SPHERE'S SURFACE AREA & DENSITY
INTENSITY OR DENSITY OF AN EXPANDING SPHERE'S
SURFACE AREA, $D \alpha$ (IS PROPORTIONAL TO) $1/r^2$

INVERSE SQUARE LAW, SPHERE
The surface area of a sphere is $A = 4\pi r^2$. It increases
exponentially as it expands. When twice the radius
from its center, its area increases four times. Three
times the distance, nine times, and so on.

So the intensity or density (they're the same
thing) of any assumed flux (a quantity at the surface)
will dissipate exponentially following the inverse
square law where D (the density at a chosen radius)
= S (the original source density) / $4\pi r^2$ (the area of a
sphere). So its density twice the radius from its
center will be one-fourth its original. Three times the
radius, one-ninth, and so on, indicated in diagram **1**.

This same principle would have to apply to our
big bang universe that's assumed to be finite and
expanding. The material at each consecutive radius
throughout its interior would be simultaneously
diffusing exponentially per the inverse square law,
as represented graphically and qualified numerically
in diagram **2**.

But we observe a homogeneous and isotropic
universe. So it can't be finite, or expanding. This
simple indisputable fact of three-dimensional
spherical geometry by itself completely undermines
the big bang.

To preserve it, we've employed Einstein's belief
that it actually expresses two-dimensionally, like a
sphere's surface. Its homogeneousness is
maintained by confining galaxies to the surface of a
spherical universe that can then dissipate uniformly
with expansion. But there's no existence in two
dimensions. So the big bang can't be real, which
reduces it to nothing more than theoretical whimsy.

A more pragmatic direction would begin with an
infinite (non-expanding) universe that suggested
realistic alternatives for astronomical redshift.

(7.1 Inverse Sq Sphere 13a)

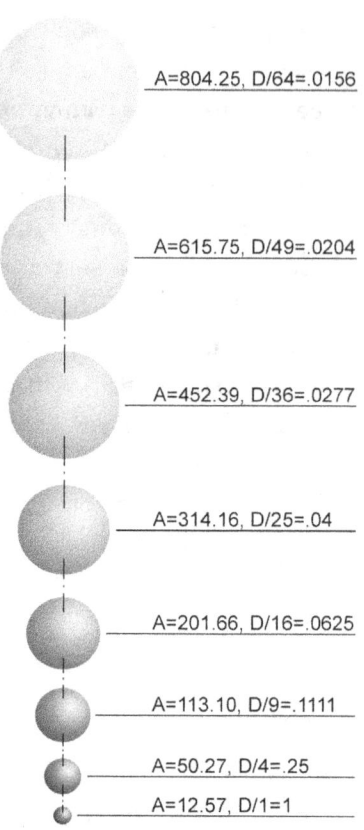

A=804.25, D/64=.0156

A=615.75, D/49=.0204

A=452.39, D/36=.0277

A=314.16, D/25=.04

A=201.66, D/16=.0625

A=113.10, D/9=.1111

A=50.27, D/4=.25

A=12.57, D/1=1

LATERAL VIEW OF AN EXPANDING
SPHERE DEPICTING THE
EXPONENTIAL DIFFUSION THAT
OCCURS AT EACH RADIUS
INCREASES FROM, r = 1, 2, 3,... 8

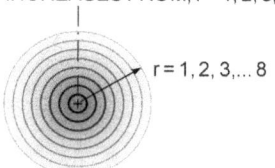

r = 1, 2, 3,... 8

SECTION VIEW THROUGH AN
EXPANDING SPHERICAL VOLUME
PORTRAYING THE EXPONENTIAL
DIFFUSION THAT OCCURS AT EACH
RADIUS SIMULTANEOUSLY AS IT
INCREASES FROM, r = 1, 2, 3,... 8

2. SPHERICAL EXPANSION
SURFACE AREA: $A = 4\pi r^2$
SURFACE DENSITY: $D \alpha 1/r^2$

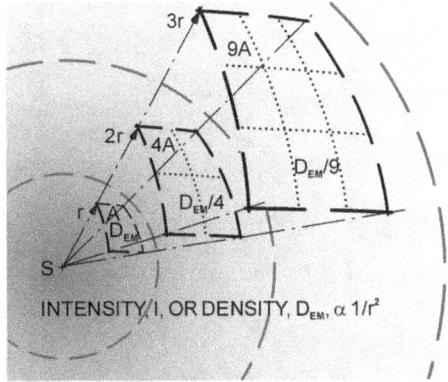

INTENSITY, I, OR DENSITY, D_{EM}, α $1/r^2$

1. ELECTROMAGNETIC FIELD
ELECTROMAGNETIC FIELD INTENSITY, WHICH IS DENSITY, DECREASES PROPORTIONAL TO THE INVERSE OF THE SQUARE OF THE RADIUS. THIS DIFFUSES ITS OUTWARD ACTING PRESSURE GRADIENT EXPONENTIALLY.

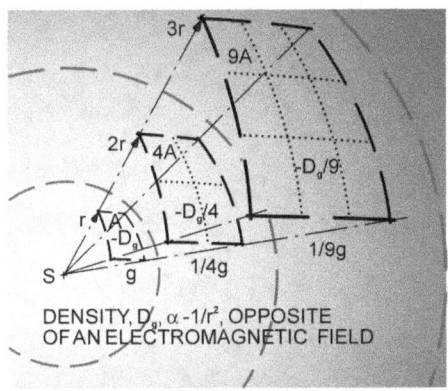

DENSITY, D_g, α $-1/r^2$, OPPOSITE OF AN ELECTROMAGNETIC FIELD

2. GRAVITY FIELD
GRAVITY FIELD INTENSITY, OR DENSITY, ALSO DECREASES PROPORTIONAL TO THE INVERSE OF THE RADIUS SQUARED. BUT ITS DIFFUSION IS INWARD. THIS INCREASES ITS OPPOSITE, NEGATIVE PRESSURE GRADIENT EXPONENTIALLY PER GRAVITY'S FORCE, g.

INVERSE SQUARE LAW, FIELD
An electromagnetic field (EM), depicted in section view as the diffusing background in diagram **1**, is subject to the inverse square law that's the product of the three-dimensional geometry of a sphere. So the field's intensity, which is the same as density that produces pressure which is force, twice the distance from its source is diluted by four times the area. This reduces its density to 1/4 the original. At three times the distance, it's spread over nine times the area, which reduces the density to 1/9 the original, and so on where D_{EM} (the density at a given radius) = S (the original density) / $4\pi r^2$ (the area of a sphere).

The big bang's radiant energy, correctly interpreted as a finite universal field, could never express uniformly as we presume. It's subject to the same physical geometry as every other field. It too would have to obey the inverse square law that'd have it diffusing exponentially with its expansion. The homogeneous isotropic expression that we observe is not physically possible for a finite, expanding universe. So in our actual three-dimensional, nontheoretical reality, the big bang is untenable.

The tangible, radiant, EM energy of our real universe that particles condense out of is all-pervasive, continuous, inseparable, and it varies in density. So the remaining ambient radiation that's not been drawn into a congealed particle has to thin inward, diffusing exponentially toward its center. This is what constitutes their, or collectively the bodies they compose, gravity field, portrayed in section view as the diffusing background in diagram **2**.

It's the opposite of an EM field. Its lowest density is reciprocal to an EM field's highest. Still bound to a sphere's inverse square law, its density, which is still intensity, which still equates to pressure and force, still has to dissipate exponentially. The gradient remains the same. It just expresses the opposite direction, diffusing inward instead of outward where D_g (the density at a given radius) = -S (the original point source strength or negative density established by a body's mass) / $4\pi r^2$ (the area of a sphere).

So at twice the distance from the center, its original negative density is diffused over four times the area, which is 1/4 the original that reduces the inward acting pressure by the same amount, decreasing gravity's force to 1/4g. At three times the distance, its negative density is spread over nine times the area, which is 1/9 as dense as its original that decreases the inward acting pressure the same, reducing gravity to 1/9g, and so on.

(7.2 Inverse Sq Field 8a)

231

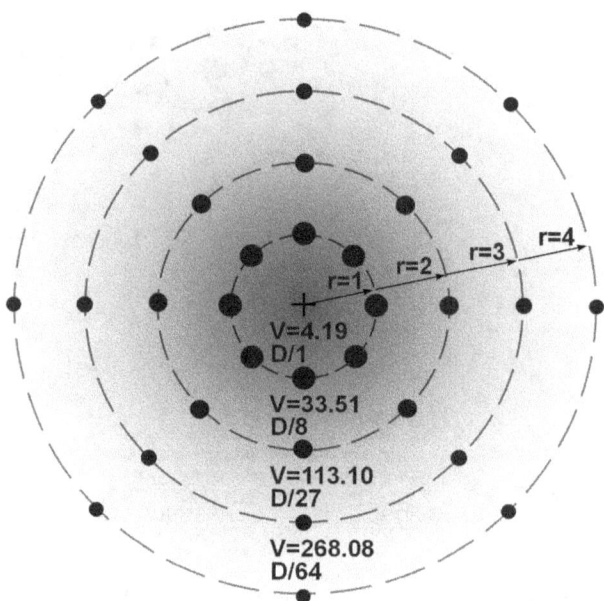

r=1
r=2
r=3
r=4

V=4.19
D/1

V=33.51
D/8

V=113.10
D/27

V=268.08
D/64

THE DENSITY OF THE VOLUME DIFFUSES EXPONENTIALLY

AN EXPANDING SPHERE
VOLUME OF A SPHERE, V = $4/3\pi r^3$
DECREASING DENSITY OF THE VOLUME, D $\propto$ $1/r^3$

This conceptual depiction of a section view through a finite universe with a uniform distribution of galaxies at its expanding perimeter, represented by the dots, shows how its spherical two-dimensional surface, beginning at r=1, would expand uniformly with every galaxy moving away from every other while remaining an equal distance apart. But the density of its three-dimensional volume has to dissipate exponentially per the cube of the radius where D (volume density at a given radius) $\propto$ $1/r^3$, as portrayed by the diffusing background and the disbursing dots.

This simply illustrates again how the big bang's uniform distribution of galaxies without some supernatural mechanism counterbalancing a sphere's innate diffusion like the constant exponential creation of new material out of nothing at its ever-expanding perimeter or a rate of expansion that somehow exponentially decreases from its center out is conceptually and physically impossible in a real three-dimensional nontheoretical world that obeys the natural laws of geometry.

(7.5 Sphere Volume 4a)

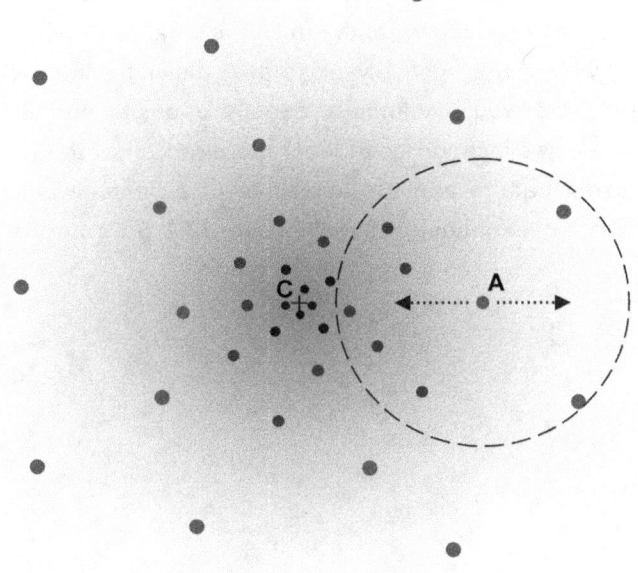

LOCATING A FINITE UNIVERSE'S CENTER
PROPERTIES OF A SPHERE & SELF-GRAVITY

If the big bang does not express metaphysically like the two-dimensional surface of a sphere but manifest normally with three tangible dimensions then it'd have to be subject to the inherent properties of a sphere and the laws of self-gravity, which would cause its material to be dispersing and condensing exponentially, portrayed in section view by the diffusing grey background and the dispersing dots that represent galaxies.

If our galaxy was located at **A**, we'd be able to quickly determine the location of the universe's center origin if it actually had one, which would confirm its finiteness and expansion. It'd be at **C**, identified easily by a pattern of increasing density of galaxies across the entire sky. The location where the density culminates to its highest would correspond to its origin.

Suggesting that the light from the galaxies near its origin is just too faint or beyond our ability to see for other reasons, the horizon implied by the dashed circle, wouldn't be a viable argument. We already insist that we're viewing back to within a few hundred thousand years of the universe's inception when we observe the cosmic microwave background radiation. But even if the origin itself was not visible, its precise direction would still be identifiable. It'd be in line with and beyond the same location where the density of galaxies peaked, exactly opposite the lowest density.

All we see, though, is an endless homogeneous uniform expanse of galaxies. So our universe can't be finite or expanding. It has to be infinitely vast and ageless, unless you believe that it can express two-dimensionally or have rashly decided that the big bang is no longer finite, but now must be infinite. An infinite universe that's expanding makes no sense whatsoever. You're faced with the impossibility of finding explanations for all of its primordial conditions that are dependent on it having a beginning and size, density, and pressure, which all become meaningless in an infinite reality.

(26.1 U Center 5a)

The expanding motion of galaxies would also generate a varying distribution of their redshifts that would reveal the universe's center. Both would confirm its expansion as well, if they actually existed. A distinct pattern of progressively increasing redshifts would culminate exactly opposite our direction of travel signifying the precise location or at least the direction of the universe's center. Both the redshift pattern and the existence of a center would verify that the universe was in fact expanding. (See Diagram **26.2** U Center 5a)

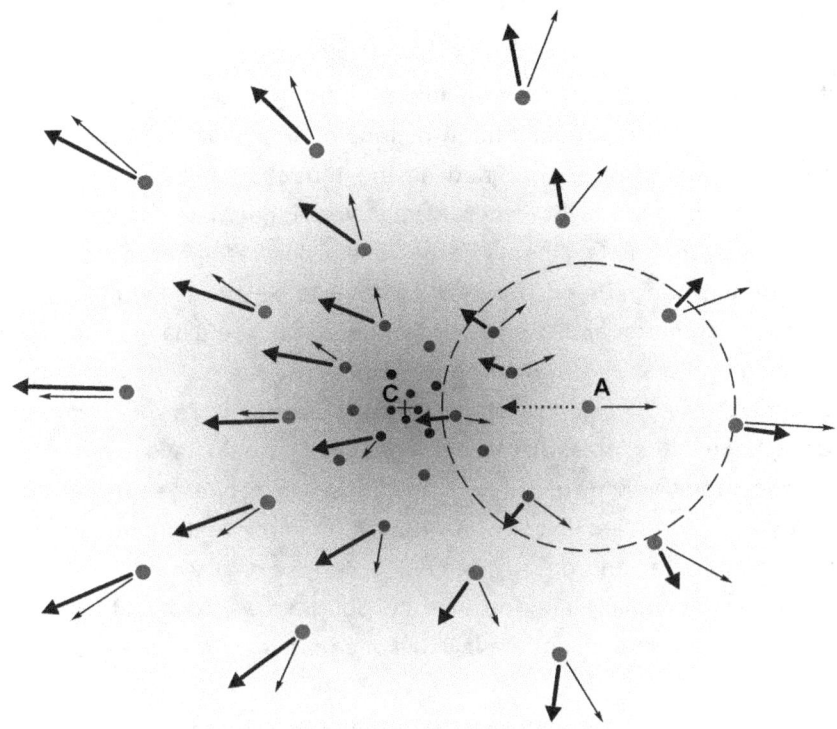

LOCATING A FINITE UNIVERSE'S CENTER
EXPANSION & REDSHIFT
In addition to the increasing density of galaxies in a three-dimensional big bang, their redshifts would also identify its center, if it actually had one, and confirm its expansion and finiteness. Looking at it in section view, the diffusing grey background represents its diffusing matter and radiation. The dots represent its dispersing galaxies. The lighter arrows indicate the direction of their outward motion with their length implying their relative rate of expansion. The heavy arrows indicate the direction of their recessional velocity from our galaxy at **A** with their length also suggesting their relative rate.

Bound to the inverse square law, a finite three-dimensional big bang would have to disperse its galaxies exponentially. This would generate a distinct pattern of increasing redshifts from increasing recessional velocity, or space's stretching, that would culminate directly opposite our direction of outward travel on the other side of its center, identifying its location that would confirm the big bang was finite and expanding.

But what we see is a uniform distribution of galaxies and all magnitudes of their redshifts. Which means the universe doesn't have a center, isn't finite, and can't be expanding. It must be infinite and redshifts must be indicative of something other than a galaxy's recessional velocity/universal stretching.

Trying to invoke faster-than-light inflation or some other gimmick to explain away the missing pattern of redshifts and a lack of an observable center still won't work. Despite a visible horizon, indicated by the dashed circle, there'd still be a distinct telltale pattern of galactic redshifts from the universe's increasing rate of expansion that would be higher in the direction of our outward motion and slower in the opposite direction directly toward its origin that would still indicate the exact direction of its location that would confirm the universe's finiteness and expansion and validate the big bang if it were real.

(26.2 U Center 5a)

It could be argued that the principle of entropy also suggests a nonuniform expression for an expanding three-dimensional big bang universe. For our discussion, entropy can be defined as the tendency of the universe's matter and/or energy to degrade, or disperse, into ever higher states of static uniformity. Higher entropy means less heat energy, less order, less compaction.

For the big bang, its entropy would have to be naturally increasing from its inception. Its matter and energy, and its galaxies, would have to be dispersing. But they cannot disburse uniformly as implied by the balloon analogy. A two-dimensional three-dimensional existence within the surface of a sphere is impossible. A real big bang that expressed three-dimensionally would have to manifest as the three-dimensional volume of a sphere and adhere to a sphere's three-dimensional geometry and properties just like all spheres do in our real environment.

So all its matter and energy, including its galaxies, would also have to be dispersing exponentially per the inverse square law. And just like the other applications of the inverse square law this observable pattern would also produce a distinct origin.

The only way an expanding spherical universe could theoretically remain uniform is if its rate of expansion increased toward its center exponentially to exactly counterbalancing a sphere's three-dimensional geometry. Or it was constantly creating new matter/energy, out of nothingness, at its expanding perimeter.

The rate of that creation would have to be increasing exponentially, coinciding exactly with the inverse square law that'd be inversely proportional to the universe's self-gravity as it enlarged. Concocting a totally implausible creation process and then having that process even more implausibly create something out of nothing is not at all a viable explanation to justify fanciful speculations of finiteness and expansion.

If you can accept that the finite big bang universe has to be three-dimensional then you have to also accept that its material could never express in a homogeneous uniform manner because of a sphere's inherent nature, self-gravity, and expansion. But isotropic uniformity is exactly what we observe. So the universe can't be finite. It must be infinite.

But even if you could somehow imagine a homogeneous expanding spherical universe in three dimensions, its existence would still be confirmed by a distinct pattern of increasing density of galaxies and their increasing redshifts. They'd still have to culminate with the highest density and redshifts exactly opposite our direction of travel that would also set the location or direction of an origin. This pattern would even hold for a finite universe that wasn't expanding. (See Diagram **26.3** U Center 5a & **50.1**, **50.2** Uniformity & Finiteness 10a)

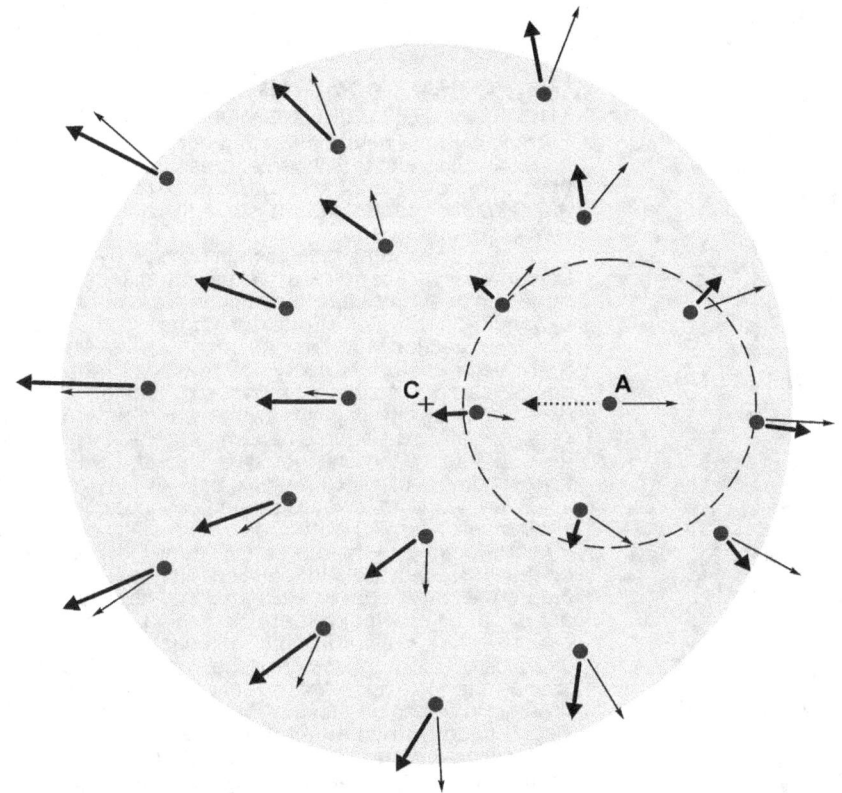

LOCATING A FINITE UNIVERSE'S CENTER
THEORETICAL HOMOGENEOUS SPHERE'S EXPANSION & REDSHIFT

Let's say you accept that the universe can't express two-dimensionally like the surface of a sphere, but you're not yet ready to accept the impossibility of a homogeneous three-dimensional spherical big bang, portrayed in section view by the uniform background and the uniform distribution of dots representing galaxies. The smaller arrows again indicate the direction and relative rate of their outward expansion. The larger arrows the direction and relative rate of their recessional velocity from our galaxy at **A**.

A three-dimensional big bang that was theoretically arrayed uniformly with galaxies would still exhibit a pattern of increasing density and higher redshifts that peaked directly opposite our direction of travel that would indicate the location of its origin and confirm its finiteness and expansion, if it actually was.

If we invoking some visible horizon for some reason, indicated by the dashed circle, to explain away this fundamental reality, it still doesn't help. The pattern of increasing redshifts toward our outward direction of travel would still hold because of the big bang's inferred accelerating rate of expansion that would again indicate the direction of the big bang's origin and confirm that's its finite and expanding.

Ultimately, any three-dimensional expansion, even if theoretically uniform, would generate a distinct pattern of increasing density and increasing galactic redshifts that would comfirm the big bang's three-dimensional existence. Without them, we're forced to realize that our real nontheoretical three-dimensional universe can't be expanding or finite. It must be infinitely vast and ageless. Galactic redshifts must indicate something other than a galaxy's recessional velocity or universal stretching like the recessional velocity of each galaxy's own continuously coalescing infalling material that's perpetually transmuting and recycling because of gravity's innate runaway nature.

(26.3 U Center 5a)

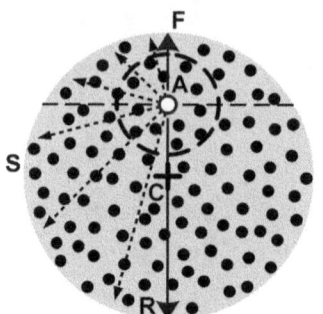

1. TOP-DN SECTION VIEW

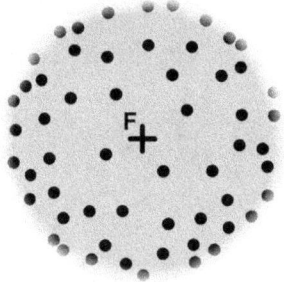

2. FORWARD VIEW

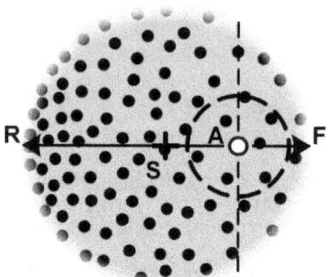

3. LEFT SIDE VIEW

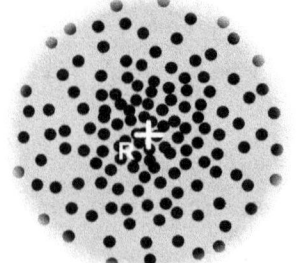

4. REAR VIEW

(50.1 Uniformity & Finiteness 10a)

UNIFORMITY & FINITENESS
THE VISUAL DISTRIBUTION OF GALAXIES

Starting with the assumption that our universe is finite and it's expanding and that it can express uniformly, which isn't physically possible for a sphere in three dimensions. But for the sake of argument, let's assume it is because uniformity is what we observe.

If we also assume that we didn't end up by chance at the universe's exact center, but were located, for convenience, about halfway between the universe's center, **C**, and its expanding perimeter, **F**, at **A** in diagram **1** that portrays a top-down section view through our universe. We'd then see a two-dimensional array of galaxies that visually condensed across the entire sky, represented by the black dots beginning in diagram **2**, that was least dense in our forward outward-bound direction of travel toward **F** where the universe's perimeter would be closest. That's where the fewest number of galaxies would be.

If we were to sweep around, as suggested by the sequence of smaller dotted arrows in diagram **1**, from **F**'s forward-looking view, diagram **2**, through **S**'s side view, diagram **3**, and look to our rear in the direction of **R**, diagram **4**, the visual two-dimensional density of galaxies across the sky would keep increasing, peaking exactly opposite our outward-bound direction of travel in the direction of **R** through the universe's origin at **C**, as depicted in diagram **4**. That's the direction where we'd find the greatest number of galaxies. We'd have to see this same visual pattern whether our (presumed) finite universe was diffusing or uniform or expanding or static.

If our universe was diffusing from expansion and condensing from gravity, as it'd have to be if it were actually finite because of a sphere's inherent geometry that's bound to the inverse square law, it'd still express the same visual array of galaxies across the sky. It'd just be more exaggerated, more dispersed in the forward direction, **F**, and more condensed in the rearward direction, **R**.

If we were to apply a cosmological redshift to galaxies from the universe's assumed expansion/stretching, whether it was diffusing or theoretically uniform, we'd get an exact correlation to the pattern. The highest redshift would be directly opposite our direction of travel where the galaxies would be at their farthest and densest and be receding the fastest. And the lowest redshift would be in front of us in the direction of our travel where the fewest, closest, slowest receding galaxies would be.

What we actually see though is a uniform isotropic/homogeneous distribution of galaxies and their redshifts. This explicitly indicates an infinitely vast, ageless cosmos where cosmological redshift has to originate from a source other than universal expansion/stretching.

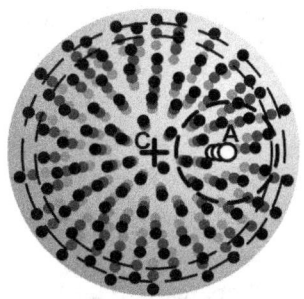

5. 3D EXPANSION
REALITY: NONTHEORETICAL
OUTWARD RADIAL DIFFUSION.

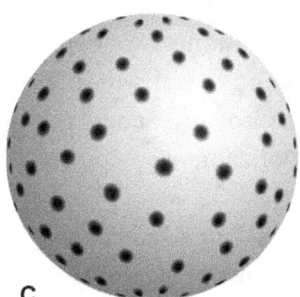

C

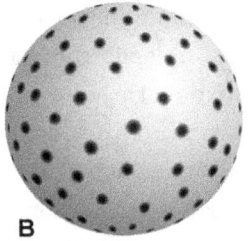

B

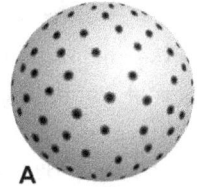

A

6. 2D EXPANSION
IMAGINED: THEORETICALLY
UNIFORM DIFFUSION WHOSE
CURVED PERPENDICULAR 2D
DISPERSION IS IMPOSSIBLE.

(50.2 Uniformity & Finiteness 10a)

UNIFORMITY & FINITENESS - CONT.
THE RADIAL & UNIFORM EXPANSION OF GALAXIES
Arguing that there must exist a visible horizon that limits our view to a certain distance, indicated by the dashed circle that represents a spherical region around our position at **A** in diagram **5** and **1** and **3** on the previous page, where all we can effectively see is uniformity doesn't work. Even if we ignore the interior diffusion that's inherent to a three-dimensional spherical volume, its expansion, or contraction, always has to occur radially. Its geometry is physically bound to the inverse square law's exponential diffusion, as portrayed in diagram **5**.

The volume of an expanding, or contracting, sphere can never express uniformly. And that diffusion would be easily perceivable whatever our location in our (presumed) finite universe. The only way to maintain its observed uniformity is to adopt Einstein's purely theoretical, curving non-Euclidean, finite yet somehow unbounded universe that with expansion has become the big bang.

It expresses two-dimensionally like the surface of a sphere, as depicted by the sequence **A-B-C** in diagram **6** that shows its galaxies' uniform dispersal with expansion. But there's no existence in two dimensions. Two dimensions can only define the location of a plane. So it doesn't work either.

No matter what we try, we're forced to return to an infinitely vast and ageless universe that requires practical, commonsense explanations for cosmological redshift and the cosmic microwave background radiation.

To maintain its static nature, Einstein adopted his cosmological constant to resist the gravity of his finite universe and prevent its ultimate collapse. But he could only achieve this mathematically in the two-dimensional realm of a sphere's surface that can express uniformly, or constantly.[156] But forced to accept that a real spherical universe could never manifest uniformly in three-dimensions, we're also forced to realize that his cosmological constant could never work in our actual three-dimensional world. The force of gravity decreases exponentially while his constant would remain just that, constant. They wouldn't coincide at all.

The same has to be said for the uniformly expanding space that he later accepted as its replacement. If space were actually something, its uniform expansion, if it was actually possible, could never resist gravity's natural coalescing in a real three-dimensional environment either. Gravity's exponential condensing and uniform expansion would be completely mismatched in the same way. The only way the universe's expansion, or some mathematical term, could resist its gravity and neutralize matter's coalescing is if its rate of expansion coincided exactly with gravity's exponentially decreasing strength.

If our existence has to be three-dimensional then you have to also accept that the expansion of a three-dimensional big bang could never resist the coalescing and eventual collapse of all its material, which is the whole reason expansion was adopted in the first place. With expansion now an unworkable and unnecessary holdover, there's no reason to keep it other than to explain galactic redshifts, which have several other more viable explanations. Without expansion, we've now lost the whole argument for the big bang, along with a finite universe.

The big bang's expansion could never proceed uniformly in three dimensions. Its self-gravity would dissipate from its center out per the inverse square law, causing its strength to decrease exponentially, allowing the velocity of its diffusing outward-bound matter and galaxies to increase, making its expansion nonuniform everywhere.

The expansion for any three-dimensional spherical universe of fixed mass would have to vary from zero at its origin to its maximum at its enlarging perimeter. So to achieve uniformity, any inverse reaction would have to neutralize all expansion, uniformity only being attainable at zero. Expansion couldn't begin or even exist if its uniformity was a prerequisite from the beginning.

The self-gravitation and expansion of a real three-dimensional big bang universe would cause the strength of gravity to be different everywhere and vary over time, further nullifying its ability to expand uniformly. Because it dissipates exponentially and compounds, every body of mass would have a different strength of gravity depending on its distance from the universe's origin. That's assuming the mass could have been formed in the first place.

156. Einstein, *Relativity*, 122-127.

Gravity's strength would also have to be continually decreasing exponentially as it expanded outward. There'd even have to be an ever-shrinking horizon around every body that would define a limit to its ability to attract other mass. The horizon's decreasing radius would depend on the body's mass and its increasing distance from the origin. (See Diagram **29** Threshold 4a)

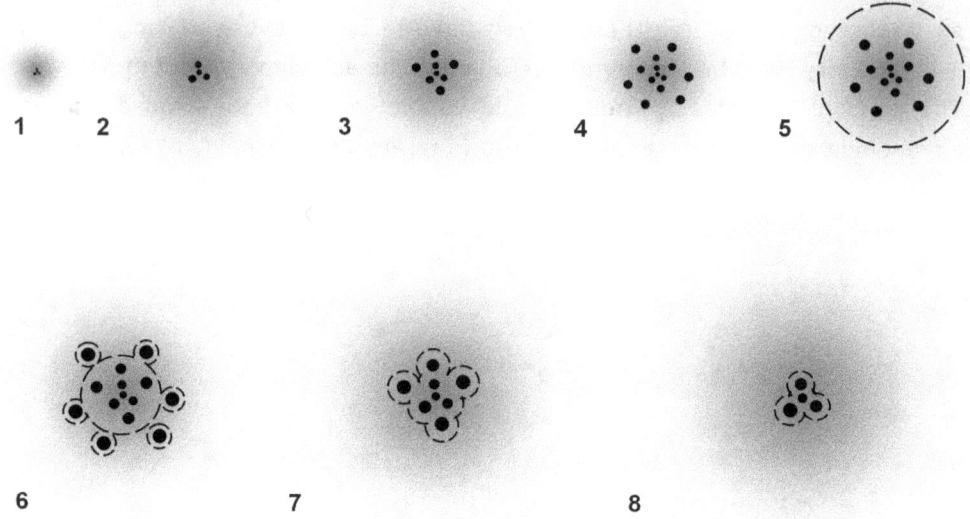

SPACE'S THREE-DIMENSIONAL EXPANSION

If expanding space was a reality in a normal Euclidean, three-dimensional, finite universe subject to the same laws of geometry we experience, its expansion could never manifest uniformly. Its matter and energy would dissipate exponentially, diffusing per the inverse square law just like any sphere, as depicted in section view in diagrams **1** & **2**.

Because expansion and gravitation are opposite effects, elemental matter would begin congealing, forming stars that collect into galaxies, represented by the dots, where and when the universe was at its most dense and its rate of expansion was at its least, its center at its inception. It could only congeal more galaxies outside its central region if its expansion were to slow, stop, or reverse, depending on its rate of expansion, as portrayed in **3** and **4**, where galaxies would be newly forming while still dispersing exponentially.

Because of the runaway nature of gravitation, the never-ending coalescing of material that would eventually lead to the collapse of the entire universe would continue unabated unless the universe started expanding again, which would restart the diffusion and dispersion of galaxies, prohibiting their formation while creating an ever-contracting threshold at its perimeter where space's outward expanding motion equaled gravitational coalescing's inward contracting motion, indicated by the dashed circle in diagram **5**. Everything outside would continue to spread while everything inside would still be subject to gravitational coalescing.

As space's expansion proceeded to overtake gravitation, this expansion-contraction boundary would continue to shrink exponentially, eventually engulfing each individual galaxy, as implied in **6**, dispersing their material at an ever-increasing rate, as suggested in **7**. Ultimately, the threshold would shrink to nothing at each galaxy at the universe's center, dispersing all material back to elemental matter, as inferred in **8**.

This simplified conceptually accurate portrayal of finite, three-dimensional, spherical expansion isn't anything like what we observe, especially the homogeneous distribution of galaxies without an identifiable center. Nor is it like what's predicted by the big bang, which delusively applies two-dimensional geometry to our three-dimensional existence to rationalize uniform expansion.

(29 Threshold 4a)

For matter to initially form and condense out of the big bang's primordial radiant energy its rate of expansion is required to fluctuate. But in a real three-dimensional universe, its expansion would already be varying everywhere. It'd always remain zero at its origin and would be constantly increasing to its maximum at its perimeter.

With no expansion at its origin, wouldn't matter had to have been congealing there from the universe's inception? And how could matter ever congeal anywhere else if expansion was increasing everywhere else? Wouldn't it essentially have to slow to a point below gravity's ability to coalesce to allow coalescing to begin?

And once expansion begins, how does it start slowing? What is the actual mechanism that would cause it? It can't be the gravity. Matter wouldn't have been created yet. That'd be putting the cart before the horse. Fluctuations in the rate of expansion is a theoretical concoction to justify the big bang.

If its expansion could actually be uniform, which it can't, and not just its rate of expansion but its expansion everywhere, how would it ever have allowed matter's initial coalescing? Wouldn't space's initial expansion have disbursed and continue to disburse primordial radiation? Wouldn't it have to remain in its most elemental form?

If matter wasn't condensed before the big bang's dispersion, how would there ever be a denser condition later that would allow its initial formation or it to begin congealing? And slowing or bringing the expansion to a complete stop would do nothing either to increase the density so that matter could then congeal into existence.

If matter formed before its expansion, wouldn't its exponentially decreasing density cause it to instantly begin condensing at an exponentially increasing rate? And what if its matter could somehow form sometime after its expansion began? Wouldn't the universe's exponentially decreasing density still cause the exponentially increasing coalescing of its matter toward its center origin? How could any of these realistic scenarios that all express exponentially because of the basic three-dimensional geometry of a sphere ever produce the homogeneous condition we observe today?

How is it that the primeval big bang could be a little unsmooth to initiate radiation's coalescing to create matter? How could minor density variations have ever developed in the first place, especially when it's so tightly compressed? What is the actual mechanism that caused the unevenness? How is it exactly that its small variation in density increases just enough to induce coalescing and creation?

But wouldn't an unforced density increase in radiation not lead to coalescing but actually have had the opposite effect? Wouldn't it push outward with greater force as it sought to find equilibrium? Why would it be any different from the presumed increased density of dark energy? That's what it's theorized to do. Those regions envisioned to be more concentrated with hotter radiation don't cool and condense into ordinary matter before the less dense regions and coalesce more rapidly into heavenly bodies and eventually galaxies. It's exactly the opposite. They push out.

There are more fundamental flaws in the whole concept of universal expansion. If we start with the assumption that the universe didn't spontaneously spring from nothing. And we accept the most popular view that it's been condensed down into a small point that's nearly infinite. And we assume that it's now composed of only matter or radiation, matter always coalesces because it has mass and radiation always disperses because it does not. Then if matter's coalescing condenses until fusion reactions are triggered that transform it back to radiation that no longer congeals but begins dispersing, how can a pre-matter, pure radiation universe even exist in a condensed state?

Wouldn't it have begun diffusing the instant its mass was transmuted to radiation? Radiation can't further condense. It no longer has mass, or mass relative to when it was matter. It has to be diffusing. That's a property inherent to radiation.

If the universe's radiation began diffusing innately from entropy, it'd continue unabated, disbursing exponentially per the inverse square law until it filled the entire void it inhabited. At that point, it'd attain an equalized state. It seems that only then could it possibly begin to condense into matter and later start coalescing. But there would be no finite volume outside of our finite universe to infill.

But let's say that space is actually a physical, three-dimensional something and its presumed expansion has nothing to do with radiation's innate diffusing, but it somehow expands separately, causing the dispersion of objects. Then because it cannot curve, we might reason that gravitation's coalescing and compaction must originate from the opposite effect, space's contraction or condensing. That seems like a sound conclusion, right? But how could space ever be both expanding and condensing at the same time when they're opposite effects?

We know things gravitate. We experience that in our everyday lives. The universe's dispersion, uniform or otherwise, is only theory and a tenuous one at that. It's based almost solely on redshift observations that we accept also originate from several sources other than universal expansion/stretching. So if universal expansion or gravitational condensing cannot coexist, which seems more likely?

There seems to be no conceivable way the universe can be expanding. Whether it's the big bang's purely theoretical, curving non-Euclidean, uniform diffusion that somehow expresses two-dimensionally like a sphere's surface or a real three-dimensional, finite, spherical universe's exponentially increasing

244

dispersion, expansion is not a viable alternative. It's truly fascinating how the fundamentally conflicted conditions of first space's actual existence and then its presumed expansion, either two or three-dimensionally, remain unaddressed, ignored by virtually the entire profession, but are instead regarded as a solid scientific certainty despite all the glaring questions.

This as much as anything else in our relativity-based cosmology exposes the extent of our indoctrination and devotion to groupthink, and how that indoctrination/groupthink has us desperately rushing headlong to manufacture and accept any kind of justification to validate it instead of rationally and objectively investigating the conceptual and real physical processes of our presumed facts.

Astronomical redshifts, when interpreted as originating from each galaxy's recessional velocity or space's stretching, actually suggest radial expansion, not uniform expansion. Uniform expansion is an assumption inferred from the universe's homogeneous nature, but more so a necessity of Einstein's two-dimensional finite yet unbounded delusions that require the isotropic properties of a sphere's curving surface to resist gravity uniformly.

Without that requirement, we would have never even considered that expansion should be uniform. Any inference of expansion would have been radial for a normal three-dimensional spherical universe. But that would have placed us at the universe's exact center. The odds of that coincidence being infinitesimal, we would have gone back to an infinite universe looking for other explanations for galactic redshifts.

To explain earlier than expected galaxy formation, we've mathematically induced unevenness into the early big bang by adopting a so-called inflationary period where we have it undergoing a brief period of faster-than-light expansion. But exceeding light's fixed velocity is a violation of relativity's founding premise.

So as a workaround, we decided that expansion could be accomplished through space's stretching that would cause light's velocity to slow while maintaining the validity of galactic redshifts. But if the velocity of an object's light actually was decreasing, slowing with space's stretching, then it'd be compounding with space's motion while varying everywhere, still violating relativity's principal tenet.

Relativity doesn't place any qualifiers on what the source of an object's velocity is. Two objects can never recede from one another faster than the speed of light no matter what's driving them. The consequence being that time would stop, dimension in the direction of travel would become infinitely small, and mass would become infinitely large.

It's a mathematical certainty of relativity, albeit contradictory and conceptually impossible. But that's not the point. The problem still exists. And explaining it away with spatial stretching or any other concocted rationalization that allows faster-than-light travel doesn't provide a viable workaround. We have to either abide by relativity or abandon it altogether. If we abandon it, light's fixed velocity isn't a restriction so there'd be no need to have space stretch.

But light's fixed velocity is a fallacy anyway. Conceptually, it has to compound with motion and its velocity does vary depending on the density of the field it's traversing. So there's no need to have space stretch to begin with. Also, space by definition is the nothingness between objects. There's nothing there to stretch. But if it actually were stretching then an object's spectral lines would have to be progressively stretching as well in a progressively stretching spectrum along with any displacement.

If we decide to keep space's stretching, the Doppler effect from a galaxy's recessional velocity wouldn't just go away. It'd still be a relevant physical effect. But it's just ignored. So is the redshift from a galaxy's gravity, gravitational redshifting, that's widely accepted. So is Einstein's, gravitational redshifting, an assumption of redshifting from relativistic time dilation from a galaxy's rotation. But most important, so is the redshift from every galaxy's constant infall of coalescing material.

There are potentially eight other redshift sources maybe more besides space's stretching, tired light that loses its velocity over distance, motion induced charge that slows an atom's natural frequency shifting its light toward the red, relativistic time dilation that shifts light toward the red, light's slower velocity in gravity fields, gravitational redshifting from loss of energy, Einstein's gravitational redshifting that's based on relativistic time dilation, Doppler effect from the recessional velocity of each entire galaxy, and the Doppler effect from the recessional velocity of each galaxy's constant infall of coalescing material. But we ignore all of them because if even one turns out to be correct, we'd never be able to know how much redshift to contribute to each and could never determine the universe's rate of expansion or even infer that it was expanding.

Why is it that astronomers only use the redshift from elliptical galaxies when attempting to establish the universe's (presumed) rate of expansion? What is it that makes the redshift from all other galaxy types less reliable than that of ellipticals? It can only be that their redshift is originating from some other source(s). Being that galaxy type is based on shape/rotation, variation in redshift is most likely caused by rotation and/or the ever-increasing infall velocity of its ever-coalescing/condensing material. The redshift from the infall velocity of an elliptical's material would be more consistent than any other.

What this tells us, though, is that astronomers are aware that other redshift sources exist and that they're affecting their readings. But they ignore the undermining implications for the big bang and continue anyway. Subjugated by pervasive confirmation bias, they can't help but cherry-pick data and shape the results to support their foregone groupthink consensus of universal expansion.

The compounding effect of space's assumed uniform expansion would instantly cause the matter at the opposite sides of the universe to exceed the speed of light with respect to one another, again violating relativity's founding premise. If the most-distant galaxy we're able to observe is more than 13 billion light-years away and if the Hubble constant (the universe's supposed rate of expansion) is about 45miles/second/ megaparsec (mega means a million and a parsec is 3.26 light-years)[157], then the recessional velocity of that galaxy would be close to the speed of light. And any galaxy the same distance away in the opposite direction would have to be separating from the other at almost twice the speed of light. This would have all objects close to the universe's expanding perimeter exceeding the speed of light, violating relativity's maximum speed limit law.

It's been calculated that space's rate of expansion would create a 16 billion light-year cosmic (event) horizon.[158] Everything beyond it wouldn't be visible because it'd be moving away faster than light, once again violating relativity's founding premise. This horizon is also equal in all directions. It doesn't take into account our own expanding motion because it's based on Einstein's unfeasible non-Euclidean solution that has our three-dimensional existence impossibly expressing two-dimensionally.[159]

It's also been calculated that the universe has an observable radius of about 46 billion light-years but that its existence is only about 14 billion years.[160] It's believed we're seeing all the way back to within a few hundred million years of its inception. First, for a real three-dimensional expanding universe, how would it even be conceptually possible to see its primeval beginnings? Its inception would have taken place 14 billion years ago, before our existence, at a single specific location and its light would have been diffusing exponentially at an ever-expanding perimeter.

With distance set by how far light travels in a year, how is it even conceptually possible for its radius to be farther than the distance to its origin? Wouldn't its minimum age have to be equal to its light-year radius? With a radius set by the speed of light that's three times farther than its age, doesn't that imply a perimeter

157. "Hubble's law," Wikipedia, last modified Mar 17, 2023, https://en.wikipedia.org/wiki/Hubble's_law.
158. "Cosmological horizon," Wikipedia, last modified Mar 14, 2023, https://en.wikipedia.org/wiki/Cosmological_horizon.
159. Einstein, *Relativity*, 125-126.
160. "Observable universe," Wikipedia, last modified Mar 22, 2023, https://en.wikipedia.org/wiki/Observable_universe.

that's expanding faster-than-light? And wouldn't that also imply light's compounding with space's expansion, both in violation of relativity?

To actually observe the light from the universe's primeval beginnings, wouldn't we have to travel faster than light to overtake its expanding perimeter to get to a location outside of the universe so we could then look back and see its approach? Even then, if we could actually exist outside the universe, the event would have dissipated to such an extent that any accurate record of it would have been lost forever.

With the farthest known galaxy more than 13 billion light-years away in a universe that's almost 14 billion years old, and if everything started out together at one location then began expanding outward, separating with the big bang, wouldn't the universe's minimum age then have to be more than 26 billion years? At a minimum, wouldn't you have to add the time it took for our galaxy to separate from the one we're observing to the time it took for its light to reach us from where it is today?

So if it were separating from us at nearly the speed of light, that'd have to be 13 billion years plus the 13 billion years they believe it's taken for its light to get to us. But wouldn't it be more likely that our separating velocity would have been very much slower than light? It could have taken hundreds of billions of years for the galaxy to get 13 billion light-years away? So really, the big bang universe could be virtually any age, but it'd have to be well over 26 billion years old.

Doesn't all this unresolvable conflict and contradiction and nonsensical reasoning really come down to just more mathematical sleight of hand that impossibly has our non-Euclidean universe's origin located everywhere on its impossible two-dimensional expanding perimeter? It's not at what it should be, a center beginning point for an actual three-dimensional, finite, spherical volume.

The cosmic microwave background radiation that we believe confirms the big bang could actually originate from any source, and it'd have to if the big bang is a misperception. It doesn't necessarily hail from the big bang's primeval beginnings. Sky surveys show that it correlates with galaxies.

So why isn't it more natural that it originates from them? The way we see it today is a two-dimensional portrait across our sky of how it looks from our vantage point on Earth. Its three-dimensional distribution would be and look totally different from every other location in the universe. And the same would be true for what it would be and look like just after a big bang, even if it correlated with the galaxies back then. So its association with galaxies now has no relevant connection to its presumed origin from the big bang other than to evidence the opposite.

Most people will easily agree that a three-dimensional environment that expresses two-dimensional properties is not possible. They don't have a problem conceding that existence on the two-dimensional surface of a sphere is fanciful. They do have a problem, though, facing the reality that homogeneousness for a three-dimensional finite expanding universe is just as impossible. The inherent properties of a sphere, a sphere's self-gravity, and its expansion would all produce an observable exponential condensing/diffusion of the universe's material.

Faced with the implication of this realization, the complete repudiation of the big bang, most acknowledge the discrepancy, conceding their inability to explain it but then put their blinders back on and place their faith back in established doctrine and its number of adherents, arguing that the big bang must be correct. All our brightest, well-educated authorities say so. So many couldn't be so wrong. No different than lemmings, people always follow the herd's prescribed groupthink regardless of where it leads them.

Beaten Track

Even though it can be easily demonstrated and understood how a two-dimensional universe is conceptually and physically impossible and how it's also conceptually and physically impossible for a normal three-dimensional finite universe that was expanding to express homogeneously, overcoming the big bang's momentum will still prove to be very difficult. It may even take generations.

But after we've seen through our conditioning and broken free of groupthink's grip and fully awakened to this elementary reality, the big bang will be embarrassingly viewed as science's "big blunder." Not so much because it was originally such a bad idea. But because we insisted on holding fast to and promoting its metaphysical nonsense even after being confronted with all of the invalidating evidence and nullifying logic.

There are several reasons. A primary one is the existing authorities' strong incentive to preserve the status quo. Aside from not wanting to appear silly, overturning their established doctrine loosens their grip on power. It undermines their control, which is what most people in authority crave. But the root cause of the big bang's persistence, or for that matter any commonly held but fictitious idea, is the fact that the mass of humanity remains hypnotized, hopelessly deluded by acquired beliefs.

The notion that nearly all of us are asleep to reality is a fundamental principle of all legitimate philosophies. Functionally impaired by indoctrination, groupthink, and what's misperceived as self-interests, we're unable to reason through and see even the most obvious fallacies. Until we're willing to admit the likelihood that we may harbor false beliefs and are controlled by groupthink, any objective investigation is virtually impossible.

Those burdened with the most always proclaim the loudest that they're objective freethinkers. They believe they're unaffected by any conditioning or influenced by so-called groupthink. But reality is the opposite. We're all largely affected.

Our pervasive big bang-relativity indoctrination demands that all contrary evidence, no matter how substantive or obvious, be forced into agreement with its dogma. For example, we insist:

- that light cannot compound with motion despite that its fixed velocity is conceptually impossible,
- that light's velocity varies in gravity fields despite that we insist that it's also constant,
- that objects contract in the direction of their motion to enforce light's fixed velocity despite that linear contraction due to motion is impossible,
- that time is an independent property of the universe when we ourselves create time and establish its rate,
- that mass increases because of an object's motion,

- that acceleration/braking and rotation create gravity fields despite that they are totally different from gravity and don't coalesce,
- that space is actually something when it's actually the nothingness between objects,
- that space can curve like a nonexistent, two-dimensional plane when it would have to vary in density if it were actually a three-dimensional something,
- that two-dimensional space dents underneath three-dimensional massive bodies when a third, more massive body does not exist below them to give them weight,
- that the universe is finite yet unbounded despite that this is a self-conflicted condition,
- that the universe is two-dimensional and non-Euclidean in nature expressing like a sphere's surface despite that there is no existence in two-dimensions,
- that the entire universe was compacted down into a size smaller than an atom despite that it's unfathomable,
- that the universe began by exploding into nothingness nor even empty space despite that the reason is unknown,
- that just after its big bang the universe underwent a brief period of faster-than-light inflation despite relativity's principle that faster-than-light travel is mathematically impossible even in a stretching universe,
- that the big bang's origin is not at its center beginning location but everywhere,
- that we can look out in any direction and almost see the universe's inception,
- that if we had a powerful enough telescope we could look out in any direction and see the backside of our galaxy,
- that if we were to set out from our galaxy straight in any direction we'd eventually return to our starting point,
- that a three-dimensional, finite, expanding universe can express homogeneously when it's gravity and expansion would condense and diffuse its material exponentially,
- that the material of galaxies rotates around their central black hole in a fixed circular orbit despite gravity's runaway nature,
- that the central material of a galaxy is endlessly falling down the infinite, two-dimensional, funnel-shaped space of its black hole despite that in three dimensions the material would instantly arrive at the galaxy's center,
- that the mass of an unseeable dark matter is affecting the distribution of galaxies, not their exponential condensing due to spherical and circular geometry,
- that an unseeable repulsive dark energy is accelerating the universe's inferred rate of expansion not a galaxy's exponential condensing due to spherical/circular geometry that'd exponentially increase their redshifts commensurate with their decreasing size,
- that the cosmic microwave background radiation evidences the big bang despite that it's associated with the location of galaxies today and could never appear as it would have at the universe's inception.

And there's even more. The list seems endless.

All of these irrational precepts seem perfectly reasonable to us because we've been conditioned to believe them, in or outside of academia. It's no different than if you were raised being taught that 2 + 2 = 5. Despite that it's unworkable, you'd still believe it's entirely logical. It's exactly the same with our uncritical acceptance of our big bang-relativity orthodoxy that we insist is rational. We don't and we can't see it for what it really is: a collection of wholly impractical, ad hoc justifications that are intended only to validate groupthink's entrenched ideology.

So we ignore, dismiss, or explain away all contradictory observations or disputing logic so we can blindly continue down the same baseless, illogical, preordained path. Our unfortunate predicament is well characterized in a poem by D. D. Palmer (born Canadian, founder of chiropractic theory, 1845-1913) reproduced here in part:

THE PATH THE CALF MADE

One day through the primeval wood,
A calf walked home as good calves should,
But made a trail all bent askew,
A crooked trail, as all calves do.

Since then two hundred years have fled,
And, I infer, the calf is dead;
But still, he left behind his trail,
And thereby hangs my moral tale.

The trail was taken up the next day
By a lone dog that passed that way;
And then a wise bell-wether sheep
Pursued the trail o'er vale and steep.

And drew his flocks behind him, too,
As good bell-wethers always do.
And from that day, o'er hill and glade,
Through those old woods a path was made.

And many men wound in and out,
And dodged and turned and bent about,
But still they followed - do not laugh -
The first migrations of that calf;

And through the winding wood way stalked,
Because he wobbled when he walked.
This forest path became a lane
That bent and turned and turned again;

This crooked lane became a road
Where many a poor horse with his load
Toiled on beneath the burning sun
And traveled some three miles in one.

And thus a century and a half,
They trod the footsteps of that calf.
The years passed on in swiftness fleet,
The road became a village street.

And this, before men were aware,
A city's crowded thoroughfare;
And soon the central street was this
Of a renowned metropolis.

And men two centuries and a half
Trod in the footsteps of that calf.
Each day a hundred thousand rout
Followed the zigzag calf about;

And o'er the crooked journey went
The traffic of a continent.
A hundred thousand men were led
By one calf near three centuries dead.

They followed still his crooked way,
And lost a hundred years a day.
For thus such reverence is lent
To well-established precedent.

A moral lessen this might teach
Were I ordained and called to preach,
For men are prone to go it blind
Along the calf paths of the mind;

They follow in the beaten track,
And out and in and forth and back,
And work away from sun to sun
To do what other men have done.

They go two miles instead of one,
To follow a path that was begun
By one lone calf; and to this day
They follow this old crooked way.

It continues ...

The point is, it must surely be evident by now that pursuing our big bang-relativity cosmology is to continue traveling the wrong path. To discover a straighter, more correct path, all that's necessary is to admit that the expert authorities who originally charted our course may have simply been wrong. There's no other way around it. We can't keep incorporating and explaining away their nonsense to preserve their reputations, and protect our egos, and expect to arrive at reality.

Coda

The only reasonable way to reconcile our observation of a homogeneous/ isotropic, uniform distribution of galaxies and their redshifts with the actual three-dimensionality of our universe is to accept that those redshifts cannot be the product of the recessional velocity of each entire galaxy being swept away by expanding or stretching space. Cosmological redshift must originate from some other source.

The most practical is the recessional velocity of each galaxy's own constant infall of material that's continuously migrating inward toward its center, ceaselessly coalescing and condensing and ultimately collapsing due to the inherent runaway nature of gravity. That collapsed material is being continuously transformed back into the radiant/plasma energy it originated from and is radiated back out.

For well-developed spirals, it's spewed out in high-velocity jets. Eventually, it slows and cools and reconstitutes back into ordinary matter. It can then begin gravitating back to its or another nearby galaxy. This extremely dynamic process of perpetual retroversion and regeneration, that's impossible to rationally deny, affirms our infinitely vast and eternal universe and how it can be continuously changing while remaining in a balanced steady-state condition.

By systematically demonstrating the flaws of our big bang-relativity cosmology, it becomes clear that it's evolved into nothing more than a belief system, no different from any other religion. It's fascinating how our insistence on preserving the big bang, with an origin that somehow manifests everywhere, reveals the hidden-most, egocentric desires of the human psyche and to what extent we'll go to keep ourselves at the center of the universe as we've done consistently throughout history.

Despite sound nullifying reasoning and an overabundance of contradictory factual evidence from all of our sophisticated technology and precision instrumentation, we still anxiously hold fast to our make-believe universe. We unwittingly rush to subjectively interpret every questionable observation or inconsistent result, forcing them to conform to our fictitious, preordained big bang-relativity dogma.

It's like we're Vernon Howard's (American philosopher, 1918-1992) ignorant cave dwellers frantically running in the dark from straining timber to straining timber desperate to shore up an impending cave in. Well, regardless of what must be the greatest fear for believers, the time has come to just quietly step aside and allow our entire big bang-relativity cosmology to completely collapse. The facts that remain intact will reveal the universe's true nature.

Part Three
Commonsense Cosmology *for the Commoner*_____

A persistent and difficult problem for astronomy is that many galaxies and quasars (presumed to be the most distant energetic objects in the universe) appear to be associated, some even sharing common material, while exhibiting very different redshifts. In an expanding big bang universe, for any two or more galaxies or quasars to be physically interacting it would mean that they have to be located approximately the same distance from us.

But if their redshifts are considerably different, it would indicate that they should be located a significant distance apart along their line of sight. This paradox it seems is not uncommon, but remains without (conventional) explanation.[161] The correct resolution of this single issue, though, is exceedingly simple. And it indicates nothing less than the true nature of the universe.

The fact that we simultaneously observe gravitation, the coalescing of objects, and the presumed uniform expansion of space, the dispersion of objects inferred from cosmological redshift, could imply an inherent conflict. Viewed in the most elemental way, one of these conditions would necessarily have to overcome and dominate the other. They're opposite effects.

Since gravity is an observational fact that we cope with in our everyday lives and space's, or the universe's, expansion/stretching is speculation, it's reasonable then that expansion/stretching might be a misperception. The redshifting of a galaxy's light may indicate something other than its recessional velocity and/or space's stretching.

But to draw this basic conclusion, we need not abandon the established precept of light's redshifting due to the Doppler effect of receding objects. We need only concede that we've been misconstruing the motion of a galaxy's material. Instead of assuming that it constantly revolves in a nearly circular orbit around a galaxy's center, especially spirals, we only have to recognize gravitation's inherent runaway nature that we already accept as a principle of gravity.

This would cause all of a galaxy's material to be constantly coalescing and condensing, continuously migrating inward, endlessly infalling toward every galaxy's center of mass. Conceptually, this would have to create a recessional velocity with an associated redshift.

---

161. Halton Arp, *Seeing Red: Redshifts, Cosmology and Academic Science* (Montreal: Aperiron, 1998). Throughout the book, Arp presents an ironclad case that cosmological redshift cannot be an indication of a galaxy's recessional velocity/distance by presenting factual evidence of companion and/or associated galaxies/quasars with dramatically different redshifts. He also reveals how groupthink's corrupting influence has perverted the profession of astronomy. It's an excellent resource with critical insight.

Incredibly, all we've been doing is simply mistaking the recessional velocity of each galaxy's own constant infall of material for the recessional velocity of each entire galaxy. This, along with our own infall velocity in our own galaxy, makes it appear as if all galaxies are receding from us and one another when in reality they remain essentially stationary, discounting any local gravitational interplay.

As unbelievable as it may seem, this plainly obvious but crucial error has given rise to the entire big bang theory. With the additional realization that this constant condensing of material ultimately leads to its collapse at the galactic center, we can see that galaxies are not close to being static. They're very active, constantly transmuting their material, perpetually recycling it back into radiant energy while their size remains nearly constant, only fluctuating slowly over long periods.

This rudimentary understanding completely resolves how associated or companion galaxies and/or quasars could even be physically interconnected, yet have any magnitude of redshift without resorting to other more exotic explanations of redshift. This simple realization allows our universe to remain infinite and eternal, as we all intuitively know it is, in a static or steady-state condition as it dynamically recycles material.

This is not to be confused with the steady-state theory advanced in 1949 by Fred Hoyle (British astronomer, 1915-2001) and others as an alternative to the big bang. It proposed the (unworkable) idea that new matter is continually being created in between galaxies to keep the density of the universe constant.[162]

Before we can fully explore this practical resolution and its relevance for cosmology, we need to first make sure we have a concise and coherent understanding of the true nature of light, time, and space and their implications for special and general relativity.

162. "Fred Hoyle," Wikipedia, last modified Mar 20, 2023, https://en.wikipedia.org/wiki/Fred_Hoyle.

c?

Imagine you're walking down the street at 5mi/hr holding a flashlight that you're pointing directly forward. If you were asked how fast is the light leaving your flashlight, you'd probably respond 186,000mi/s. That'd be correct. But according to Einstein, and conventional wisdom, it's only because your time has "slowed" and your dimension in the direction of travel has contracted to compensate for your 5mi/hr speed.[163] It's assumed that because light's velocity is fixed, its speed leaving your flashlight is 186,000mi/s minus your 5mi/hr walking speed.

Now let's say you had a second flashlight and were pointing it perpendicular to your direction of travel and were asked the same question. You'd probably respond the same, 186,000mi/s. And you'd be right again. But there's a problem. For the light pointed perpendicular there's no motion in that direction so there's contraction but time's rate is still being "slowed." Its decreased rate would have to apply equally everywhere over the same moving reference frame, which is you. It can't read differently in different directions. This makes for two conflicting velocities for light with the perpendicular exceeding 186,000mi/s by 5mi/hr. (See Diagram 3.1 Light's Constancy 7a & 3.2 Light's Compounding 4a)

---

163. Einstein, *Relativity*, 25-46.

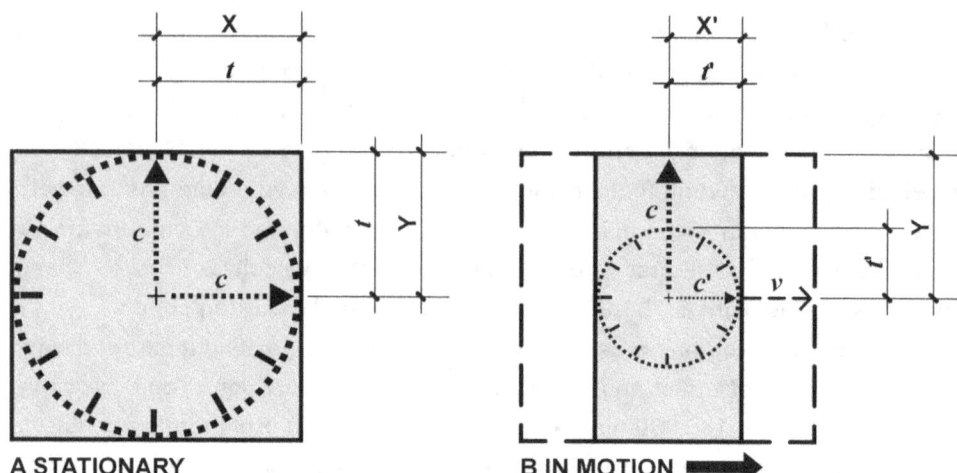

A STATIONARY　　　　　　**B IN MOTION** ➡

LIGHT'S CONSTANCY

A simple way to illustrate the impossibility of light's fixed velocity (relativity's founding premise) is by establishing a two-dimensional square reference frame that could be of any size, as depicted in diagram **A**. When theoretically stationary, its **X** and **Y** dimensions from its center would correspond to light's constant velocity, indicated by the arrows at c, and time's constant rate, symbolized by the clock-like circle that fills the entire reference frame equally that equates to t.

Einstein would contend that when the reference frame is put in motion, let's say moving from left to right at velocity v, as depicted in diagram **B**, for light's velocity to maintain its constancy in the direction of motion, it would have to slow in that direction by the amount of the reference frame's velocity to c'. This would require the reference frame to contract correspondingly in the direction of motion to the distance **X'** while its rate of time also contracted ("slowed/dilated") equivalently to t', as implied by the smaller clock-like circle.

But since there's no motion in the perpendicular direction, the reference frame's **Y** dimension and light's velocity, c, are not required to contract to maintain its constancy. And since time's contracted rate, t', has to apply equally over the entire reference frame, this creates an unresolvable conflict in every other direction, as indicated by the smaller clock-like circle. Its contracted time, t', corresponds to the contracted **X'** dimension and light's contracted velocity, c', in the direction of motion. But in the perpendicular direction, its contracted rate conflicts with the noncontracted dimension at **Y** and light's noncontracted velocity at c, which would cause it to exceed 186,000mi/s by v.

This unresolvable conflict clearly shows how light's velocity can only remain fixed, theoretically, in the one abstract dimension of linear motion. Even if time had a real existence and was a viable constituent of the universe, light's constancy would still be conceptually impossible in two or the three real dimensions of our physical environment. Which means, its velocity has to compound with the motion of its source and that of other reference frames. Left without the possibility of light's velocity ever being fixed, relativity loses its underlying premise and becomes altogether untenable.

(3.1 Light's Constancy 7a)

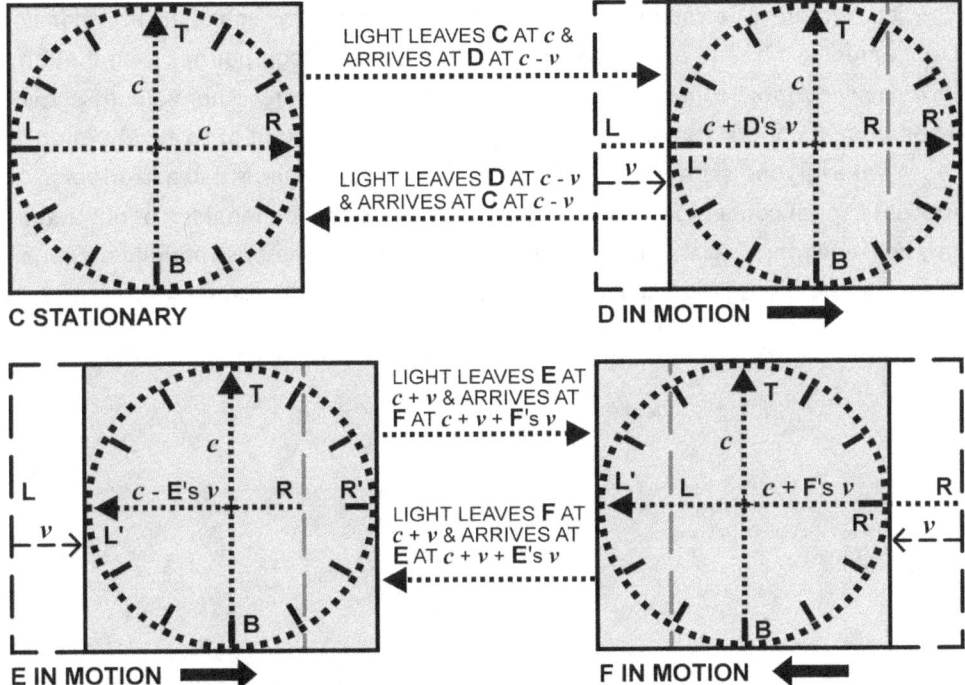

C STATIONARY

LIGHT LEAVES **C** AT c & ARRIVES AT **D** AT $c - v$

LIGHT LEAVES **D** AT $c - v$ & ARRIVES AT **C** AT $c - v$

D IN MOTION ➡

LIGHT LEAVES **E** AT $c + v$ & ARRIVES AT **F** AT $c + v + $ **F's** v

LIGHT LEAVES **F** AT $c + v$ & ARRIVES AT **E** AT $c + v + $ **E's** v

E IN MOTION ➡   **F IN MOTION** ⬅

LIGHT'S COMPOUNDING

Light compounds with the motion of its source and that of other reference frames just as we'd naturally infer. The four conditions represent generic reference frames theoretically free of gravity fields to avoid light's variability. The clock-like circle in each symbolizes time's theoretical rate that remains constant throughout the entire reference frame. The dotted arrowed lines denoted with c indicate light's velocity. The dashed grey lines indicate the reference frame's original location prior to motion.

Reference frame **C** is portrayed as theoretically stationary. For its observers, light moves from left to right and from the bottom to the top at c. An outside observer also theoretically stationary would record the same.

For reference frame **D**, it's depicted as moving from left to right with velocity v. For those observers, light moves normally as if it were stationary from **L** to **R** and **B** to **T** at c. This is what all of the Michelson-Morley and Sagnac type experiments show. Light always leaves its source at the same time in all directions at the same rate regardless of motion.

For an outside observer that's theoretically stationary, light begins from its initial position at **L** and arrives at **R'**. Light travels a longer distance in the same amount of time, leaving **L** at $c + $ **D's** v. This is a compounding of velocities. Light projected between **C** and **D** also indicates its compounding as noted for the different directions.

For **E**, this time light is shown as projected from right to left, opposite the direction of its motion. Its observers again record the light's progress but this time from **R** to **L** as if stationary. But a stationary, outside observer records it traversing a shorter overall distance from **R** to **L'** in the same amount of time. This compounded velocity would be slower than c by **E's** v because the light leaves **R** opposite the direction of **E's** motion at $c - $ **E's** v.

F is the same as **D**, just in the opposite direction. The light projected between **F** and **E** indicates the compounding conditions for the other circumstances of relative motion.

(3.2 Light's Compounding 4a)

This simple illustration clearly reveals the fallacy of light's fixed velocity. Conceptually, it has to compound with motion. And if it compounds with motion, its velocity cannot be limited to 186,000mi/s. It has to be any other velocity above 186,000mi/s up to instantaneous. Everything's in some kind of relative motion.

What everyone seems to have missed, including Einstein, is that a fixed velocity for light cannot conceptually work in two or the three real dimensions of our actual physical environment. It only works theoretically in the one abstract dimension of linear motion. (See Diagram **47.1**, **47.2**, **47.3**, **47.4**, **47.5** Carriage 6a)

SAME RELATIVE MOTION
Using the same props and circumstances that Einstein liked to use in his thought experiments, we can envision ourselves positioned on an embankment alongside a railway carriage where we can see a passenger inside timing a "ray" or a quanta of light as it travels from the ceiling at **A** to the floor at **B**. If we also timed the same ray of light, compensating for our distance, we'd obviously get the same reading as the passenger, 186,000mi/s, as long as we were both stationary or our relative motion remained the same. Few would disagree. Problems only arise when there's relative motion between objects.

(47.1 Carriage 6a)

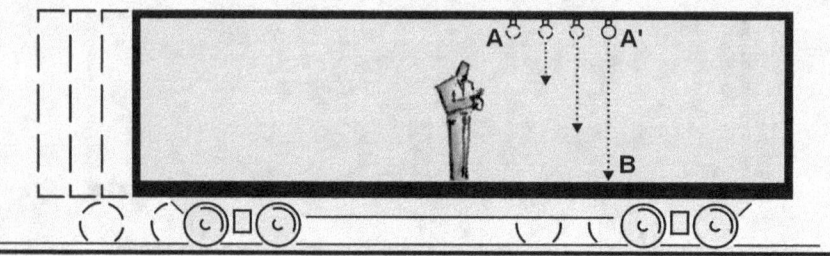

DIFFERENT RELATIVE MOTION ➡

Let's say we're again at our position on the embankment, but this time we're stationary and observing the railway carriage moving with velocity v from left to right, set into motion by any force other than that of gravity. The passenger again times a quanta of light beginning from the ceiling at **A** that travels to the floor at **B**. But unable to perceive his motion, he again measures its vertically from **A'** to **B**, recording the same time despite that it actually traversed the longer diagonal distance of **A** to **B**, which reveals that light's velocity actually compounds with the motion of its source.

From the embankment, we too measure the ray of light from the ceiling at **A** to the floor at **B** and record the same time as the passenger. But because we can perceive the carriage's motion, we measure the traveled distance from **A** to **B**, the longer diagonal distance. The longer distance covered in the same time also gives us a velocity for the speed of light that's 186,000mi/s + v, confirming its compounding.

Conceptually similar diagrams attempt to validate relativity by projecting a ray of light in the diagonal direction. In this case, it'd be from **A** toward **B**. They're incorrect. The stop-action nature of this diagram shows the true path of the quanta or ray of light would always remain vertical. A diagonal ray of light would be a completely different ray.

Einstein would have us believe that the carriage must be contracting the distance **A** to **A'**, and that time and light's velocity in the direction of motion must be slowing correspondingly so that both the passenger and an observer on the embankment would record a speed for light that remains constant. His assertion that time "slows" is also an error. An object's necessary contraction to enforce light's constancy would require a corresponding smaller, faster running time, not a larger, slower running time. But for this example, we'll ignore his incorrect assumption. We'll also ignore the more fundamental fact that in reality there's no such thing as time. For now, we need only assume that time changes with motion to follow the argument.

The error Einstein made was that he addresses light's presumed constancy only in the theoretical, one-dimensional condition of motion. He ignores the other two dimensions of our real three-dimensional environment. If the carriage's motion resulted in its contraction in the direction of motion and a slowing of its rate of time to maintain light's constancy, then because the entire carriage is of the same three-dimensional reference frame, all subject to the same slower rate of time, this would create an unresolvable conflict with any ray of light projected in any other direction than directly forward.

If the distance between the sidewalls and floor and roof could also somehow contract to maintain light's constancy, then its velocity for an outside observer would be less than 186,000mi/s in all directions except in the direction of motion. The only other option is to have time vary in all directions simultaneously for the passenger, which of course is not possible either. All of this presents us with unresolvable conflicts that cannot be reconciled, which unequivocally establishes the fallacy of light's constancy.

(47.2 Carriage 6a)

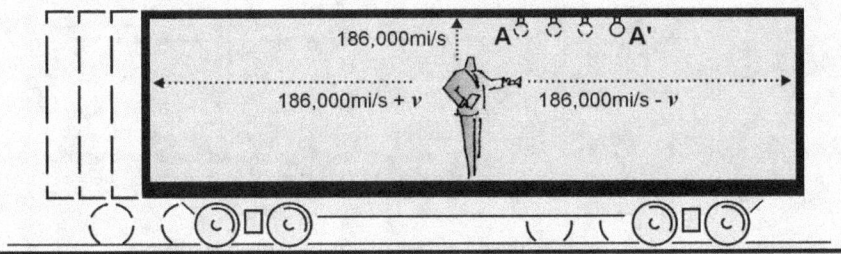

PROJECTED LIGHT IN MOTION ➡

What would happen if our passenger, located at the carriage's midpoint, was holding three flashlights, one pointed directly forward, one to the ceiling, and one to the rear while it was in motion. For his speed of light to remain constant, wouldn't it have to vary in every direction? In the forward direction, wouldn't it have to be light's velocity minus the carriage's velocity? To the ceiling or sidewalls perpendicular to its direction of motion, wouldn't it have to remain 186,000mi/s? And to the rear, wouldn't it have to be light's velocity plus the carriage's? Wouldn't all these conflicting velocities also conflict with light's velocity measured across the entire carriage from end to end in the direction of motion?

If the carriage's size could also change to enforce light's constancy, wouldn't it have to impossibly vary simultaneously in every direction, where it'd be contracting ahead of the passenger, not changing in the perpendicular directions between the ceiling and floor and between the sides, and expanding behind the passenger? And would all these conflicting dimensions conflict with the contraction from **A** to **A'** for the light measured across the entire carriage in the direction of motion?

And if the carriage's time could change to enforce light's constancy, wouldn't it also have to impossibly vary simultaneously in every direction, running slower for the forward directed light where the carriage would have to be contracting, normal for the light pointed to the ceiling and sides where its size would remain the same, and faster for the light directed to the rear where it'd have to be expanding? And wouldn't all these conflicting times also conflict with the contraction from **A** to **A'** for the light measured across the entire carriage in the direction of motion?

Also, if time and dimension were actually changing with motion to enforce light's constancy, then from our passenger's perspective wouldn't all of the stationary objects outside of the moving carriage have to also be contracting with his increasing speed to maintain light's fixed velocity? If not, wouldn't he be measuring the speed of light everywhere else as increasing, exceeding 186,000mi/s by the velocity of the carriage?

Of course, there's no way to reconcile any of these fundamental conflicts. We just have to concede the impossibility of light's fixed velocity and accept that it simply must compound with motion.

(47.3 Carriage 6a)

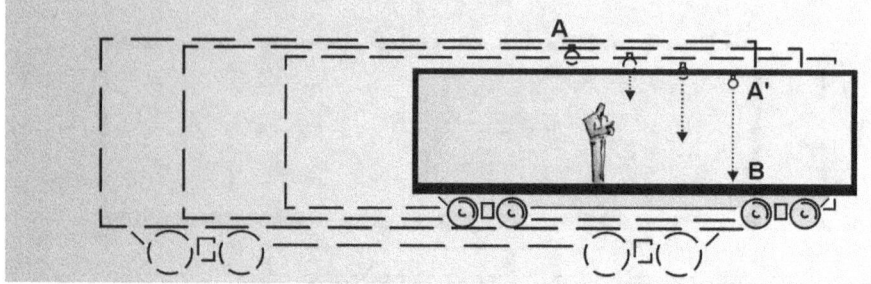

RELATIVE MOTION DUE TO GRAVITY ➡

Diagrams **47.1**, **47.2** & **47.3** all assume a hypothetical existence in a gravitational field of constant density, which in an absolute sense is impossible. All gravity fields, no matter the scale, vary in density even if it's only slightly that impart some relative motion. So once again, let's view our railway carriage from our stationary position on an embankment in a gravity field that's still theoretically of constant density. But now let's imagine that the carriage is not bound to its rails but is in outer space somewhere in a real gravity field. Its pursuit of equilibrium in gravity's exponentially decreasing field density naturally causes its accelerating free-fall with an increasing velocity v from left to right while contracting in a near equal omnidirectional manner.

Our passenger again times a ray of light traveling from the ceiling at **A** to the floor at **B**, also measuring the distance from **A'** to **B** just as in diagram **2**. But this time, it's a somewhat shorter distance. Even though he's unaware of the shorter distance because of the carriage's omnidirectional contraction, he still calculates the same 186,000mi/s. This is because light's velocity decreases in the gravity field's decreasing density while time's assigned rate increases in a nearly synchronized way because all the objects within the carriage's reference frame are also contracting in gravity's decreasing field density, which is increasing their natural frequency, which is how we establish time's rate. All of which compensate for the reducing distance between **A'** and **B**.

For us, though, as outside observers, we can again not only conclude a compounding of velocities just as in diagram **2** because we're able to perceive the light ray's path from **A** to **B**. But because our rate of time is constant, we can also perceive that the distance from **A'** to **B** is decreasing and that the carriage may not necessarily be moving away into the distance as its omnidirectional contraction would suggest. But if we were to travel to the carriage, we would find that it is in fact farther away because distance is only a function of a gravity field's decreasing density, which decreases light's velocity while contracting objects, which increases their natural frequency or their assigned rate of time.

With the fact of light's variability due to its refraction through the varying density of a gravity field, its compounded velocity has to be based on the velocity of the light determined by the density of the gravity field where the measurement is taken. Compounding velocities implies an unlimited speed for light, but only up to instantaneous. Velocities faster than instantaneous would mean that information was traveling forward or ahead of its assigned rate of time, which would be time travel and would violate laws of cause and effect.

With the reality of light's compounding and the fact that an object's mass does not increase because of motion, becoming infinite at the speed of light, we have to also accept that there can be no restriction on an object's velocity either. It must also have unlimited potential up to instantaneous.

(47.4 Carriage 6a)

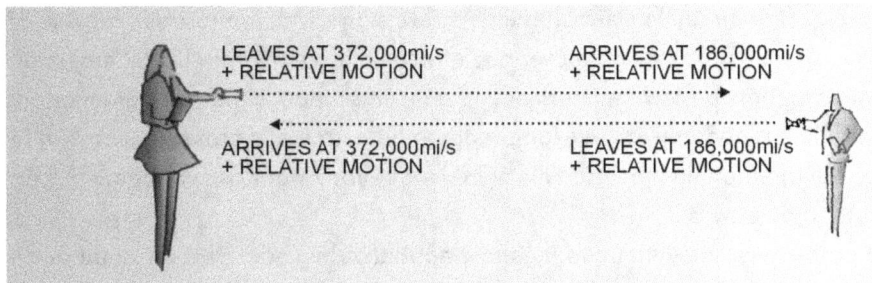

LEAVES AT 372,000mi/s
+ RELATIVE MOTION

ARRIVES AT 186,000mi/s
+ RELATIVE MOTION

ARRIVES AT 372,000mi/s
+ RELATIVE MOTION

LEAVES AT 186,000mi/s
+ RELATIVE MOTION

LIGHT'S VARYING & COMPOUNDING VELOCITY

For a simple example of how light behaves in the decreasing density of a gravity field, let's say that we've placed the person on the left at a location outside of our galaxy where the field density, represented in section view by the diffusing background, is twice that of another person that we've placed near the Earth's location. Let's assume that their relative motion is exactly the same and that each's field density remains constant and that we're observing from a third location equal distance from both. Notice that the size of the person in denser field is twice that of the other.

Let's also give them extremely powerful flashlights that could reach the other. If the person on the right near the Earth's location was to turn on his light, it would leave him at 186,000mi/s but might arrive many thousands of years later at the other person whose field is twice as dense at a much higher velocity. Let's say 372,000mi/s. Conversely, if the person on the left in the denser field was to turn on her light, it might leave her at 372,000mi/s but arrive at the other person at 186,000mi/s.

Now let's say that she was to begin walking toward the other person at 5mi/hr. The speed of her light would now arrive at 186,000mi/s + 5mi/hr. If the person on the right in the thinner field was to shine his light while walking away from the other person at 5mi/hr, his light would arrive at 372,000mi/s - 5mi/hr. You get the idea. Velocities would always have to compound, but compound with a velocity of light that's established by the field density where the measurement is taken.

47.5 Carriage 6a)

Some justify Einstein's claim that time's rate slows with motion with the use of a so-called "light clock." It would theoretically have a pulse of vertically projected light oscillating between two mirrors acting as a timepiece. When put into motion, its light would supposedly travel a longer diagonal path that's projected forward to keep up with the moving mirrors. This would make for longer oscillations, causing time's rate to be slower.

The problem is, the light traveling the diagonal path would be a different pulse or quanta of light, a photon if you will. The original vertical pulse would always remain vertical from its origin. But its origin is in motion. So it would have to be moving in unison with it. As long as the light clock's speed remains constant, they'd remain in the same reference frame. This reveals, once again, how light conceptually has to compound with the motion of its source, and that of other sources. It cannot remain fixed. (See Diagram **30.1**, **30.2** Light Clock 2a)

1. LIGHT'S CONSTANT OSCILLATION - STATIONARY (REALITY)

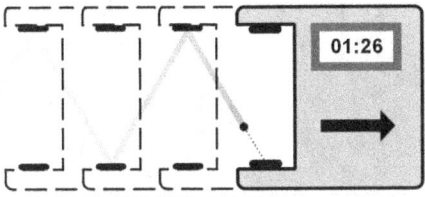

2. LIGHT'S "SLOWER" OSCILLATION - WITH MOTION (THEORIZED)

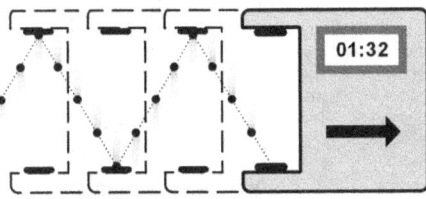

3. LIGHT'S CONSTANT OSCILLATION - WITH MOTION (REALITY)

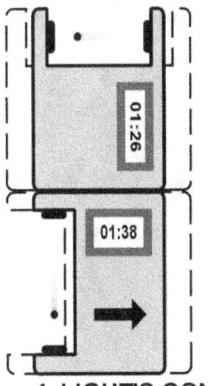

4. LIGHT'S CONFLICTED OSCILLATION - WITH MOTION (RELATIVITY)

(30.1 Light Clock 2a)

LIGHT CLOCK

This depiction of a so-called light clock is similar to those commonly used to shore up Einstein's assertion that time slows with motion. At diagram **1**, it's first shown as stationary where a photon or a pulse of light oscillates vertically between mirrors acting as a clock that runs at a certain fixed rate. When it's in motion, as shown in diagram **2**, its time is imagined to run slower because the light's diagonal path is longer. This is represented as a smaller time, as its reading indicates.

The problems with this modeling quickly become apparent with diagram **3**. First, the photon wouldn't travel a diagonal path. Its path would remain vertical, always moving with the clock. So its rate of time would remain constant. It would take a different photon projected at a forward angle to travel a diagonal path. But the lateral motion of the vertically oscillating photon would cause it to traverse a diagonal path. The distance of the diagonal path being longer but traveled in the same amount of time reveals that light's velocity has to compound with the velocity of its source. If it is compounding with the motion then it can't be restricted to 186,000mi/s but has to be any velocity up to instantaneous.

Second, the clock when it's in motion is not shown to contract in the direction of motion. Without its synchronized contraction but a contracting rate of time and light's slower velocity, this would have the clock expanding in the direction of motion. Third, with the speed of light fixed, traveling a longer diagonal path, this would produce a larger time than shown in diagram **1**.

Besides all of these conceptual failures, this model also provides another opportunity to reveal the fallacy of relativity's founding premise. Referring to diagram **4**, we see two light clocks attached to one another but arranged so that their photons travel in perpendicular directions. When the clocks are put in motion, the necessary contraction that would have to occur in the direction of motion to maintain light's constancy would affect the distance between the mirrors in the upper horizontally-oriented clock only. But since time's smaller rate that would be required maintain light's constancy in the direction of motion has to apply equally to both clocks being that they're of the same reference frame, this would create an unresolvable conflict between their rates of operation, as indicated by their different readings. This plainly demonstrates one more time that light's constancy is a conceptual impossibly and that it must compound with the motion of its source while its assigned rate of time and dimension remain constant.

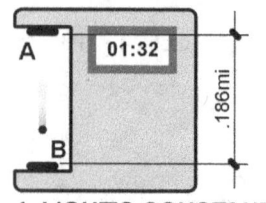

1. LIGHT'S CONSTANT VELOCITY - STATIONARY

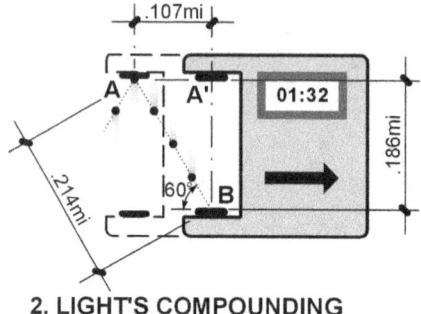

2. LIGHT'S COMPOUNDING VELOCITY - WITH MOTION

LIGHT CLOCK'S COMPOUNDING

Just for fun, let's look at how the compounding of light with its source might occur for our light clock example if we were to apply some numbers. To start with, let's assume for convenience that it's very large, more than .2 of a mile across, and that we're able to place it out in space somewhere, stationary, where the density of the universal field would theoretically remain constant along with light's non compounded velocity that we'll assume is 186,000mi/s. Let's also say for numerical convenience that the vertical distance between the mirrors, from **A** to **B**, is .186 miles so that the oscillating photon will cover the distance in one microsecond (one millionth of a second), as indicated in diagram **1**.

For diagram **2**, to keep the calculation simple, let's set the clock's velocity very fast so that in one microsecond it covers a distance of .107 miles, from **A** to **A'**, so that the photon's path scribes a diagonal that's 60 degrees to its vertical oscillation that establishes a 30-60-90 degree right-angled triangle whose sides are proportionally 2, 1 & 1.73 with 2 being the hypotenuse.

Because the light pulse still originates from the clock when it's in motion, it moves with it, and its oscillation always remains vertical, covering the same .186 miles in a microsecond. So the clock's rate of operation remains constant whether it's in motion or stationary. But when it is in motion, the photon traverses a diagonal, from **A** to **B**, that's a longer distance, about .214 miles, covered in the same microsecond. That equates to a speed of light that would be 214,000mi/s.

With the light pulse always remaining vertical but covering the increased distance of the diagonal in the same amount of time, this confirms the compounding of its velocity with the velocity of its source. It's not an illusion or some kind of trick. It's just a natural consequence of two and our real three-dimensional geometry. And it's exactly how we intuit it. But it's completely contrary to our big bang-relativity conditioning.

(30.2 Light Clock 2a)

272

The other quality of light that academia refuses to face is its variability, even though Einstein fully accepts it. It refracts through any field, including gravity fields. Its velocity and position changes depending on the density of the field.[164]

The problem is, its refracted variability, even aside from its compounding, completely invalidates all of relativity. It's founded and is wholly dependent on light's velocity remaining fixed. When that's lost, every aspect of it fails. Einstein concedes this as well.[165] (See Diagram 15.1.1, 15.1.2 Refraction 5a & 15.2.1, 15.2.2, 15.2.3, 15.2.4 Refraction 3a)

164. Einstein, *Relativity*, 84-85, 104, 109,171.
165. Einstein, *Relativity*, 85, 151.

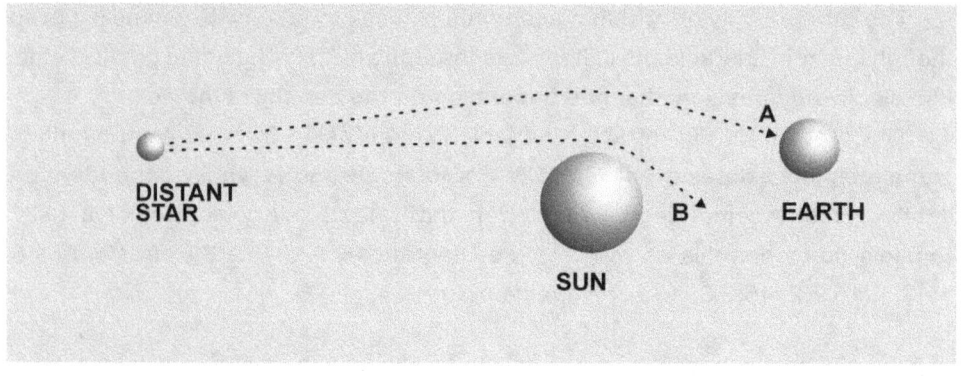

DISTANT STAR

SUN

A

B

EARTH

LIGHT'S BENDING

Our current belief is that a "ray" of light from a star or any distant object passing near a massive body like our sun is being pulled by gravity, that it's being bent from its otherwise straight path in the direction of **B** as it follows space's geodesic that somehow curves two-dimensionally in the vicinity of mass. And when viewed from Earth, its position is distorted in the direction of **A** due to an optical illusion.

Even though Einstein contends that light's distortion is actually due to its slowing through gravity fields, which is nothing more than refraction, which contradicts relativity's founding premise, light's fixed velocity, we reject his explanation. Instead, we hold to our belief that a photon, which remember is only a hypothetical quantum of massless energy, is subject to gravity's influence. We first mistakenly assumed that a photon is a particle. And then we incorrectly reason that because it's in motion it must have momentum. If it has momentum, it must have inertial mass. And then because of relativity's principle of equivalence, if it has inertial mass it must also have gravitational mass. And if it has gravitational mass, it must then be affected by gravity.

We're highly motivated to retain this convoluted logic because if we use light's refracted slowing like Einstein, we're abruptly confronted with the total collapse of relativity, which is wholly dependent on light's constancy. Incredibly, Einstein actually agrees that relativity would completely unravel if it were found that light's velocity was not fixed but variable.

(15.1.1 Refraction 5a)

274

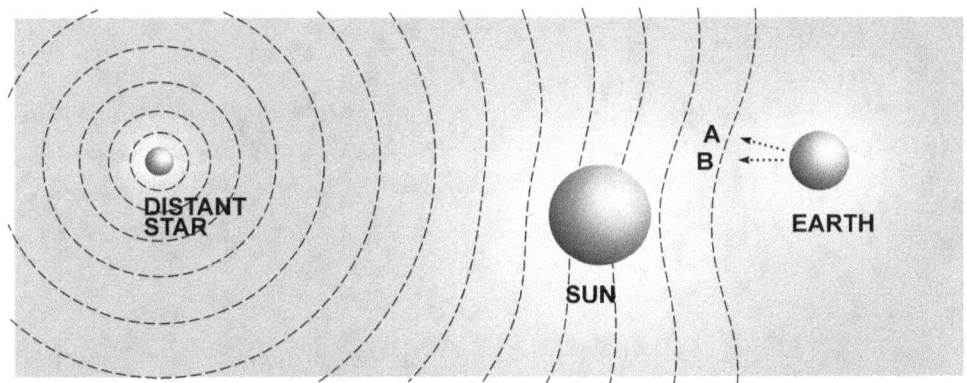

LIGHT'S REFRACTION

Light refracts through gravity fields. The distant star appears displaced in the direction of **A** not because light rays follow the impossible curvature of two-dimensional space or a resultant optical illusion but because the light in that direction reaches us slightly before the light coming directly straight from the star in the direction of **B**. Light's velocity slows through the decreasing density of the Sun's gravitational field, depicted in section as the diffusing background, just as any wave travels slower through a less dense medium, as portrayed by the series of circular and waving dashed lines that indicate the varying velocity of the incoming light emanating from the distant star.

It's also light's refracted slowing that's responsible for the gravitational lensing of distant galaxies or quasars that are split into two or more images that are assumed to be the product of the mass of some unseen foreground galaxy that's closer but fainter. But more often than not, it's just the common center of mass of any number of galaxies or galaxy clusters that is located between us and the object along its line of sight that's responsible for the lensing effect, which is why the refracting mass is so often never identified.

(15.1.2 Refraction 5a)

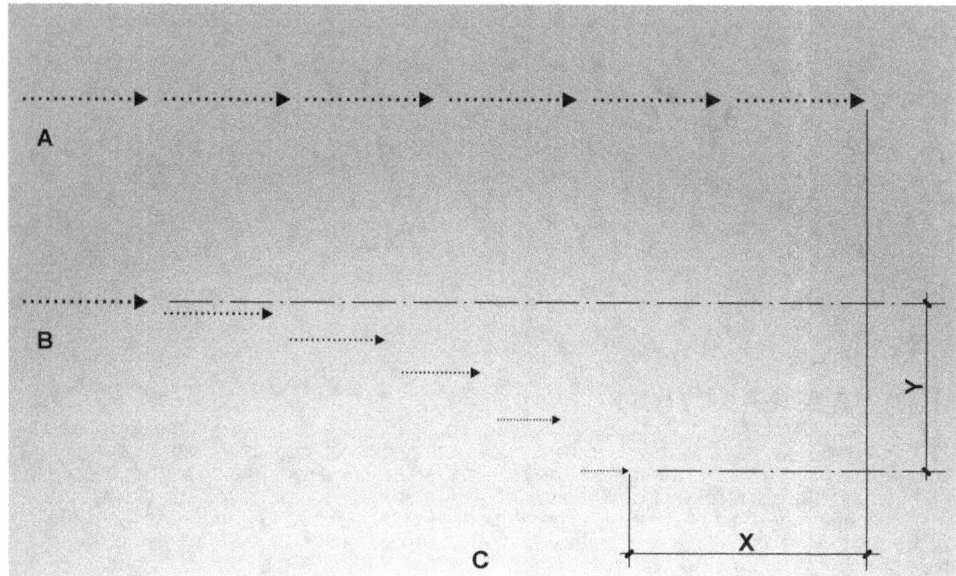

REFRACTION: LIGHT

To conceptually portray light's refracted slowing, let's say we could theoretically quantize it into a "ray" and represent it with a dotted arrow of fixed length. And then let's set two parallel rays out in space somewhere and project them so that the ray in row **A** would pass through a region where gravity essentially remained constant. Its velocity and direction would remain constant as well, as indicated by the sequence of unchanging arrows.

Let's set the other ray in row **B** to simultaneously pass through the gravity field of a nearby massive body like our Sun or a galaxy located in the direction of **C**. That field represented in section view by the diffusing background. As that ray's quantized light traverses the field, it would begin to slow in its exponentially decreasing density, as suggested by the decreasing length of the arrow. The difference expressed by the distance **X**. The light's course would also begin to deviate from its parallel path toward the body as it followed the field's decreasing density. That difference expressed as **Y**. Any other distortion is omitted for clarity. Anyone at the receiving end of our hypothetical rays would see the light in row **A** first because of its faster velocity.

(15.2.1 Refraction 3a)

276

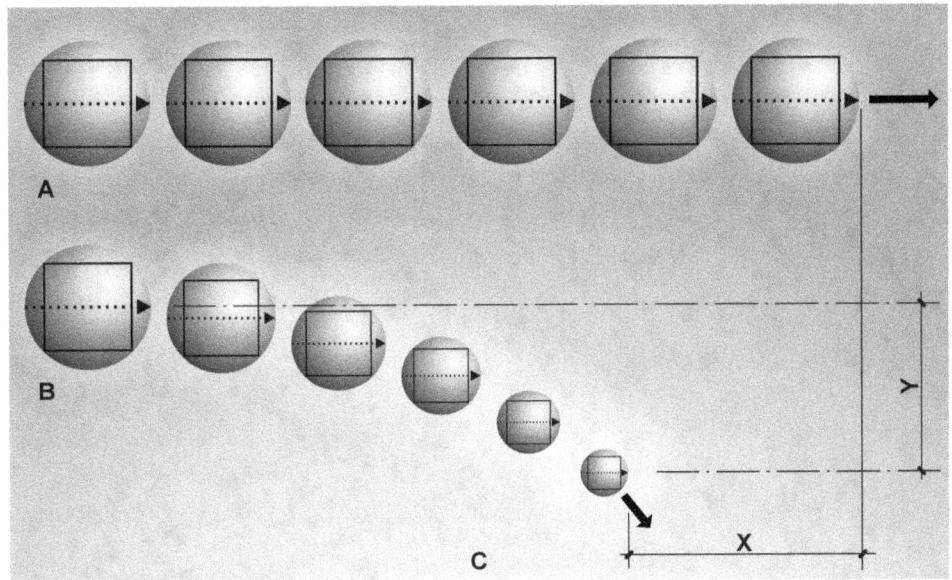

REFRACTION: OBJECTS

Objects themselves can also be interpreted as refracting in the decreasing density of a gravity field. Let's say we could associate a massive body, the shaded sphere, with our theoretical quantized light ray. The body in row **A** would remain unchanged just like its ray as it passes through gravity fields that we're assuming are uniform.

But in row **B**, as it enters **C**'s gravity field, it also begins to slow in the forward direction, indicated by **X**, while accelerating in **C**'s direction, indicated by **Y**. Because the universal field of radiant energy that comprises the electromagnetic spectrum is essentially the support structure for matter, its inherent exponential thinning at any mass, which is a gravity field, causes the body's synchronized contraction, and distortion, just the same as the ray's, as both our body and the body at **C** naturally pursue equilibrium in the decreasing density of their combining gravity fields. Any other distortion is still omitted for simplicity's sake.

If we were to put an object on our moving bodies like a box, symbolically represented by the corresponding rectangle, it too would theoretically contract, and distort, the same as the body and the same as the light ray.

(15.2.2 Refraction 3a)

277

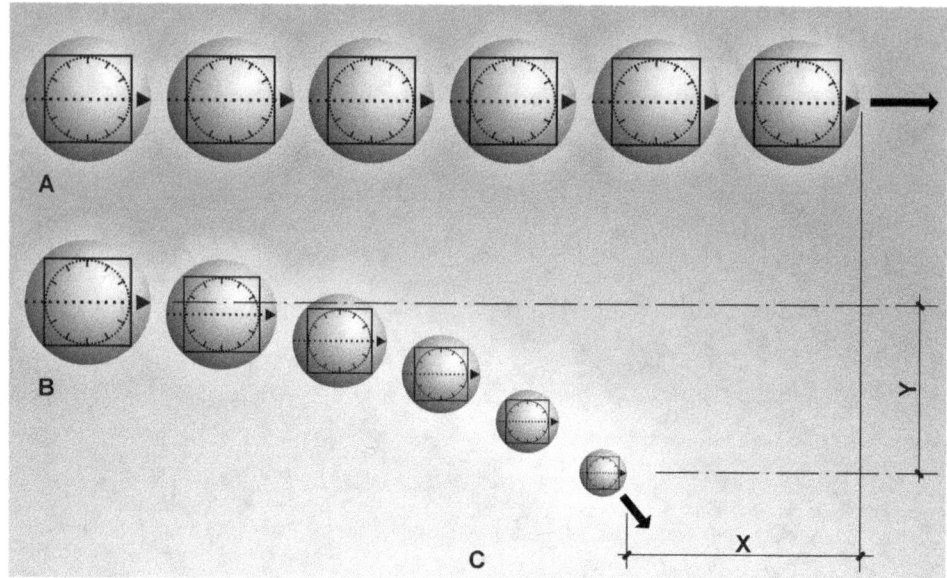

REFRACTION: TIME

Now let's establish a rate of time for our moving bodies by setting it to the natural frequency of their rotation just like we do for ourselves with the Earth's rotation, and let's represent that with a clock-like circle within the box on the bodies so that it too is associated with our body and light's velocity where its changing rate is represented by the changing size of the clock. In row **A**, the clock, the box, the body, and light's velocity all remain consistent in our assumed uniform gravity fields.

In row **B**, they all contract, and distort, in unison while both our body and that of **C**'s naturally draw together because of their reactive search for equilibrium in their combining gravity fields' decreasing density, which is always less between them.

As our body condenses, its mass is redistributed inward, which increases its rate of rotation, which in turn increases time's rate, as signified by the clock's smaller size. A smaller time being a faster running time that theoretically exactly compensates for light's decreasing velocity, as represented by the synchronized contraction of its arrow.

(15.2.3 Refraction 3a)

278

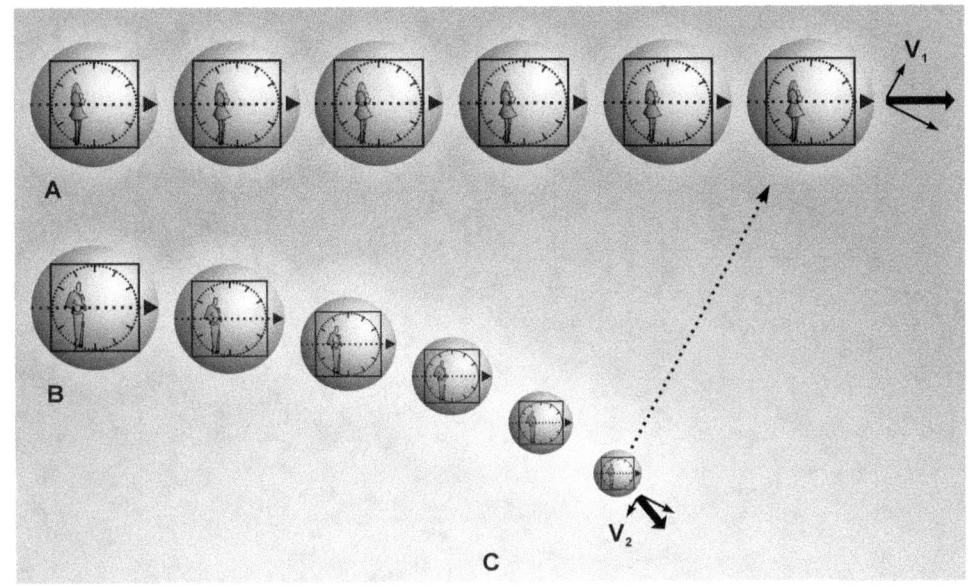

REFRACTION: PERCEPTION

With light, objects, and time all in place, let's put someone inside our box, or we'll call it a crate like Einstein. But we'll make ours transparent. For row **A**, everything still remains the same, just as the person perceives.

But it's also the same for row **B**. The person perceives no change despite that everything would be condensing. That's because the contraction, and any other distortion, omitted for clarity, all occur in a synchronized way. With faster time compensating for light's slower velocity and all measuring devices condensing, and distorting, the same as the objects they're measuring, there's no way, theoretically, to compare a previous measurement to a current measurement to perceive a difference. Everything's contracting in unison. So the surroundings appear to remain unchanged. And for all practical purposes, row **B**'s Euclidean environment remains intact, no different from **A**'s.

If row **B**'s person were to look out at the object in **A**, it might appear to be maintaining a consistent distance or even enlarging while higher redshift observations confirmed an increasing recessional velocity. That increasing recessional velocity would express as a function of the three-dimensional line-of-sight vectors associated with each object's motion, diagramed two-dimensionally as V_1 & V_2. A perpendicular perspective could confirm its farther distance and ever-growing separation.

We could apply these same circumstances as if our objects were in separate galaxies that were a fixed distance apart. Let's say the object in **B** was infalling towards its galaxy's center in the direction of **C** while **A** was at the outer reaches of its own galaxy so its trajectory was nearly circular. Both objects' condensing would occur in the same manner, they'd have the same recessional velocity based on the same line-of-sight vectors, and they'd record the same redshift.

The problem is for us here on Earth, we misinterpret our observations of recessional velocity as every entire galaxy moving away from every other because of the big bang's presumed expansion. When in reality, it's simply the recessional velocity of each galaxy's continuous infall of material along with our own infall velocity that's a natural consequence of gravity's inherent runaway nature.

(15.2.4 Refraction 3a)

Real Time Observations

There is no actual property of the universe that can be labeled as time. The idea of time is a notion we've devised that establishes a convention for reference. We determine time's rate by adopting the periodic motion of an object as a baseline and use that natural frequency as a unit of measure for comparison.[166] For us, it's usually the Earth's rotation and its orbit around the Sun. But it could just as well be the oscillating motion of a swinging pendulum or the natural frequency of an atom.

Time does not exist outside of the physical process that we select for reference. It cannot subjectively vary with motion. It cannot change with the variables of an equation. And it cannot originate from an assumed beginning of the universe and vary with the universe's assumed varying rate of expansion. It can only be defined by us.

The prospect of it being linear with a past or future somehow and somewhere coexisting with the present, where we presume the potential of time travel or its varying rate of passage, is only a construct of the imagination and cannot exist anywhere in reality. Some truly objective and mature reflection along with a willingness to outgrow the fanciful illusion of time travel will confirm this essential fact.

Since time is only the product of the periodic motion or natural frequency of the object that we've selected to establish its rate, there can never be an independent change in time's rate by itself. It simply cannot vary independent of that object's periodic motion. Any change in time's rate can only come from a change in the rate of the object's periodic motion.

This can only occur when the object's size or the distribution of its mass has been altered like changing the length of a pendulum or repositioning the arms of a spinning ice skater. Shortening a pendulum's length would increase its natural frequency or the rate of its periodic motion, which would cause its rate of time to run faster.

Time being only a unit expression of the periodic motion of an object's natural frequency also precludes it from running in reverse, as has been naively proposed for the "big crunch" (the tenuous theory that an expanding finite universe would begin collapsing in on itself after it has been completely slowed by gravity).

Of course, the direction of the motion of the object we've selected to represent time, either forward or backward or outward or inward, can have no influence on the object's natural frequency. So it can't influence its assigned rate of time. But any reduction in the density of space's field would condense the objects in it,

166. "Time," Merriam-Webster, last modified Mar 28, 2023, https://www.merriam-webster.com/dictionary/time.

causing the rate of their cyclic motion, their natural frequency, to increase, which would in turn cause an increase in their assigned rate of time.

When this elemental concept of time is understood and embraced, and then merged with the realization that the size of objects vary and is dependent only on the varying density of the universal field, we can then begin to provide realistic examples for the phenomena of contracting objects in motion and time's varying rate with motion. (See Diagram 4 Time's Rate 7a)

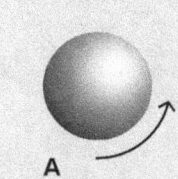

1. HOMOGENEOUS FIELD

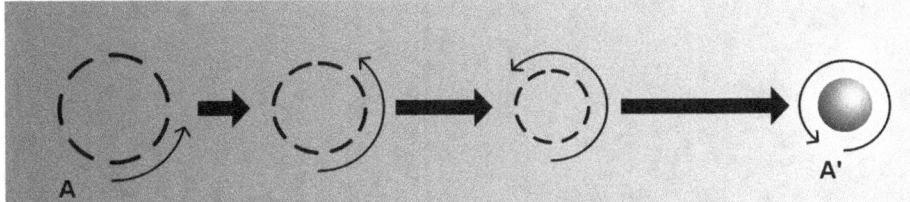

2. GRAVITY FIELD DECREASES IN DENSITY EXPONENTIALLY

TIME'S CHANGING RATE

If we could view two identical heavenly bodies, one in a field that was theoretically uniform in density and the other in a gravity field that naturally decreases in density exponentially per the inverse square law, as represented in section by the diffusing background, the body at **1.A** in the homogeneous field would remain constant, the same without motion, or contraction, or increase in its rate of rotation. This is how we perceive our environment.

Reality, though, is the condition in row **2**. The body in the gravity field would be mechanically seeking equilibrium in the diffusing density, moving in the direction of the least resistance, which is always directly toward other objects, or more accurately a common center of mass, increasing in velocity from position **A** to **A'**, as indicated by the heavy arrows, while contracting in size in a near omnidirectional manner due to the field's decreasing density. As it condenses, its natural frequency increases, increasing its rate of rotation, as implied by the increasing length of the radiused arrows.

If time's rate were established by the rate of the periodic motion of a celestial body's rotation, as ours is, then as outside observers we could see that the assigned rate of time quickens for objects traversing fields of lesser density due to their condensing which increases their rate of rotation. But for anyone inhabiting this heavenly body, there'd be no perceptible change. Without an outside reference, the length of a day would still appear to be the same despite its perpetual shortening. Those inhabitants might look out at other heavenly bodies in regions of space with higher field densities and it'd appear as if the rotation rate for those bodies was slowing.

Altering the naturally frequency of the object we've selected to establish time's rate by either condensing its mass through the natural coalescing process of gravitation or by increasing its mass by subjecting it to a charge is the only way time's assigned rate can change with or without motion.

(4 Time's Rate 7a)

Spaced Out

Grasping the true nature of space is simple. It's the nothingness between objects.[167] So there's nothing there to comprehend. If space were actually something, it'd still have to be three-dimensional to exist. So it couldn't bend, warp, or curve to facilitate gravitation. Those are properties of one and two dimensions that don't exist. But it could vary in density. (See Diagram 12 3-D Space 13a & 17 Curvature 6a)

167. "Space," Merriam-Webster, last modified Mar 26, 2023, https://www.merriamwebster.com/dictionary/space.

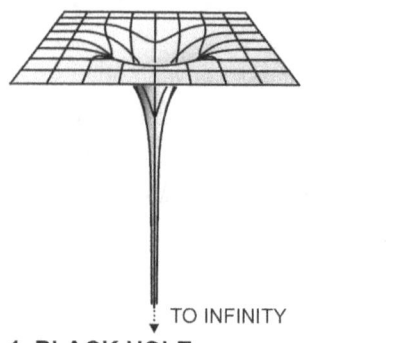

TO INFINITY

1. BLACK HOLE

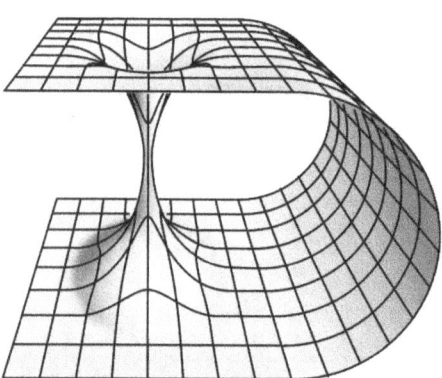

2. WORMHOLE

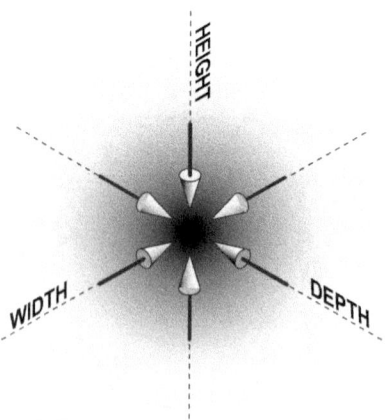

HEIGHT

WIDTH

DEPTH

3. REALITY

(12 3-D Space 13a)

THREE-DIMENSIONAL SPACE

The notion that space is something and that it expresses two-dimensionally has led us to mistakenly reason black holes. Their gravity is presumed to be so strong they create a two-dimensional funnel-shaped curvature in their inconceivable four-dimensional space-time, usually portrayed in diagrams similar to **1**. It's imagined that any object, or light, that strays too close will be drawn over a theorized event horizon, never to be heard from again, where it endlessly infalls while stretching, or "spaghettifying" as it's called, as it accelerates without end toward the infinite (mathematical) condition of a singularity.

The same mistaken logic is applied to so-called wormholes. Space's fictional two-dimensional composition, commonly depicted in diagrams similar to **2**, would supposedly allow us to shortcut through a black hole that's not infinite but tied to another black hole at another location in our curving two-dimensional universe.

In reality, space is the nothingness between objects. There's nothing there. But if it really were something, it'd still have to be three-dimensional, as indicated in diagram **3**. It'd have to have width, height, and depth. That's it. Additional dimensions are inconceivable. And two-dimensions, despite being incorrectly justified as merely an analogy, only describe the location of a plane. Without the third dimension of height, existence isn't possible.

Whether you insist that space is actually something or accept the universe as a field of radiant electromagnetic energy that corresponds to all space, in our reality of three actual dimensions, coalescing infalling material, represented by the condensing grey three-dimensional sphere in **3**, would quickly arrive at a common center of mass. There's nowhere else for it to go. It cannot slip over the edge of a nonexistent two-dimensional funnel-shaped space on an endless journey to infinity. That's a make-believe science fiction fantasy that's justified through two-dimensional mathematical gimmickery.

Gravity's exponentially decreasing field density continuously coalesces and condenses matter together in a runaway recycling process where it's increasingly compressed until fusion reactions are triggered that perpetually convert it back to its original state of radiant energy.

286

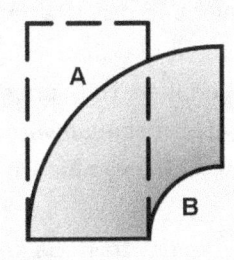 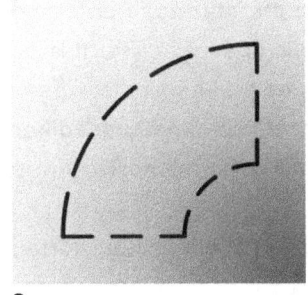 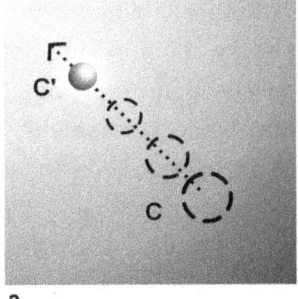

1   2   3

CURVATURE & DENSITY

If we were to envision the shape in diagram **1** as three-dimensional, giving it the same depth as its width, we can see how if it were bent over where its two-dimensional surfaces, at **A** & **B**, now curved, its interior would have to vary in density, as represented in section view by the diffused shading that dissipates toward the outer curved side with a greater radius, **A**. Note how the density along both the outer and inner curved surfaces would remain constant as **A** diffused and **B** condensed when bent.

If we were to imagine that shape as part of a larger volume of the same material, as depicted in diagram **2**, we can more easily see how any three-dimensional volume of something can never curve but can only vary in density. Curvature is a property of only one and two dimensions, which don't actually exist in reality. Without the third dimension, they only define a location of a line or a plane.

These principles of curvature and density would also have to hold true for the three-dimensional volume that's commonly refer to as space. It could never curve. It could only vary in density. But space is the nothingness between objects. It's nonexistent by definition, and it would also be nonexistent if it were a curving two-dimensional plane. So there's nothing there to vary or curve. But there is something that does correspond exactly to all space: radiant electromagnetic energy. Everywhere that space is, so is the all-pervasive universal field of radiation that comprises the electromagnetic spectrum. This field is continuous, its continuity uninterruptible, and it's average volume is fixed per any given quantity. And its density varies.

Because of its innate continuity, the density of the radiation that's been drawn together to form a particle has to diffuse. But it diffuses inward toward the particle's center, still dissipating exponentially per the inverse square law because of the three-dimensional geometry of a sphere. It's the compounding of those inward diffusing fields of the particles themselves, and the objects they compose, that's responsible for gravitation.

Referring to diagram **3**, if we were to place a small object into the decreasing density of a gravity field of a much larger object, it would begin moving with acceleration in the direction of least resistance directly toward the larger object as it mechanically sought to find equilibrium in the field's decreasing density. It'd naturally move from the highest density toward the lowest. But it would also be contracting in size. Its atoms also naturally coalesce and condense within the field's decreasing density, as indicated by the sequence of dashed circles representing the object's motion from **C** to **C'**. In reality, each object would move toward the other in proportion to the size of their gravity fields toward a common center of mass because of the compounding effect of their gradient.

(17 Curvature 6a)

There is something, though, that does correspond to all space: the radiant energy of the electromagnetic spectrum. But this universal field of radiation not only occupies the emptiness between objects, when congealed it also comprises the objects themselves. There's no constituent difference between the universal field and the objects that inhabit it. So the entire universe consists of only one thing, electromagnetic radiation.

Electrons and protons don't have an actual surface where radiation stops and matter begins. The congealed radiation they're composed of just gets more dense toward their center until it reaches a maximum that can never be considered solid.

With electrons and protons being aware of each other's presence through their gravity and electromagnetic fields, this suggests that this all-pervasive universal field is also self-aware, which could be interpreted as consciousness. Since it would be aware of everything, some might go further and say it's all-knowing.

With it also being fundamentally energetic and creative, "spontaneously" bringing into existence these particles out of itself, this might also be construed as all-powerful. Those with a religious bent, but still somewhat open-minded, might find themselves leaping to the conclusion that electromagnetic radiation might be synonymous with the God of their Bible. It'd have the same omnipresent, omniscient, and omnipotent qualities.

Since this omnipresent electromagnetic radiation that coincides with all space has an actual physical existence, it has to be three-dimensional. So it can't curve underneath objects. Curvature again being a property of one and two dimensions. It can only vary in density. Because there's no place where it's not, it has to remain continuous and uninterrupted. It can never be terminated, discontinued, separated, or split into pieces.

When it's condensed into a particle, the ambient radiation remaining outside the particle that's not drawn into it has to thin. That thinning has to diffuse inward toward the particle's center. It decreases in density exponentially per the inverse square law because of the immutable, three-dimensional geometry of a sphere. This defines every particle's, and the objects they compose, gravity field.

So gravity's attracting effect can only be the product of every object's mechanical reactive search for equilibrium in this universal electromagnetic field's varying density that innately decreases around them. With the compounding of their gravity field, they're naturally pushed toward one another by the highest density toward to lowest density that's always directly between them toward their common center of mass, or center of gravity, same thing.

Because an object's decreasing field density can't just stop at its surface but has to continue on through it, in between its molecules and even in between the subatomic particles of its atoms, that decreasing field also has to promote its overall contraction/condensing along with that of other nearby gravitating objects.

So every gravity field varies in density from its least dense at the center of every object to an ever-increasing density with distance. And each object free falls toward every other, accelerating directly opposite the direction of the most amount of pressure as they naturally pursue equilibrium, condensing and continually contracting in size until they merge. (See Diagram 28 Gravitation 12.1a)

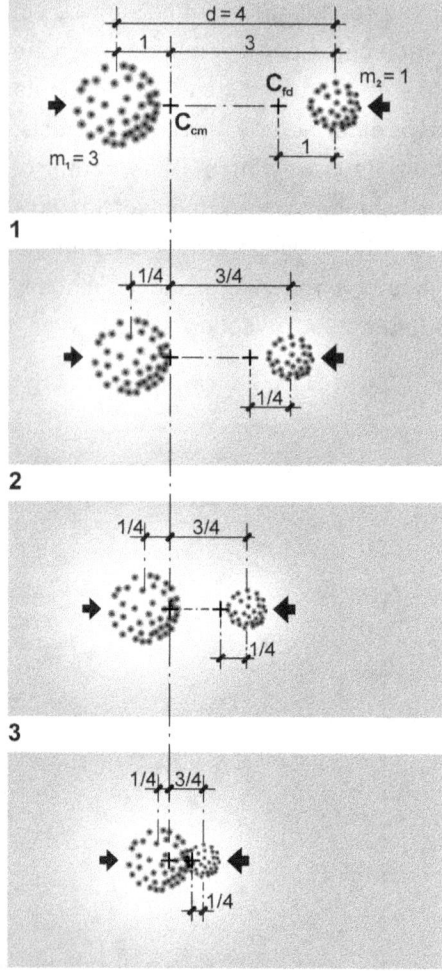

1

2

3

4

COALESCING BODIES
BODIES AREN'T ATTRACTED TO ONE ANOTHER. THEY'RE CONSTANTLY PRESSED TOGETHER BY HIGHER FIELD DENSITY TOWARD LOWER FIELD DENSITY AS THEY MECHANICALLY REACTIVELY SEARCH FOR EQUILIBRIUM IN THE EVER-DECREASINGDENSITY OF THEIR EVER-MERGING GRAVITY FIELDS.

(28 Gravitation 12.1a)

GRAVITATION
A natural consequence of a particle's emergence, gravity fields, depicted in section view by the diffused background, necessarily diffuse inward exponentially because of basic spherical geometry and the uninterruptible continuity of radiant electromagnetic energy.

Gravity fields' innate compounding causes that inward diffusion to always be at its least directly in between the particles and the bodies they surround at their common center of mass, C_{cm}, which is the same as their common center of gravity.

Mechanically pursuing equilibrium in the ever-decreasing density of their ever-compounding gravity fields, all bodies, be it particles or galaxies, are constantly pushed by the highest field density toward the lowest. This inexorably leads to runaway coalescing that ultimately ends with fusion reactions transmuting all matter back into the radiant energy it originated from.

Because gravity fields not only surround but also permeate all bodies, including atoms, depicted as the small spheres comprising the spherical bodies, their compounding simultaneously causes both coalescing and condensing at all scales consistent with Newton's law of gravitation: $F = G(m_1 m_2) / d^2$, where F is the "attractive" force, G is the gravitational constant, m the mass, and d is the distance between their centers.

The distance to their C_{cm} from m_1 is $d_{cm} = m_1 d_1 + m_2 d_2 / m_1 + m_2$, where $d_{cm} = 3(0) + 1(4) / 3+1$ or 1. From m_2, it'd be $1(0) + 3(4) / 3+1$ or 3.

C_{fd} indicates the location in between them where they share a common field density. The distance to their C_{fd} is opposite of or naturally reciprocal to their C_{cm}. Both their C_{cm} and C_{fd} could be interpreted as non-centrifugal Lagrange points where the gravitational influence remains in equilibrium.

Actual Lagrange points incorporate orbital motion's centrifugal force. It's not included in this example for clarity. If it were, their C_{fd} would become the L_1 Lagrange point that'd have to be closer to m_1 to compensate for the outward centrifugal force.

The distance to their C_{cm} and C_{fd}, their relative rate of motion toward each other, and their relative condensing, all remain proportional to their masses as they relentlessly gravitate in the ever-thinning density of their ever-compounding gravity fields, conceptually portrayed in the sequence of diagrams **1-4**.

Eventually, this process leads to every object's collapse and transformation to radiant energy as they migrate toward their galaxy's center where the field density is at its least. Because their contraction corresponds with their persistent inward migration, their natural frequency, or their rate of rotation, or any periodic unit of motion, is constantly increasing until enough matter is condensed and the pressure becomes great enough that fusion reactions are triggered that convert it back to pure radiation.

For someone experiencing this ongoing contraction, their increasing velocity, their decreasing size, and the increasing natural frequency, which we could establish as their rate of time, would theoretically always appear normal to them, completely the same without change because the coalescing-collapsing effect of gravitation is omnipresent and omnidirectional. This means that they and all of the objects around them would be essentially contracting in unison, making it virtually impossible for them to ever perceive any of these changes because there's no physical way to maintain a record of their previous size or rate of time for comparison to their current size and rate of time. (See Diagram **31** Condensing 7a)

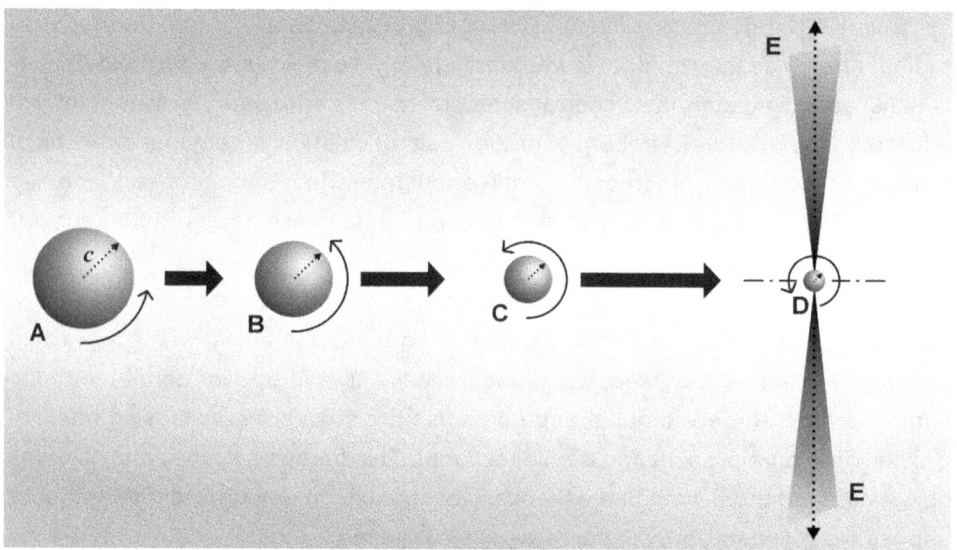

SECTION VIEW THROUGH A MATURE SPIRAL GALAXY

CONDENSING BODIES

The sphere depicted as moving in the sequence from **A** to **D** represents the condensing and ultimate collapse of an infalling star or any heavenly body as it whirlpools in toward a spiral galaxy's center. It is accelerating and contracting while its rotation rate increases, implied by the increasing length of the radiused arrows, as it mechanically pursues equilibrium in the decreasing density of the spiral's gravity field. Light's velocity decreases correspondingly, as indicated by the dotted arrows of decreasing length at c.

Galaxies naturally coalesce from ellipticals into the flat rotating plane of mature spirals because the slower rotation rate nearer their developing poles resists gravity less. That's where the downward migration of its material occurs the most, which produces the disk's overall flattening. So the impetus of its coalescing transitions from the geometry of a full sphere more into the two-dimensional geometry of a circle. With its area, $A = \pi r^2$, it's now condensing exponentially more toward the disk's center within its planar vortex. The diffused background portrays the spiral's flattening, disk-shaped gravity field.

Because the diffusing density of a gravity field causes the omnidirectional contraction of all infalling bodies and a corresponding decrease in light's velocity, all measurements are shrinking while time's rate quickens with the body's increasing rotation, assuming time's rate is established by that rotation. This causes any measurement to always read the same for anyone at the body, making it virtually impossible for them to ever directly perceive their contraction or increasing rate of time.

As infalling bodies approach the galactic center, instead of disappearing over the presumed two-dimensional event horizon of a supermassive black hole that's believed to lurk at the center of almost all galaxies devouring anything that strays to close, the increasing pressure created by the galaxy's exponentially decreasing field density rapidly increases the infalling body's fusion reactions near the spiral's center, at **D**, hastening its collapse. This essentially squeezes its material back into radiant/plasma energy that's then spewed out in a bipolar manner as high velocity jets, as represented at **E**, which act as the vortex's drain. As the expelled radiation slows and cools, it reconstitutes back into ordinary matter. It can then begin gravitating back to its or another nearby galaxy to be perpetually recycled in a ceaseless process of degeneration, transmutation, and rejuvenation.

(31 Condensing 7a)

292

Special Considerations

By acknowledging that space is the nothingness between objects, so doesn't have an actual existence and couldn't curve even if it did, the elemental flaws of special relativity, light's constancy, time's slowing with motion, the contraction of objects in the direction of motion, and increasing mass with increasing motion, become clear.

Because of light's "bending" in space, often referred to as gravitational lensing, we're led to reason, contrary to orthodoxy, that light's varying position has to be a product of its refraction. Its velocity changes as it passes through the decreasing density of a massive object's gravity field. Its path is not bent by gravity's assumed pull on its photons. Nor is it following space's geodesic as it dents underneath those objects.

Photons, not possessing any mass, can't be pulled into a curved path by gravitational "attraction." Even if they did or could possess mass, gravity doesn't attract. It pushes objects together. Light's refraction, by definition, requires the apparent repositioning of its source due to its reduced velocity. So light's velocity isn't and can't be fixed. It must be variable. Which makes its velocity different everywhere because the density of the universal field is different everywhere. It innately decreases exponentially around any quantity of mass, particle, or galaxy, or even galaxy clusters.

The problem with attempting to directly measure light's variability is essentially the same as trying to measure its compounding. The difference between source velocities is so small as compared to light's velocity that direct measurements are next to impossible to record reliably.

But even if the difficulty could be overcome, we could never measure in one field density and then another to compare light's velocity. All the conditions, including the size of any measuring device, change correspondingly with the density of the field it inhabits. This would cause light's velocity to essentially always read the same.

Light's variability can be recorded from an outside location. And many times measurements do in fact document velocities that deviate far beyond the speed of light here on Earth. But instead of accepting these results at face value, we ruse to explain them away as an illusion, mostly through and because of relativity.

One example of factual evidence that demonstrates light's variability in gravitational fields is the so-called "Pioneer anomaly." Ever since the Pioneer 10 and 11 spacecraft passed Saturn they appear to be continually slowing, falling behind their expected positions by about 8,000 miles a year. The accepted explanation for this discrepancy is that some of its equipment is generating thermal radiation in the direction of travel that's creating a repulsive effect which is slightly slowing its velocity.[168]

The actual reason is that light's velocity is variable. Its rate increases as the field density increases, analogous to sound's increasing velocity in denser mediums. As the density of the solar system's gravity field increases with distance, light's velocity increases correspondingly. So the spacecrafts' signals are reaching us sooner than expected, causing them to appear closer than they actually are.

As repeatedly noted, Einstein actually agrees with and asserts that light's bending at massive bodies occurs because its velocity varies with position, which is refraction.[169] But its refraction occurs through the diffusing density of a gravity field, not the impossible two-dimensional curvature of a nonexistent space of constant density, or because a massive body's gravity is pulling on a (massless) photon.

He maintains, though, that its fixed velocity still holds true when disregarding the influences of gravitational fields.[170] But ignoring gravity fields is impossible. Every object has its own self-gravity. And its field extends indefinitely. So there's no place where they are not. For any practical, real world interpretation, they have to be accounted for.

And again, even he affirms, contradictorily, that relativity would completely unravel if the speed of light were found to be variable.[171] His mass-energy relation along with his special and general principles, which has all motion considered only as relative motion and where the laws of nature, like light's presumed constancy, remain the same regardless of motion, would all become invalid. They, and in fact all of relativity, originate from the premise that light's velocity remains fixed.

If we're willing to concede light's variability then faster-than-light travel must be a fact. There will always be some place where the universal field's density will be greater than ours where the speed of light would be faster. There'd also always be some place where its density would be less than ours where the speed of light would be slower.

With relativity's limiting precepts, like its mass-energy relation that has mass becoming infinite at the speed of light, now no longer valid, how can there be any limit on an object's, or light's velocity? Either can be any rate up to instantaneous.

The unlimited velocity allowed by its variability also affirms its compounding with motion. But it's a compounding with a velocity determined by the field density where the measurement is taken. If light's velocity wasn't variable and unlimited, it wouldn't be able to compound with the motion of its source or that of other sources.

168. "Pioneer anomaly," Wikipedia, last modified Feb 6, 2023, https://en.wikipedia.org/wiki/Pioneer_anomaly.
169. Einstein, *Relativity*, 84-85, 104, 109, 171.
170. Einstein, *Relativity*, 85, 109, 171.
171. Einstein, *Relativity*, 85, 151.

All of the Michelson-Morley type experiments find that the speed of light remains constant regardless of the Earth's motion. It always leaves its source at the same rate in all directions at once no matter its velocity. This by itself confirms compounding. They also confirm that objects do not contract in the direction of their motion and that their rate of time doesn't change.

The construct of these experiments, being of a single reference frame arranged two-dimensionally, have a beam of light projecting in the direction of the Earth's rotational and orbital motion that's split 90 degrees into another beam projected perpendicular to the motion then reflected back again to be recombined. They should have produced an interference pattern in the light if its velocity were fixed. But they didn't.

If the reference frame had contracted in the direction of motion, or its light's velocity slowed in the direction of motion to maintain its constancy, or its rate of time "slowed," this would have produced the expected interference pattern. It didn't happen because light's velocity didn't slow in the direction of motion, the distance between the mirrors didn't contract in the direction of motion, and time's rate didn't change. This clearly indicates that light's velocity cannot be absolute but must compound with motion.

Sagnac's similar experiment demonstrates the same result but more definitively. When the light is split, the platform's continuous rotation separates it into different beams traveling in opposite directions that become different reference frames. They return to their origin at the same time because their velocities compound with the platform's angular velocity. This produces an interference pattern because their different velocities cause them to be out of phase. (See Diagram 1.1, 1.2, 1.3, 1.4 M-M Exp 6a & 16.1, 16.2 MM 7a & 49.1, 49.2, 49.3, 49.4 Sagnac Exp 5a)

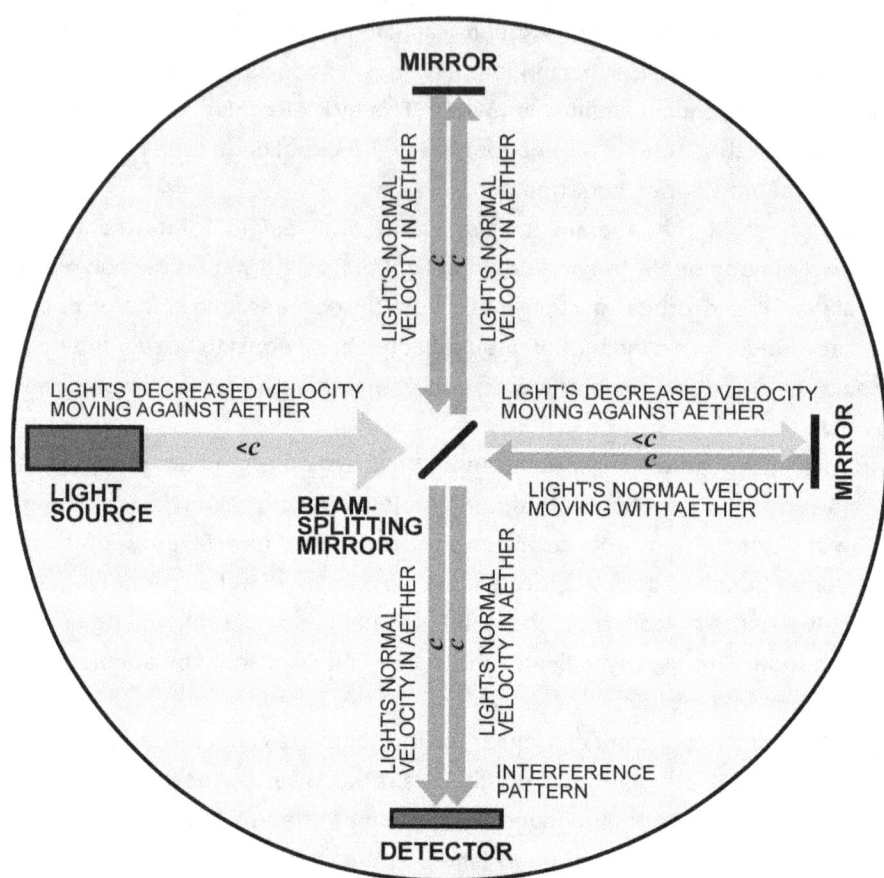

DIRECTION OF EARTH'S ROTATION & ORBITAL MOTION ➡️

MICHELSON-MORLEY - CONCEPTUAL DIAGRAM
EXPECTED RESULT

The experiment essentially consisted of a light source projected onto a series of mirrors arranged perpendicular equal distances from a central beamsplitter mounted on a rotating table oriented with one beam projected in the direction of the Earth's orbital motion and the other perpendicular. When the light was recombined, it was expected to produce an interference pattern due to its decreased velocity from the theorized aether "headwind." This would confirm the aether's existence. But no interference pattern was found.

(1.1 M-M Exp 6a)

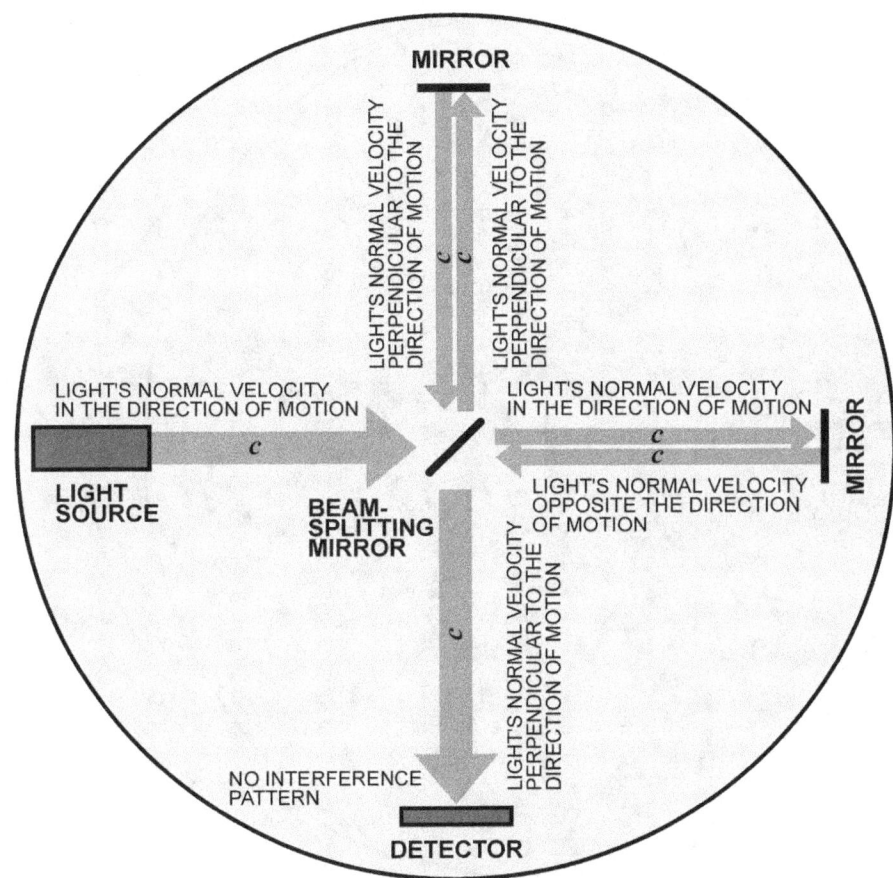

MIRROR

LIGHT'S NORMAL VELOCITY PERPENDICULAR TO THE DIRECTION OF MOTION

LIGHT'S NORMAL VELOCITY PERPENDICULAR TO THE DIRECTION OF MOTION

c
c

LIGHT'S NORMAL VELOCITY IN THE DIRECTION OF MOTION

LIGHT'S NORMAL VELOCITY IN THE DIRECTION OF MOTION

MIRROR

c

c
c

LIGHT SOURCE

BEAM-SPLITTING MIRROR

LIGHT'S NORMAL VELOCITY OPPOSITE THE DIRECTION OF MOTION

LIGHT'S NORMAL VELOCITY PERPENDICULAR TO THE DIRECTION OF MOTION

c

NO INTERFERENCE PATTERN

DETECTOR

DIRECTION OF EARTH'S ROTATION & ORBITAL MOTION ➡

MICHELSON-MORLEY - CONCEPTUAL DIAGRAM
ACTUAL RESULT

What the experiment actually showed is that light always leaves its source at 186,000mi/s in every direction at the same time as we'd naturally expect. This indicates its compounding with the motion of its source and implies its compounding with other reference frames. Which means that because everything's in motion, its velocity can never be fixed at 186,000mi/s but will always be some slower or faster rate that can be any velocity up to instantaneous.

If someone was out in space stationary with respect to the solar system, they'd be in a different reference frame recording a compounding of light's varying velocity, which is determined by the field density at their location, plus/minus the Earth's rotational and orbital velocity or some vector angle of it.

(1.2 M-M Exp 6a)

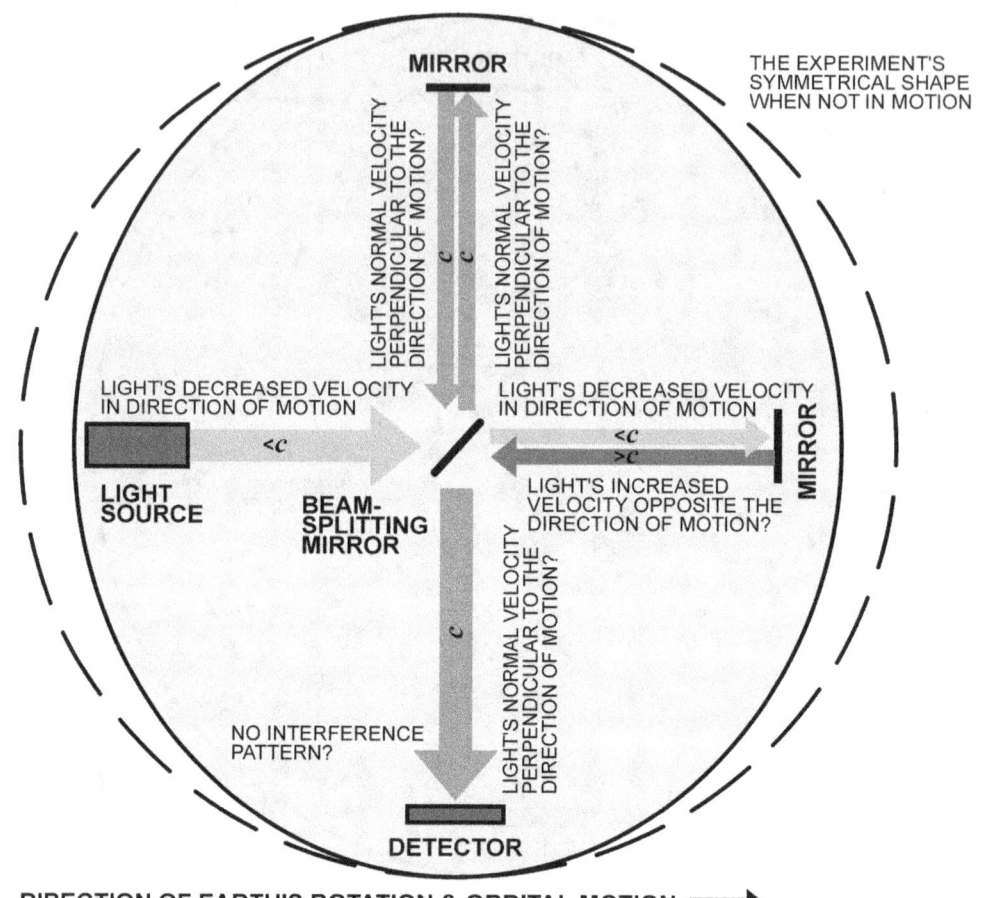

THE EXPERIMENT'S
SYMMETRICAL SHAPE
WHEN NOT IN MOTION

MIRROR

LIGHT'S NORMAL VELOCITY
PERPENDICULAR TO THE
DIRECTION OF MOTION?

LIGHT'S NORMAL VELOCITY
PERPENDICULAR TO THE
DIRECTION OF MOTION?

LIGHT'S DECREASED VELOCITY
IN DIRECTION OF MOTION

LIGHT'S DECREASED VELOCITY
IN DIRECTION OF MOTION

<c

<c
>c

LIGHT
SOURCE

BEAM-
SPLITTING
MIRROR

MIRROR

LIGHT'S INCREASED
VELOCITY OPPOSITE THE
DIRECTION OF MOTION?

LIGHT'S NORMAL VELOCITY
PERPENDICULAR TO THE
DIRECTION OF MOTION?

NO INTERFERENCE
PATTERN?

DETECTOR

DIRECTION OF EARTH'S ROTATION & ORBITAL MOTION ➡

MICHELSON-MORLEY - CONCEPTUAL DIAGRAM
LORENTZ'S EXPLANATION ADOPTED BY EINSTEIN

What Lorentz proposed to explain the result, and Einstein later adopted for relativity, was that objects contract in the direction of their motion while time slows to maintain light's fixed velocity so that an interference pattern is not produced. But what happens to the experiment for the reflected light moving opposite the direction of motion? To maintain light's constancy wouldn't it have to expand while time's rate increased? And how about in the perpendicular direction? Without any contraction wouldn't time's "slower" rate cause light's velocity to exceed 186,000mi/s? These irresolvable conflicts confirm that light's velocity cannot remain fixed in our real nontheoretical environment of three actual dimensions but must compound with motion, which invalidates any Lorentz contraction and undermines nearly all of relativity.

(1.3 M-M Exp 6a)

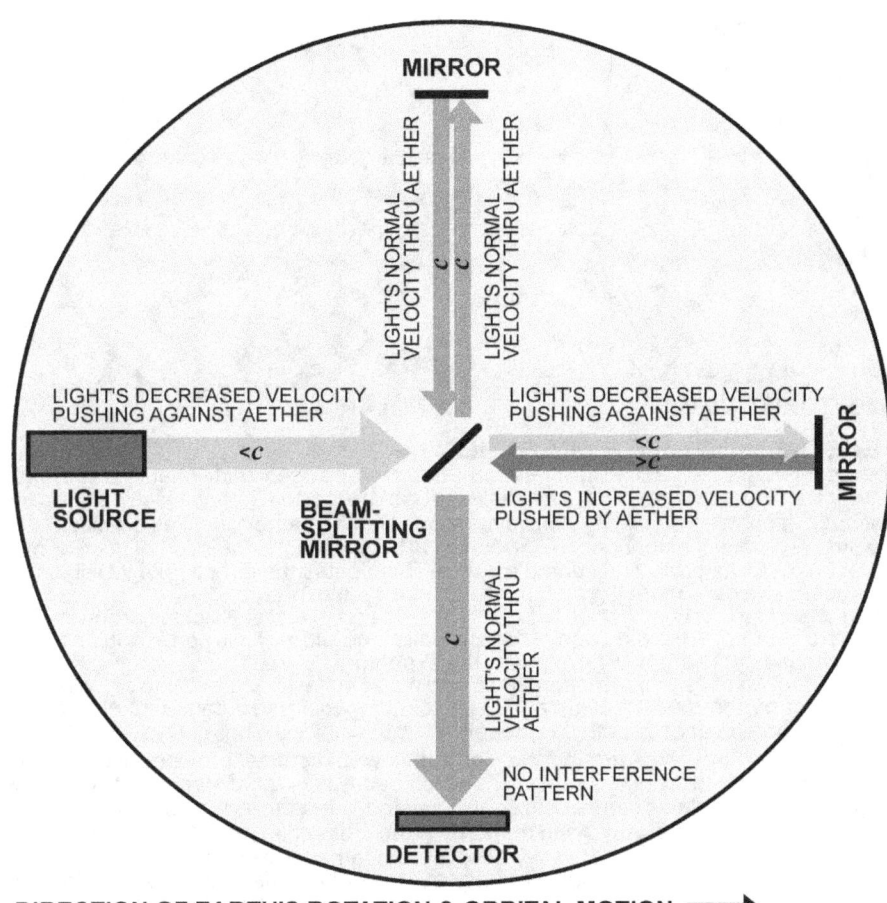

DIRECTION OF EARTH'S ROTATION & ORBITAL MOTION ➡️

MICHELSON-MORLEY - CONCEPTUAL DIAGRAM
AETHER EXPLANATION

Instead of contriving the fantastic, self-conflicted notion that objects physically contract but only in the direction of motion while time's rate slows to maintain light's constancy, why wouldn't you simply reason that light's velocity is first slowed by the aether's "headwind" then is increased by the same amount from its "tailwind" after it's been reflected backward? It's not the correct explanation. But at least it's rational.

(1.4 M-M Exp 6a)

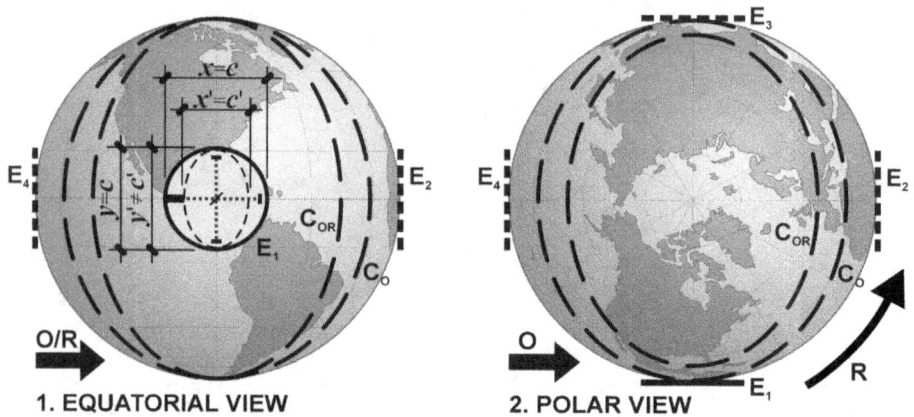

1. EQUATORIAL VIEW   **2. POLAR VIEW**

MICHELSON-MORLEY EXPERIMENT - 1

The experiment essentially consists of several mirrors arranged rectilinearly in a cross pattern around a central beamsplitter on a table that can be rotated. For simplicity, ours is at the equator, E_1. It showed that light, c, always radiates at the same rate in all directions at the same time regardless of motion. The arrow at **O/R** indicates the direction of the Earth's orbit & rotation. At **O**, its orbit. The curved arrow at **R** indicates the direction of its rotation. For our purposes, we can ignore the motion of our solar system and galaxy.

Einstein assumes that light's velocity is fixed. It's (special) relativity's basal premise. So objects have to contract in the direction of motion while time's rate slows correspondingly to maintain its constancy. That's how he explains the experiment's result.

So in the direction of its orbital motion, the Earth would have to contract into an ellipsoid shape, indicated by the dashed ovals at C_O. Its rotation would also cause its ellipsoidal contraction. The combination of both is represented by the dashed ovals at C_{OR}.

As the experiment revolves around the Earth, this would cause it to contract in the x direction, as implied by the dashed oval at E_1, which decreases the distance between the mirrors while time's rate slows to maintain light's constancy (assuming "time" is something that actually exists and that it slows with motion). In the y direction, there is no motion. So there's no contraction. The distance between the mirrors remains the same.

Since time's slower rate has to apply equally over the experiment (and over the entire Earth, they're of the same reference frame), it affects light's velocity, c', the same in both the x and y directions. But with contraction only in the x direction, not in the y, this creates different velocities for light. y' will always be faster than x'. This would cause different arrival times, which should have produced a negative interference pattern. But it didn't.

For the Earth's contraction from its orbital motion, the distance between the mirrors in the y direction would also remain unchanged, while the distance in the x direction would be constantly fluctuating. The Earth's rotation would pass the experiment through its orbital tangents, causing maximum contraction at E_1 & E_3, with no contraction at its perpendicular positions, E_2 & E_4, which would compound with its constant contraction from its rotation. The result would be C_{OR}. This should have also produced a negative interference pattern, but one that was expanding and contracting every twelve hours. But this didn't happen either.

Einstein also asserts that special relativity's effects can't occur within gravity fields where light's velocity varies. But the experiment rests in the Earth's gravity field, and all others. They extend indefinitely. So how can it even be considered to explain the results?

The experiment clearly demonstrates that objects do not contract in the direction of motion. Nor does (nonexistent) time slow with motion. So its velocity can't be fixed. It compounds with the motion of its source and that of other reference frames, and that's in addition to its variability. With a founding premise that's untenable, relativity has no viability.

(16.1 MM 7a)

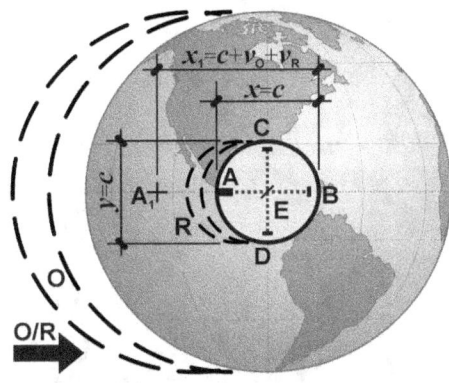

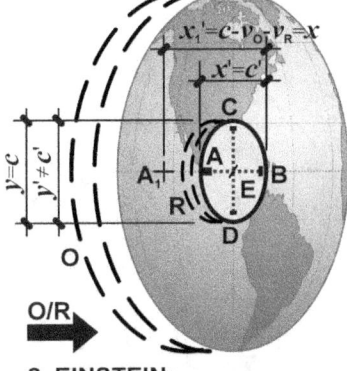

1. REALITY:
LIGHT'S COMPOUNDING
TIME IS NONEXISTENT

2. EINSTEIN:
LENGTH CONTRACTION
TIME DILATION

MICHELSON-MORLEY EXPERIMENT - 2

Light always radiates at the same rate in all directions at once regardless of the motion of its source. So someone at the experiment, **E** in diagram **1**, measuring light's velocity from **A** to **B**, distance x, and also from **C** to **D**, distance y, would get 186,000mi/s in each direction, $x=c$ and $y=c$, despite their motion. They're of the same reference frame, moving in unison with the Earth's orbit and rotation, implied by the dashed ovals at **O** and **R**.

If we positioned ourselves just outside the Earth's orbit, stationary with respect to the solar system, which is a different reference frame, we'd see that the measurement they made in the x direction actually covered a longer distance from A_1 to **B**, distance x_1. But for them, they still measured from **A** to **B**, covering the shorter distance x due to their motion.

If we measured light's velocity from **A** to **B** as the Earth sped by, our measurement would also begin at A_1 and end at **B**. But light would traverse the longer distance, x_1, in the same time it took for it to go from **A** to **B**, distance x, for the person at **E** because we're not moving with the Earth. So for us, light's velocity would have to include the Earth's orbital velocity, v_0 (66,000mph), and its rotational velocity, v_R (1,000mph). Their compounded velocities over distance $x_1=c+v_0+v_R$. That works out to be about 186,020mi/s.

With his underlying assumption that light's velocity is fixed regardless of motion, Einstein reasons that time slows for moving objects while their length contracts to preserve light's constancy. So for the Earth, light's velocity has to decrease by its orbital and rotational velocity, v_0 and v_R (excluding our solar system and galaxy's motion), while time's rate has to slow and the Earth has to contract correspondingly, along with everything else in its reference frame, in the direction of those motions, as portrayed in diagram **2**. The distance x_1' has to contract to distance x to maintain light's constancy where $c'=c-v_0-v_R$.

But "time" doesn't actually exist. It's not a property of the universe. So there's nothing there to slow with motion. And even if it did exist, its rate wouldn't slow. It'd increase. A unit of time, like a second, that corresponded to distance x that was theoretically contracted to the shorter distance x' would be a faster second. Compressed "time" is a faster running "time."

Einstein also "ignores" light and time's innate three-dimensionality. Time's changing rate and length's contraction only work in the one abstract dimension of linear motion. In the three actual dimensions of our real environment, they're inherently conflicted. There's no motion in the y direction. So there's no contraction. y' and x' can never be the same length. But time's "slower" rate still has to apply in the y direction. It's the same reference frame. So light's velocity can never be the same. y' 's will always exceed x' 's. This invalidates Einstein's basal premise, which undermines relativity in its entirety, and that's without even considering his contradictory but correct assertion of light's variability that does so as well.

(16.2 MM 7a)

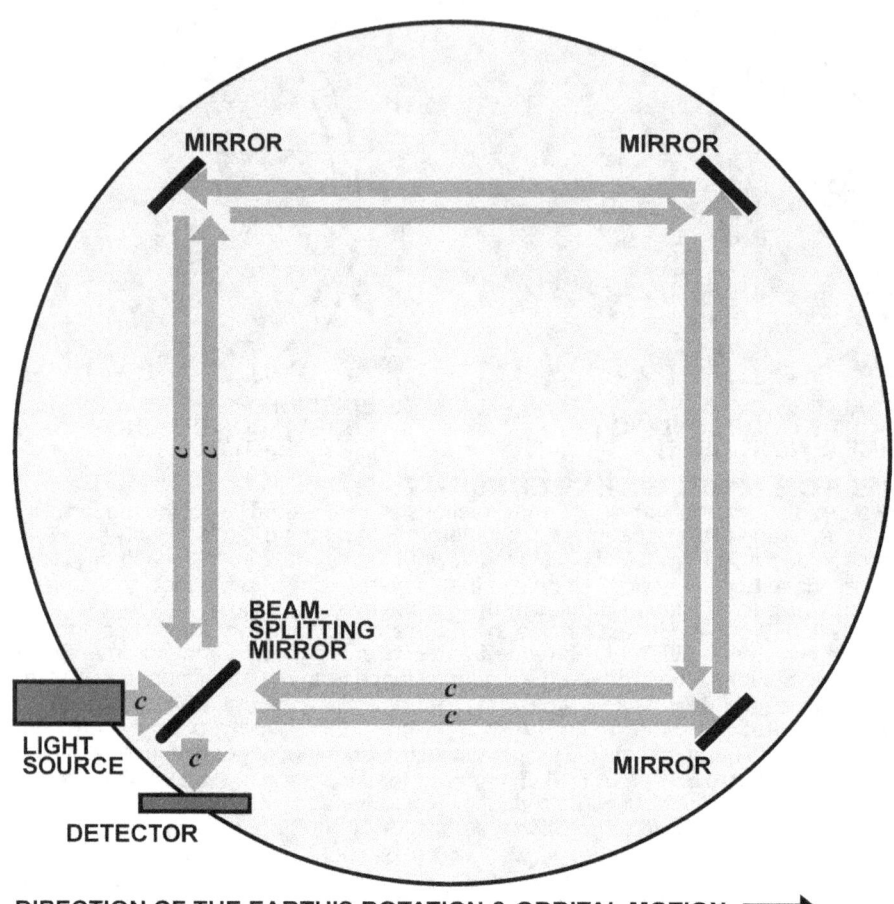

DIRECTION OF THE EARTH'S ROTATION & ORBITAL MOTION ➡️

SAGNAC EFFECT - CONCEPTUAL DIAGRAM
NO ROTATION

Sagnac's experiment essentially consists of a source of light that's projected onto a beamsplitter that sends it in opposite directions around a series of mirrors arranged in a closed loop at the perimeter of a platform form that can be spun that recombines the beams back at a detector, as suggested by the grey linear arrows labeled as c that indicates light's velocity. The inside row of arrows indicates light's clockwise path. The outside counterclockwise.

When the platform is not rotating, no interference pattern is produced. This is essentially the same result as the Michelson-Morley experiment. Both show light always leaving its source at 186,000mi/s in all directions at the same time. Light's independent motion could qualify it as a separate reference frame. Because the platform and the light it's emitting move with all of the Earth's motions, rotational and orbital, our solar system's motion through our galaxy, and our galaxy's motion through the universe, this suggests light compounds with the motion of its source and that of other reference frames.

(49.1 Sagnac Exp 5a)

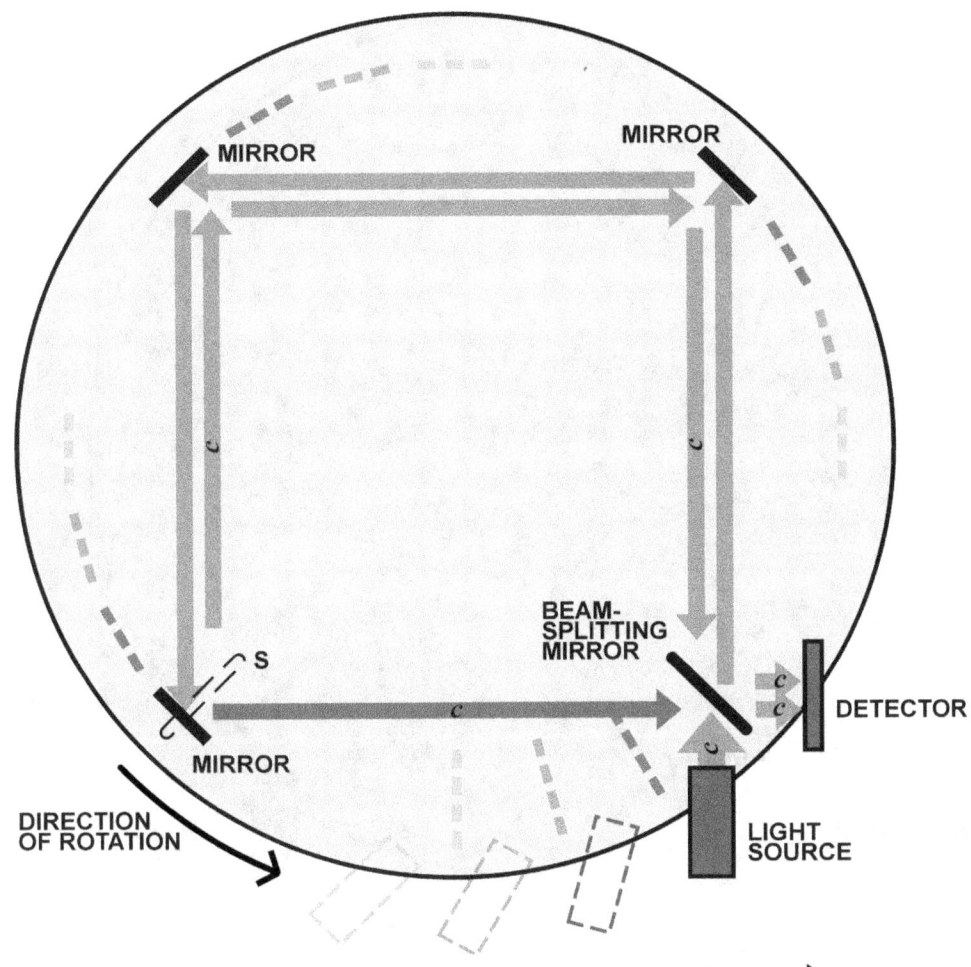

DIRECTION OF THE EARTH'S ROTATION & ORBITAL MOTION ➡️

SAGNAC EFFECT - CONCEPTUAL DIAGRAM
WITH ROTATION - ORTHODOXY

The interference pattern produced when the platform is spun has several possible interpretations. For convenience, let's establish the platform's rotation as constant with an angular velocity that completes one-quarter rotation for three-quarters of light's.

The conventional explanation assumes that the light still leaves its source at 186,000mi/s and is split from **S** at the same velocity in both directions. It's thought that because the platform is rotating into it, the rearward split light arrives at the detector first, which for our diagram is 3/4 of one revolution. While the forward split light arrives later, in 11/4 revolutions, the overlap indicated by the darker arrow. The difference in arrival times would produce a phase shift that creates an interference pattern.

Sounds reasonable enough, but it's inherently flawed. It fails to account for the platform's constant rotation. It departs from it at 186,000mi/s as if there were none.

(49.2 Sagnac Exp 5a)

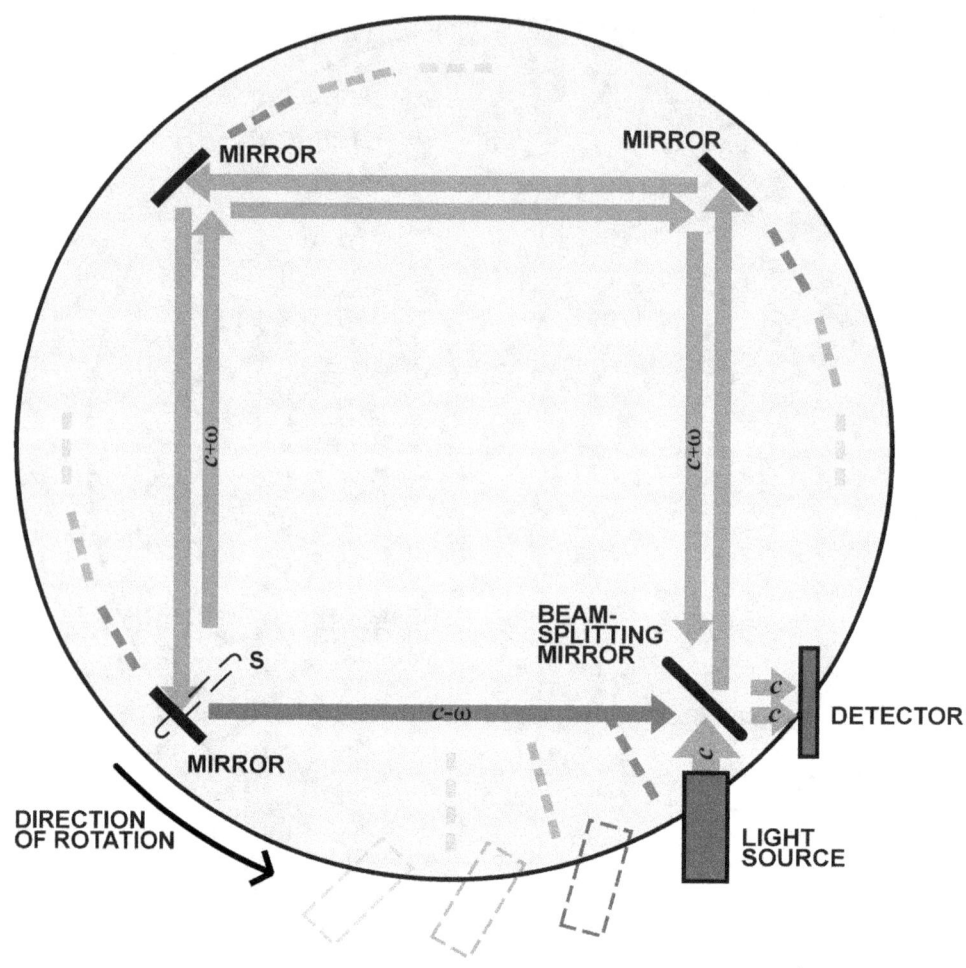

DIRECTION OF THE EARTH'S ROTATION & ORBITAL MOTION ➡

SAGNAC EFFECT - CONCEPTUAL DIAGRAM
WITH ROTATION - SPECIAL RELATIVITY

Special relativity does account for the platform's rotation. But it doesn't work either. The split light would leave at c-ω in the direction of rotation and c+ω opposite the direction of rotation to maintain light's fixed velocity. The light would arrive at the detector at different times because of the platform's rotation, creating an interference pattern.

But special relativity is inherently self-conflicted. It would have the platform's perimeter contracting while its radius remains constant and its time dilates for the entire platform. That's not even remotely feasible.

It would also conflict with the results when the platform is not rotating. It would have to be contracting in the direction of the Earth's motions to enforce light's constancy but not in the perpendicular direction while time's slowing would again have to be applied equally over the entire platform. It's one reference frame. So it fails in every respects.

(49.3 Sagnac Exp 5a)

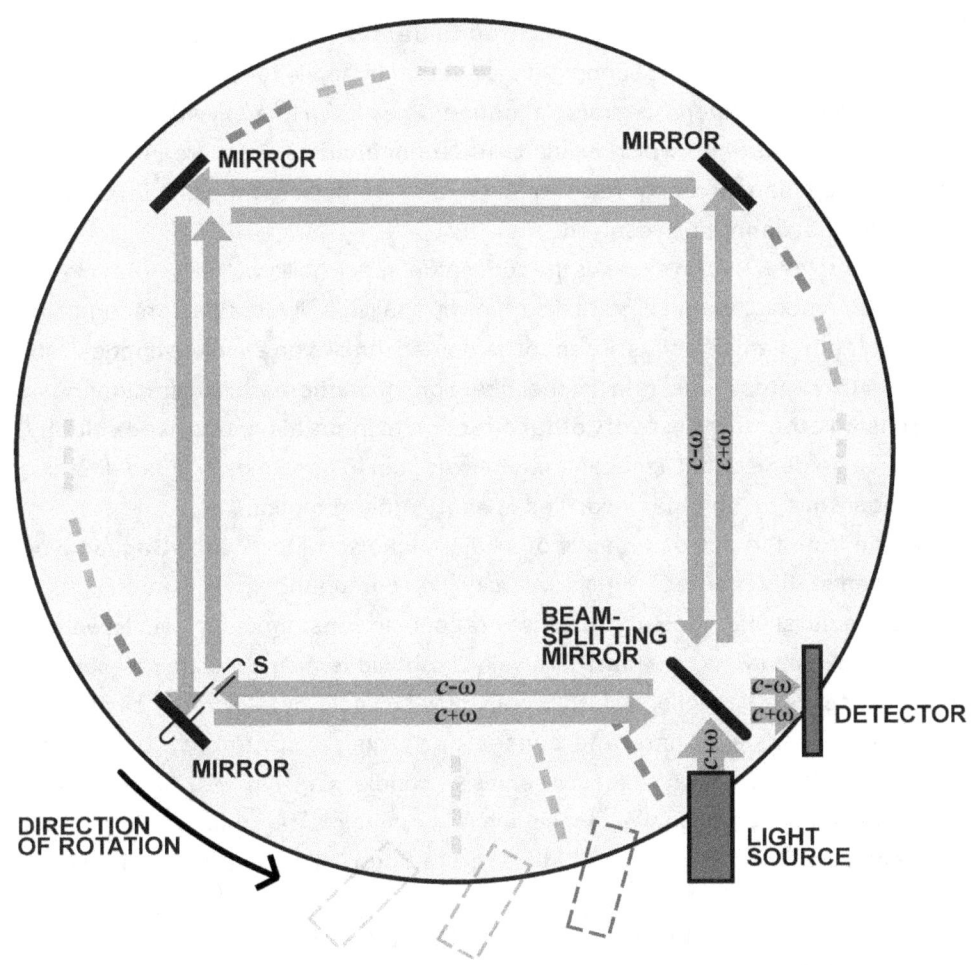

DIRECTION OF THE EARTH'S ROTATION & ORBITAL MOTION ➡

SAGNAC EFFECT - CONCEPTUAL DIAGRAM
WITH ROTATION - REALITY

The only consistent way to explain the effect is if light compounds with the motion of other reference frames. It always leaves its source at 186,000mi/s in all directions at the same time. The non-rotating condition and Michelson-Morley clearly confirm that.

The platform's angular velocity, ω, (or some vector angle of it, which for this diagram would be a 45°, or .707) has to be imparted to the light. Its emitter is moving with it. So it has to be added/subtracted to light's velocity: $c+\omega$ for the forward split light and $c-\omega$ for the rearward split light. The result is that both beams reach the detector at the same time. But it's their different velocities that puts them out of phase and produces the interference pattern, not their different arrival times.

Sagnac's experiment unequivocally establishes light's compounding with the relative motion of other reference frames.

(49.4 Sagnac Exp 5a)

305

These experiments reveal the fundamental flaw in special relativity. Time's presumed "slowing" and light's dispersal is applied in only one direction, the direction of the reference frame's motion. They have to apply equally over the entire reference frame, which exists in three dimensions. This creates inherent conflicts between differing rates of time and velocities for light in different directions that cannot be resolved.

This simple realization exposes the conceptual impossibility of light's fixed velocity. But like so much else that pertains to relativity's failings, it remains unrecognized. No matter how hard one tries, it cannot be denied that it's an absolute impossibility for time to express one rate in the direction of motion while simultaneously expressing other rates in every other direction to maintain light's fixed velocity. Nor can it be denied that light can remain fixed, being one velocity in the direction of motion while being other velocities in every other direction.

Aside from the obvious results of all the Michelson-Morley and Sagnac type experiments that address light's velocity, its compounding with motion is a perfectly natural inference. It's in no way rational to presumptively decide without any verifiable measurement that light's velocity should remain constant for everyone no matter our relative motion (which is somehow subjectively decided by each of us) that alters our rate of time and causes our physical contraction only in the direction of motion. If it wasn't for our pervasive conditioning, this would sound crazy.

Given the observational evidence, even in Einstein's day, you have to wonder how any sensible individual could be led to such an outlandish conclusion, and then use that tenuous conclusion as the fundamental premise for the defining principles of his entire cosmology. But what's even more baffling, we've completely accepted this nonsense as fact without critical evaluation and still continue to build on it.

Despite the outright physical impossibility of objects contracting only in the direction of their motion and our misconception that the three-dimensional nothingness between objects, space, has a curving, two-dimensional, non-Euclidean geometry, the decreasing density of gravity's field does cause a body's omnidirectional, albeit slightly uneven, oblate spheroidal contraction as it's pushed toward other bodies.

This does increase its natural frequency, which increases its rate of rotation, if it is rotating, which would increase any assigned rate of time, if its rate of time were established by its rotation. So time's rate could be understood as changing with motion. But it'd be running faster with the natural motion resulting from gravitation. It would not be slowing with any (subjectively decided) motion to (mathematically) enforce light's (assumed) constancy. (See Diagram 4 Time's Rate 7a)

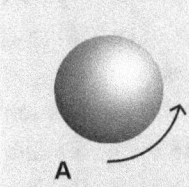

1. HOMOGENEOUS FIELD

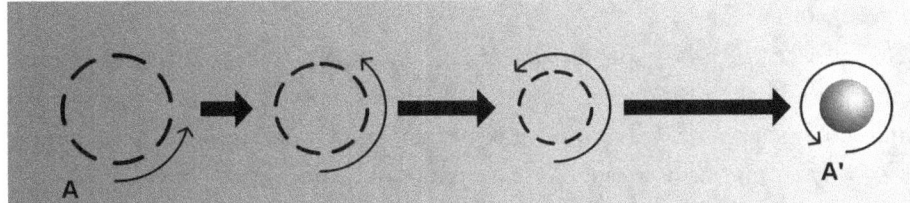

2. GRAVITY FIELD DECREASES IN DENSITY EXPONENTIALLY

TIME'S CHANGING RATE

If we could view two identical heavenly bodies, one in a field that was theoretically uniform in density and the other in a gravity field that naturally decreases in density exponentially per the inverse square law, as represented in section by the diffusing background, the body at **1.A** in the homogeneous field would remain constant, the same without motion, or contraction, or increase in its rate of rotation. This is how we perceive our environment.

Reality, though, is the condition in row **2**. The body in the gravity field would be mechanically seeking equilibrium in the diffusing density, moving in the direction of the least resistance, which is always directly toward other objects, or more accurately a common center of mass, increasing in velocity from position **A** to **A'**, as indicated by the heavy arrows, while contracting in size in a near omnidirectional manner due to the field's decreasing density. As it condenses, its natural frequency increases, increasing its rate of rotation, as implied by the increasing length of the radiused arrows.

If time's rate were established by the rate of·the periodic motion of a celestial body's rotation, as ours is, then as outside observers we could see that the assigned rate of time quickens for objects traversing fields of lesser density due to their condensing which increases their rate of rotation. But for anyone inhabiting this heavenly body, there'd be no perceptible change. Without an outside reference, the length of a day would still appear to be the same despite its perpetual shortening. Those inhabitants might look out at other heavenly bodies in regions of space with higher field densities and it'd appear as if the rotation rate for those bodies was slowing.

Altering the naturally frequency of the object we've selected to establish time's rate by either condensing its mass through the natural coalescing process of gravitation or by increasing its mass by subjecting it to a charge is the only way time's assigned rate can change with or without motion.

(4 Time's Rate 7a)

For light's velocity to appear to remain constant, as we presume to observe here on Earth, excluding its compounding with motion, it'd be necessary for the object used to establish time's rate, in our case the Earth's rotation and orbit, to contract in the ever-decreasing density of our galaxy's gravity field at a rate such that the natural frequency of its increasing rotation, which would be an increasing rate of time that'd be shortened time increments, would be exactly, or very closely, synchronized to light's slowing velocity. This correlation would cause it to always read nearly the same.

Still, we could never be aware of our increasing rate of time. All our measuring devices, including ourselves, would be contracting in unison in a near omnidirectional manner. This would make any change imperceptible. A day would still be a day no matter how much it changes or how quickly, or slowly, it passes.

Given that a relationship should exist between a gravity field's decreasing density and the increasing density of infalling objects of a given material, a condensing body's increasing rate of rotation, which is set by its natural frequency that's dependent on the size and the distribution of its mass, should also be proportional to light's decreasing velocity in the same field density. But objects composed of different material would condense at different but consistent rates while gravity's decreasing field density and light's decreasing velocity would always correspondingly change at the same rate. (See Diagram 31 Condensing 7a)

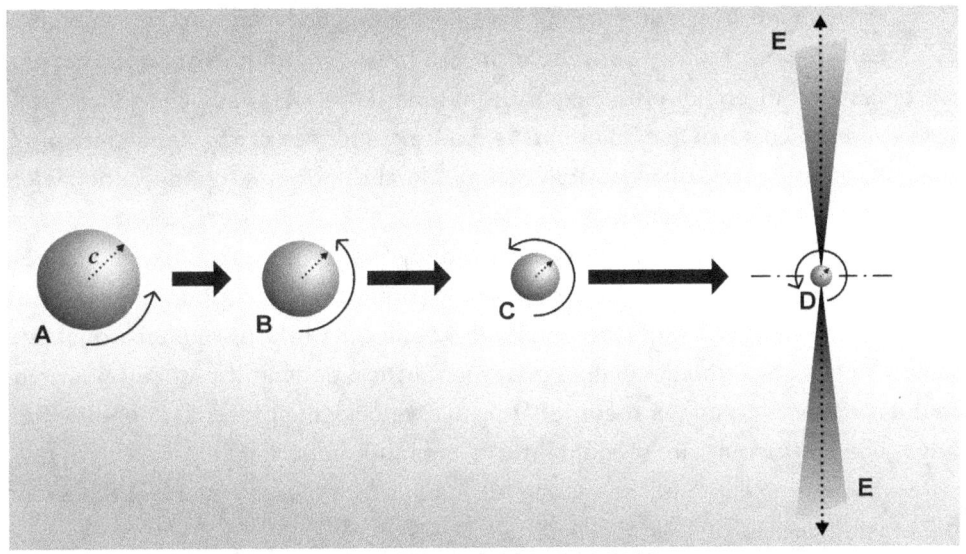

SECTION VIEW THROUGH A MATURE SPIRAL GALAXY

CONDENSING BODIES

The sphere depicted as moving in the sequence from **A** to **D** represents the condensing and ultimate collapse of an infalling star or any heavenly body as it whirlpools in toward a spiral galaxy's center. It is accelerating and contracting while its rotation rate increases, implied by the increasing length of the radiused arrows, as it mechanically pursues equilibrium in the decreasing density of the spiral's gravity field. Light's velocity decreases correspondingly, as indicated by the dotted arrows of decreasing length at c.

Galaxies naturally coalesce from ellipticals into the flat rotating plane of mature spirals because the slower rotation rate nearer their developing poles resists gravity less. That's where the downward migration of its material occurs the most, which produces the disk's overall flattening. So the impetus of its coalescing transitions from the geometry of a full sphere more into the two-dimensional geometry of a circle. With its area, $A = \pi r^2$, it's now condensing exponentially more toward the disk's center within its planar vortex. The diffused background portrays the spiral's flattening, disk-shaped gravity field.

Because the diffusing density of a gravity field causes the omnidirectional contraction of all infalling bodies and a corresponding decrease in light's velocity, all measurements are shrinking while time's rate quickens with the body's increasing rotation, assuming time's rate is established by that rotation. This causes any measurement to always read the same for anyone at the body, making it virtually impossible for them to ever directly perceive their contraction or increasing rate of time.

As infalling bodies approach the galactic center, instead of disappearing over the presumed two-dimensional event horizon of a supermassive black hole that's believed to lurk at the center of almost all galaxies devouring anything that strays to close, the increasing pressure created by the galaxy's exponentially decreasing field density rapidly increases the infalling body's fusion reactions near the spiral's center, at **D**, hastening its collapse. This essentially squeezes its material back into radiant/plasma energy that's then spewed out in a bipolar manner as high velocity jets, as represented at **E**, which act as the vortex's drain. As the expelled radiation slows and cools, it reconstitutes back into ordinary matter. It can then begin gravitating back to its or another nearby galaxy to be perpetually recycled in a ceaseless process of degeneration, transmutation, and rejuvenation.

(31 Condensing 7a)

Also, the quantity of matter composing any infalling body does not remain static. Gravity has it constantly coalescing more material, acquiring more mass. It's believed that about 44,000 tons of material fall from space onto the Earth every year.[172] This has to also affect the natural frequency of its rotation and orbit that would, even though very slightly, have to alter our rate of time, along with the density of its gravity field.

So any assumption of light's constancy here on Earth, excluding its compounding with motion, can only be theoretical. The amount of material composing it is assumed to be constant, keeping its rate of rotation constant, maintaining a constant rate of time in a constant field density. It's entirely possible that the minor differences recorded in light's velocity over the past century that have been attributed to refinements in measurement actually reflect a real difference due to an actual change in the Earth's rate of rotation and in the ever-decreasing density of its gravity field.

In the thinnest regions of a gravity field, presumably located at the center of massive galaxies or a cluster of galaxies, it's theoretically conceivable that light's velocity could approach zero. But it could never stop entirely. At the other end of the spectrum of field density, we can also infer that there exists a maximum velocity for light in a field of maximum density where the speed of light could conceivably approach instantaneous. The universal field would be most dense in the greatest voids between galaxy clusters or maybe in the ejecta of spiral galaxies or maybe even within matter itself, if it's interpreted as a congealed field. All are possible locations for the highest field density.

With light's variability that theoretically allows its velocity to range from close to zero to almost instantaneous, and because of gravitation's potential to coalesce objects at unlimited velocities in fields of ever-thinning density, we have to conclude that there simply can be no upper limit on an object's velocity. And because light conceptually has to compound with motion then for any object that's traveling any velocity, including faster than light, we have to also conclude that its sphere of influence would be compounding, still expanding from it at the speed of light, but a speed of light that corresponds to the density of its surrounding field.

So, the final velocity of light, or that of any object, can only be understood as unlimited in all respects up to the point of instantaneous. This being the case, all of relativity becomes defunct. It's all premised on light's nonexistent and impossible constancy.

172. "How much dust falls on Earth each year? Does it affect our planet's gravity?" Astronomy, last accessed Mar 25, 2023, https://www.astromomy.com/magazine/ask-astro/2014/07/space-debris.

Einstein's mass-energy relation, along with Lorentz transformations, are valid only in restricted conditions. Those that do not involve the kinetic energy of motion, like when subatomic particles are induced with a charge from an electromagnetic field.

In accelerators where the increasing energy of the electromagnetic field accelerating the particle corresponds to its increasing velocity and its increasing mass, the relation does hold true. But only because the increasing strength of the electromagnetic field pushing the particle is inducing it with an increasing charge, which is increasing the particle's mass.

The kinetic energy associated with the particle's increasing velocity lacks a physical mechanism or process by which it could impart an increase to its mass. The amount of matter it contains cannot just magically appear out of nowhere simply because it's moving. Being just congealed radiation, it's easy to comprehend how an infusion of more radiation would enlarge a particle that would in turn cause its natural frequency to decrease. But this is not a decrease in its rate of time.

The reason accelerated particles can only approach the speed of light but never attain or exceed it is because the electromagnetic field pushing them can never surpass the (uncompounded) speed of light, 186,000mi/s. Which for us is regulated by the compounded density of Earth's gravity and to a lesser degree its magnetic field.

When any chemical or mechanical force is driving a particle, or any body of mass, or if their motion is the result of gravitation, there simply can be no limit put on their velocity. And the kinetic energy of their motion has no way of inducing a charge that would in turn increase their mass. It would essentially remain constant while its inertia increased indefinitely.

It would not become infinite at the speed of light. That's a mathematical byproduct of the formula that has nothing to do with reality. So for the particular circumstances of electromagnetic energy accelerating particles, Einstein's mass-energy relation is valid. It's not valid for the kinetic energy resulting from motion, which is how it was incorrectly conceived.[173]

Specifically, for special relativity's time dilation, time's slowing with motion, which is a larger time, Einstein contends that to enforce light's constancy an object's motion causes its rate of time to slow. The faster its velocity, the slower its rate of time. This coincides with its contraction and light's slower velocity in the direction of motion that maintains its constancy.

But time's rate is solely dependent on the periodic motion of the object we've selected to use as reference. It's not an independent property of the universe. So its rate cannot be arbitrarily decided and then manipulated by a variable in an equation. It's wholly dependent on some physically real object.

173. Einstein, *Relativity*, 49-54.

Everything is in some relative motion because of gravity. Their ongoing condensing in gravity's ever-decreasing field density causes their natural frequency or periodic motion to increase. For planetary bodies, their contraction increases the periodic motion of their rotation and orbit. So if their rate of time were established the same way ours is, which is set by the Earth's rotation and orbit, then the rate of their time would be naturally increasing just as ours must be.

It actually appears, though, that the Earth's rotation is not continuously increasing from its contraction but is fluctuating by a few hundredths of a millisecond in a day.[174] This is thought to be due to other forces like the evaporation cycle of the Earth's water that gets redistributed to the poles as ice, tidal forces of the Moon, and inertial effects due to the motion of the Earth's outer molten core.

Even so, its slight overall contraction and ever-increasing rate of rotation are still a fact. It's just not observable while the redistribution of its mass is. With light's decreasing velocity and the nearly synchronized contraction of all measuring and timing devices, like atoms in atomic clocks, in the ever-decreasing density of our gravity field, time's ever-increasing rate isn't perceptible.

A day would still appear to be a day. It doesn't matter how fast it passed. Other physical phenomena that may not involve synchronized contraction like those mentioned are perceptible and can take place concurrently.

For light to maintain its constancy at any moving body, as Einstein contends, its velocity would be required to slow the same velocity as its source. This would require a matching smaller contracted time period. A smaller time is a faster running time, not a slower, longer running, dilated time. But this obvious, elemental, and nullifying conceptual error that also runs contrary to Lorentz's equations remains unrecognized.

If an object were to experience nongravitational velocity, which we can take to mean any motion resulting from a force other than that naturally produced by the decreasing density of a gravitational field, and if its trajectory is such that the density of the field it was traversing remained constant then the size of that object would remain constant. There'd be no change in the rate of its natural frequency or its assigned rate of time. This would exclude any charged acquired from its motion through the field that would increase its mass and size and reduce its natural frequency.

If its trajectory had it passing through a region of a field with an increasing density then its expansion would reduce its natural frequency and decrease its periodicity. Any charge acquired from its motion through the field would further reduce its natural frequency. If its rate of time were established by its rate of periodic motion then its time would slow correspondingly.

174. "Earth's rotation," Wikipedia, last modified Mar 23, 2023, https://en.wikipedia.org/wiki/Earth's_rotation.

This simple effect is misjudged as a changing rate of time for subatomic particles. When analyzing their motion through electromagnetic fields, they're induced with a charge, which increases their size and mass, causing their natural frequency to slow. This is misinterpreted as a slowing in the rate of their time. This would be consistent with Lorentz's transformations (one-dimensionally), giving undue credence to Einstein's incorrect theory.

With time's rate varying due to a body's varying periodicity resulting from its varying size that's due to its natural gravitational motion in the varying density of the universal field that innately decreases at every object, we now have a practical interpretation of how time's rate may change for objects in motion.

This rational, common sense deduction exposes the fantasy of time travel and completely undermines the popular twin paradox inferred from relativity. It has one twin flying off into outer space at nearly the speed of light. When he returns, he's younger than his brother. His time was passing more slowly because of his faster motion.

Well, let's assume that we actually had two identical twin Earths. But let's place them in gravitational fields of different densities. The Earth in the field of lesser density would free fall faster and experience more condensing that produced a quicker rotational rate, which would cause its assigned rate of time to run faster. So, the people on that Earth would be numerically ageing at a faster rate than their counterparts with a slower rotation. But they wouldn't actually live shorter lives. They'd just have more numerical years while remaining the same age.

The well-known clock experiments where very accurate timepieces were flown aboard commercial airliners around the world in opposite directions that are interpreted as a confirmation of time dilation are completely misconstrued.[175] The clocks did experience a change in the rate of their operation. But this is mistaken as a change in the rate of their time, which could in no way have been caused by the plane's relative motion to the ground-based clocks or gravitational time dilation (time's presumed slowing in gravity fields, the higher the altitude, the faster clocks run).[176]

Their changing rate of operation could only have been produced in two ways: The clocks' cesium atoms were induced with a charge from their motion through the Earth's magnetic and gravitational fields that decreased their natural frequency. And their elevation in those fields resulted in their contraction that increased their natural frequency.

175. "Hafele–Keating experiment," Wikipedia, last modified Aug 24, 2022, https://en.wikipedia.org/wiki/Hafele-Keating_experiment.
176. "Gravitational time dilation," Wikipedia, last modified Nov 7, 2022, https://en.wikipedia.org/wiki/Gravitational_time_dilation.

These combined effects produced an increase in the size and mass for the clocks' atoms traveling faster through the Earth's fields in the eastward direction that decreased their natural oscillation frequency, causing their slower operation, which we've incorrectly inferred as a slowing in their rate of time. These same conditions brought about a smaller relative charge and a reduction in the size of the clocks' atoms traveling slower through the Earth's fields in the westward direction. This increased their natural frequency, causing their faster operation, which we've mistaken as a quickening in the rate of their time. (See Diagram 5 Density 10a & 6 Clocks 10a)

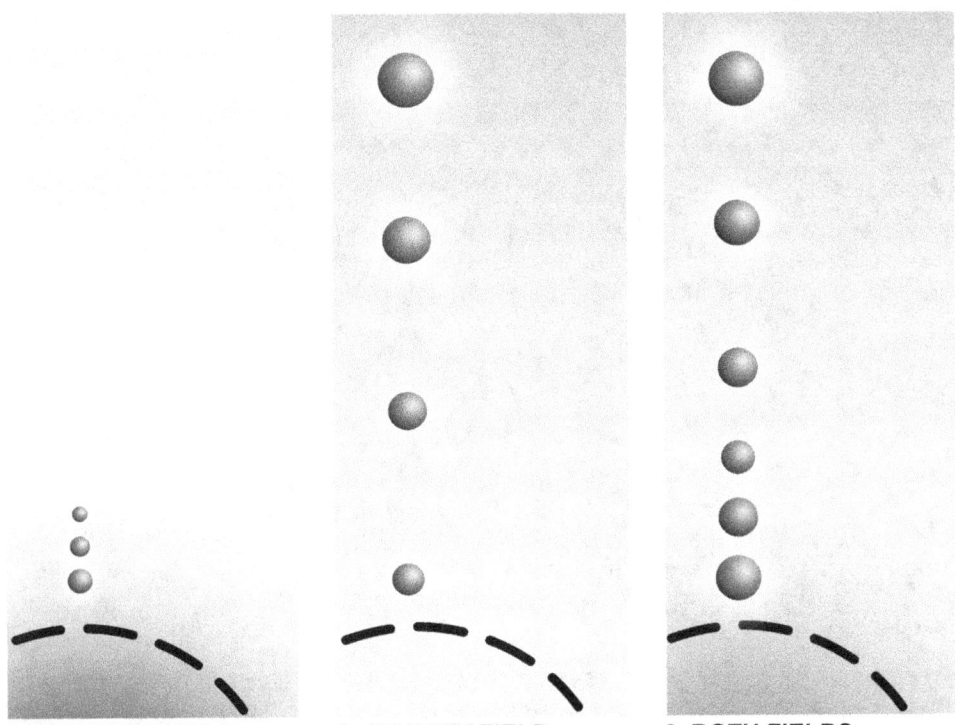

1. MAGNETIC FIELD   **2. GRAVITY FIELD**   **3. BOTH FIELDS**

FIELD'S EFFECT ON SIZE

Subatomic particles are congealed out of the universal field of radiant energy. There are no particles per se, or the objects they compose. Ultimately, there are only condensed fields that are an inseparable extension of the infinitely continuous universal field from which they arose. So when its density decreases or increases like when it expresses as a magnetic or gravity field, the size of any nearby object in it has to vary correspondingly.

In diagram **1**, imagine the Earth, omitted for clarity but where its surface is represented by the curving dashed line, without a gravity field but left with only its magnetic field. Its density dissipates from its center out exponentially, as depicted in section by the diffusing background. So all objects, including the cesium atoms of an atomic clock, would have to decrease in size correspondingly with altitude, as indicated by the sequence of spheres, which in turn causes their natural frequency to increase, making the clock run faster.

In diagram **2**, now imagine the Earth without its magnetic field but left with only its gravity field. Its density increases with altitude exponentially. So any object, or again the cesium atoms of an atomic clock, would have to increase in size correspondingly as they move farther away, causing their natural frequency to decrease and the clock to run slower.

In **3**, the compounded effect of both fields is portrayed. Objects first contract then slowly begin to enlarge as they move farther away. The gradient in the magnetic field is greater over a shorter distance because of its much smaller size while the gradient is much smaller for the gravity field because of its much larger size, which yields little change over the same distance. The sizes and effects suggested have been greatly exaggerated for clarity.

(5 Density 10a)

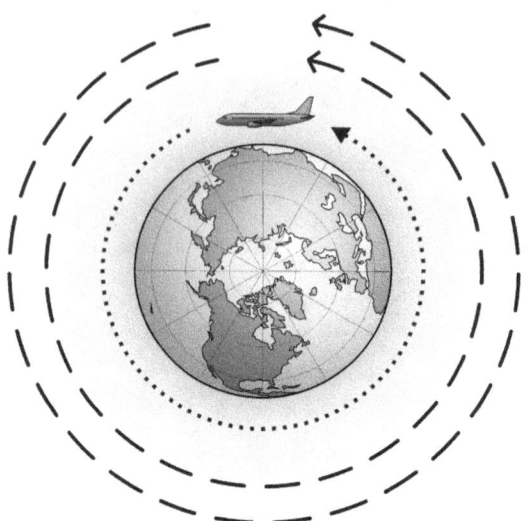
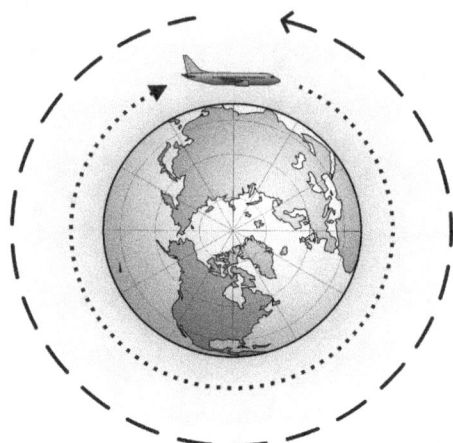

1. EASTWARD (-59ns)
THE CLOCKS' MOTION THROUGH OUR
GRAVITATIONAL & MAGNETIC FIELDS
DECREASED THEIR CESIUM ATOMS' NATURAL
FREQUENCY MORE THAN THEIR ALTITUDE IN
OUR MAGNETIC FIELD INCREASED IT

2. WESTWARD (+273ns)
THE CLOCKS' ALTITUDE IN OUR
MAGNETIC FIELD INCREASED THEIR
CESIUM ATOMS' NATURAL FREQUENCY
MORE THAN THEIR MOTION THROUGH
OUR GRAVITATIONAL & MAGNETIC
FIELDS DECREASED IT

1971 HAFELE AND KEATING AIRBORNE CLOCK EXPERIMENT

There are practical commonsense explanations for the results of all those airborne clock experiments that don't rely on special relativity's self-conflicted, metaphysical effects, length's one-dimensional contraction that's bound to nonexistent time's three-dimensional dilation that impossibly enforces light's presumed constancy.

If we assume for convenience that the speed of the jet airliners carrying the clocks is roughly 500mph and about half the speed of the Earth's rotation, we can then see how when traveling in the eastward direction, with the Earth's rotation, the airliner would complete two revolutions in the time it takes to fly one. This would induce a charge to the clocks' cesium atoms from one revolution through the Earth's magnetic field and two revolutions through its gravitational field, which would increase the cesium atoms' size and mass that would slow their natural frequency, making their clocks run slower.

In the westward direction, the clocks' motion through our magnetic field would remain the same, one revolution. But because they're traveling in the opposite direction of the Earth's rotation, they're only traveling half the distance at half the speed through its gravitational field. So their cesium atoms' acquired charge would be much less than in the eastward direction, still slowing their clocks' rate, but not nearly as much.

When the effect is compounded with the increase in the cesium atoms' natural frequency due to the aircrafts' altitude in our magnetic field where the atoms contract because of the decrease in its density, coupled with only a very slight increase in density from our gravitational field, the eastward clocks end up with a greater mass and slower natural frequency that causes them to run slightly slower than the ground-based clocks.

Conversely, for the westward clocks, not having acquired nearly the same charge, the decrease in their cesium atoms' natural frequency is less than the increase resulting from their altitude. This leaves them with a higher natural frequency than the ground-based clocks that causes them to run faster. Any effect from our orbital motion through our Sun and galaxy's gravitational field can be excluded. It's essentially the same either way.

(6 Clocks 10a)

The question that should immediately come to mind is, how is it that the clocks in the westward direction didn't slow but ran faster? Motion always causes time to run slower, right? You'd think that this factual result by itself is enough to invalidate the whole premise of light's constancy. If time isn't slowing with motion then light's velocity can't be fixed. Any number of stationary altitude experiments with multiple clocks can be imagined though that would reveal that it's actually a small decrease in the density of the Earth's magnetic field with altitude that slightly reduces the size of the clocks' atoms that quickens their periodicity, which causes them to run a little faster.

When applying a changing rate of time to any object whose natural frequency is varying due to a change in its size, there's an important distinction to be made. A contracting object, whose spin would be increasing, would have a compressed or smaller rate of time. This is a faster running time.

For example, one second for that contracted object that let's say is 3/4 of one of our seconds would really be a quicker, faster running time. It's not a slower or dilated time. This applies to objects in motion whose time is supposed to dilate, corresponding to its length contraction. It doesn't work. They're mismatched. But still, the only time we actually experience is the rate of time that we ourselves have established. All else is pure fantasy.

And any change in time's rate can only be a result of a change in the natural frequency of the object we've selected to represent its rate. And again, any variation in the object's natural frequency can only occur as the result of a variation in its size and/or mass, which can only occur by altering the amount of material it has or with the redistribution of that material. This can only occur by changing the density of the field that it inhabits.

We all have a good innate feel for how the natural frequency of objects varies when their size or mass varies. The period changes for pendulums of different lengths. The rate of a spinning ice skater's rotation changes depending on the position of their arms. The location and number of children playing on a merry-go-round affect its rate of rotation. There's no mystery here. This is the only real way time's rate can change.

It's the accumulation of its elemental errors, which are rooted in Einstein's assumption of light's constancy, that we acquired from our indoctrination that's responsible for our widely held misunderstanding that time runs slower for objects in motion while their length contracts despite that it's plainly impossible in our real physical world of three actual dimensions. And it becomes even more absurd with the realization that he's actually proposing that it's a subjective choice for each observer which clocks are in motion and the rate of their motion despite the violation of the laws of gravitation and the inherent conflict with the choices of other observers.

But just a little common sense can take us a long way toward resolving these otherwise bewildering issues. Light's variability in a field's varying density along with its compounding with motion, both conceptually indisputable, and gravitation's ever-decreasing field density that causes an object's omnidirectional condensing that increases their natural frequency, not their rate of time, and the charge that's acquired by objects like subatomic particles when subjected to electromagnetic fields that increases their mass, which in turn decreases their natural frequency, not their rate of time, finally provide us with nonmagical, consistent, realistic solutions to the phenomena attributed to relativistic effects. This rudimentary realization exposes just how nonsensical special relativity actually is.

Generally Speaking

If we're truly honest, we have to also concede that Einstein's gravity doesn't make any sense either. It's also contradictory, inherently conflicted, and physically impossible.

He wants us to believe that space is something that has a physical existence.[177] We uncritically just accept his pronouncement as a given. When the truth is, there's nothing there. Space does not exist. By definition, it's the nothingness between objects.[178] If there's nothing there, then there's nothing that can be curving (or expanding, or stretching, or cause light's redshift from stretching) to facilitate gravitation.

He melds that nonexistent space with a nonexistent time into an inconceivable four-dimensional "space-time." Time also does not exist. It's not a property of the universe. It's only the periodic motion of an object we select, like the Earth's rotation and orbit or the cesium atoms of an atomic clock, that we use to reference motion.[179] So its rate cannot arbitrarily change with location.

Then he has that purely theoretical, four-dimensional, nonexistent abstraction curve two-dimensionally as a nonexistent plane. A plane by definition doesn't exist either. Its two-dimensionality can only define a location that's planar. Curvature is a property limited to one or two dimensions. In three dimensions, any change in a substance can only express as a variation in its density. Its curvature is a conceptual impossibility.

He then takes that two-dimensional, nonexistent, curving plane of four-dimensional, nonexistent, inconceivable spacetime and somehow has it dent underneath three-dimensional, massive bodies as if it were affected by the weight of the bodies above pressing down on it due to the pull of the gravity of a third much more massive body positioned below them.

The dented, two-dimensional, nonexistent spacetime then somehow induces their attraction by somehow causing them to roll downhill toward one another, despite that they don't and can't actually roll and aren't uphill. If this were physically possible, it'd be a mechanical reaction, which act instantaneously.[180]

All of this conflicted nonsense is at odds with his contradictory assertion that gravity is propagated by a force similar to electromagnetism that somehow pulls bodies together, acting at the speed of light via waves.[181] But we maintain that

177. Einstein, *Relativity*, 155-157.
178. "Space," Merriam-Webster, last modified Mar 26, 2023, https://www.merriamwebster.com/dictionary/space.
179. "Time," Merriam-Webster, last modified Mar 28, 2023, https://www.merriam-webster.com/dictionary/time.
180. Einstein, *Relativity*, 71-73, 92-107, 112-116; "Gravity," Wikipedia, last modified Jan 1, 2023, https://en.wikipedia.org/wiki/Gravity.
181. Einstein, *Relativity*, 72.

attraction is also somehow mitigated at the same time by unobservable massless graviton particles that somehow exist physically without mass.[182] Which, if they actually were particles wouldn't be able to act at the speed of light either. They'd relativistically become infinitely large. None of this has any chance of actually working in our real nontheoretical universe of three actual dimensions.

Einstein represents his curved spacetime of a gravitational field with a Gaussian coordinate system. He acknowledges that his field is three-dimensional plus one dimension of time, but contradictorily has it expressing the two-dimensional property of curvature so it can be non-Euclidean in nature. Its grid curves two-dimensionally corresponding to his spacetime's imagined curvature.[183]

We're all familiar with the popular but misguided representation of his gravity theory that exemplifies his and our misunderstanding of gravitation. It consists of two different sized balls placed on a thin rubber sheet or any kind of flexible material that's stretched and supported around the edges.

The balls represent the Earth and our Moon, or any other pair of massive objects. The material represents the fabric his four-dimensional spacetime. When released, each ball rolls downhill toward the other because of the dents made in the material by their weight.

If one or both balls were imparted with some motion in a direction not aligned with one another, they'd tend to orbit or spiral in around each other. The dents represent spacetime's curvature caused by the bodies' mass. The greater the mass, the larger the indentation, the more curvature, and the faster the balls roll or "fall" toward one another. (See Diagram **32** Space's Curvature 5a)

182. "Graviton," Wikipedia, last modified Nov 29, 2022, https://en.wikipedia.org/wiki/Graviton.
183. Einstein, *Relativity*, 96-100.

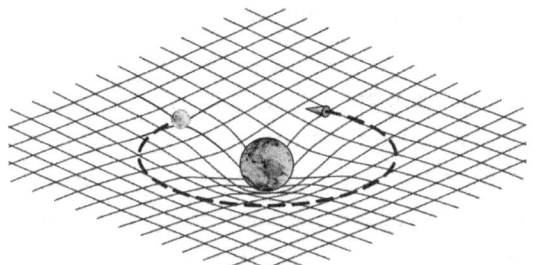

1. CREDIT: NASA (MODIFIED)

2. CREDIT: NASA/GFSC/J. FRIEDLANDER

SPACE'S CURVATURE

Despite our three-dimensional reality, we credulously accept Einstein's assertion that space is a physical something, that it manifests two-dimensionally, and that it curves underneath massive bodies to promote their attraction. It's the "dent" they make in space that causes them to "fall" toward one another or orbit if a lateral force is introduced, which he portrays with a curving, two-dimensional Gaussian coordinate system, like for the Earth and Moon in diagram **1**, that accommodates his conclusion that the universe is non-Euclidean in nature expressing curving two-dimensional properties.

We persist in modeling his misguided notion with balls placed on a sheet of stretched fabric, which only serves to further mislead and confuse. The heavier the ball, the greater the dent. Two or more balls would roll toward one another in an apparent attraction, as implied in diagram **2**. But the dent made in the fabric is due to the ball's weight in Earth's gravity. So where exactly is the source of gravity below massive bodies that would give them weight that would cause their surrounding two-dimensional space to dent?

If massive bodies actually fell, or rolled, toward one another because of a dent they made in curved space, that'd be a mechanical reaction. Mechanical reactions occur essentially instantaneously. But we believe, and Einstein agrees, that gravity is also mediated by hypothetical massless graviton particles that act at the speed of light. So, how can gravitation be simultaneously caused by both space's curvature that acts instantaneously and a particle acting at the speed of light?

Ultimately, we have to come to the realization that space is actually nothingness. It has no physical properties by definition. So there's nothing there to curve. Also, curvature is a property of only one or two dimensions, which don't really exist either. They only define the location of a line or a plane. A three-dimensional environment can never express the one or two-dimensional property of curvature. Any variation in any three-dimensional medium has to be a variation in density. Its surface can curve, but its interior volume can only vary in density. So what is it that is a physical something that can actually exist in three dimensions that might correlate with the empty space between objects that can vary in density and could somehow cause objects to accelerate toward one another?

The answer isn't difficult given that the universe is comprised of only one thing, radiant electromagnetic energy. When congealed, it's matter. When it's not, it remains the universal field of radiant energy of the electromagnetic spectrum. To induce gravitation, it must innately decrease in density exponentially around any quantity of matter, subatomic particle or galaxy, as per the three-dimensional geometry of a sphere, which yields the inverse square law (Intensity, which is really Density, $= 1/r^2$) that governs gravity's force.

So, objects move toward one another not because of nonexistent space's impossible two-dimensional curvature but because of the compounding of the universal field's decreasing density around every object, which is always at its least directly between them. Their natural tendency to seeking equilibrium in its varying density pushes them together in a ceaseless, runaway coalescing process that eventually, after enough pressure has accumulated, culminates with fusion reactions that convert them back into the pure radiation they originated from.

32 Space's Curvature 5a)

It may take another moment of objective reflection, but surely we could agree that it's not conceptually or physically possible to curve a portion of any three-dimensional volume of something. Its interior substance can only vary in density.

We might imagine bending a three-dimensional object like as a malleable cube so that it curves into sort of a partial arch shape. One side would compress inward while the opposite would stretch outward having a greater radius. The cube's surfaces can curve. They're two-dimensional. But its three-dimensional interior can't. Its density has to vary.

Now try visualizing that same bent cube when it's part of or embedded in a larger volume. That volume would be varying in density as well, both inside and outside the defined cube. (See Diagram 17 Curvature 6a)

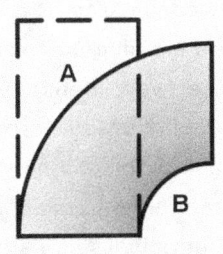
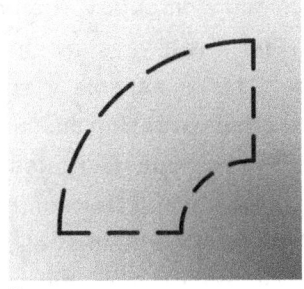
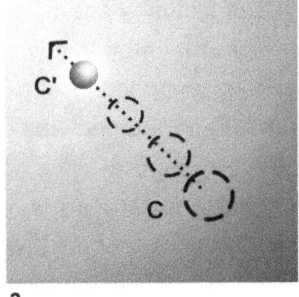

1   2   3

CURVATURE & DENSITY

If we were to envision the shape in diagram **1** as three-dimensional, giving it the same depth as its width, we can see how if it were bent over where its two-dimensional surfaces, at **A** & **B**, now curved, its interior would have to vary in density, as represented in section view by the diffused shading that dissipates toward the outer curved side with a greater radius, **A**. Note how the density along both the outer and inner curved surfaces would remain constant as **A** diffused and **B** condensed when bent.

If we were to imagine that shape as part of a larger volume of the same material, as depicted in diagram **2**, we can more easily see how any three-dimensional volume of something can never curve but can only vary in density. Curvature is a property of only one and two dimensions, which don't actually exist in reality. Without the third dimension, they only define a location of a line or a plane.

These principles of curvature and density would also have to hold true for the three-dimensional volume that's commonly refer to as space. It could never curve. It could only vary in density. But space is the nothingness between objects. It's nonexistent by definition, and it would also be nonexistent if it were a curving two-dimensional plane. So there's nothing there to vary or curve. But there is something that does correspond exactly to all space: radiant electromagnetic energy. Everywhere that space is, so is the all-pervasive universal field of radiation that comprises the electromagnetic spectrum. This field is continuous, its continuity uninterruptible, and it's average volume is fixed per any given quantity. And its density varies.

Because of its innate continuity, the density of the radiation that's been drawn together to form a particle has to diffuse. But it diffuses inward toward the particle's center, still dissipating exponentially per the inverse square law because of the three-dimensional geometry of a sphere. It's the compounding of those inward diffusing fields of the particles themselves, and the objects they compose, that's responsible for gravitation.

Referring to diagram **3**, if we were to place a small object into the decreasing density of a gravity field of a much larger object, it would begin moving with acceleration in the direction of least resistance directly toward the larger object as it mechanically sought to find equilibrium in the field's decreasing density. It'd naturally move from the highest density toward the lowest. But it would also be contracting in size. Its atoms also naturally coalesce and condense within the field's decreasing density, as indicated by the sequence of dashed circles representing the object's motion from **C** to **C'**. In reality, each object would move toward the other in proportion to the size of their gravity fields toward a common center of mass because of the compounding effect of their gradient.

(17 Curvature 6a)

323

All this may seem obvious and elementary when actually considered. But it's never presented in this way. Space is always misrepresented two-dimensionally with its impossible curvature. This is especially true for the two-dimensional, funnel-shaped grid we see so often portraying the curving, gravitationally warped space around a black hole. It supposedly facilitates an infalling object's never-ending, accelerating journey toward infinity as it somehow continuously collapses while ever-stretching toward the black hole's theoretical/mathematical singularity.

When space is correctly envisioned with its three complete dimensions, width, height, and depth, where gravitating objects quickly coalesce together at their common center of mass/gravity, a single point suspended in three-dimensional space, the failings of Einstein's two-dimensional, curving, non-Euclidean gravity become obvious, which out of necessity evokes a totally different solution. (See Diagram 12 3-D Space 13a)

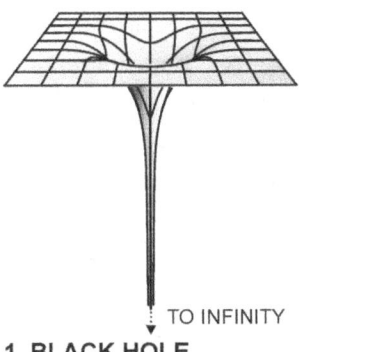

TO INFINITY

1. BLACK HOLE

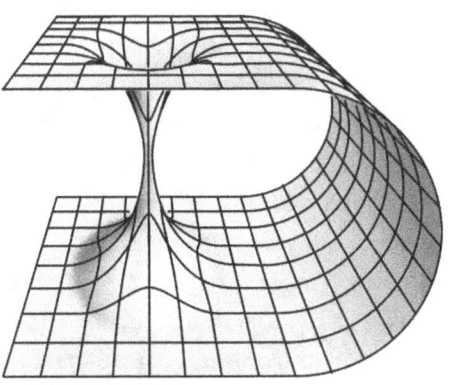

2. WORMHOLE

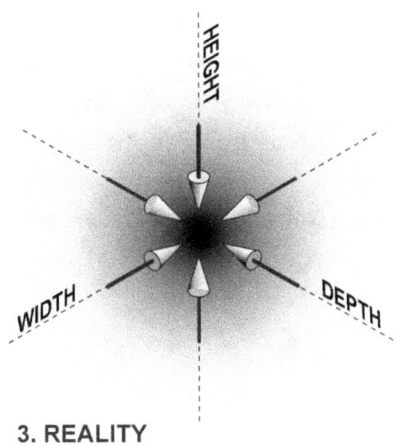

HEIGHT

WIDTH

DEPTH

3. REALITY

(12 3-D Space 13a)

THREE-DIMENSIONAL SPACE

The notion that space is something and that it expresses two-dimensionally has led us to mistakenly reason black holes. Their gravity is presumed to be so strong they create a two-dimensional funnel-shaped curvature in their inconceivable four-dimensional space-time, usually portrayed in diagrams similar to **1**. It's imagined that any object, or light, that strays too close will be drawn over a theorized event horizon, never to be heard from again, where it endlessly infalls while stretching, or "spaghettifying" as it's called, as it accelerates without end toward the infinite (mathematical) condition of a singularity.

The same mistaken logic is applied to so-called wormholes. Space's fictional two-dimensional composition, commonly depicted in diagrams similar to **2**, would supposedly allow us to shortcut through a black hole that's not infinite but tied to another black hole at another location in our curving two-dimensional universe.

In reality, space is the nothingness between objects. There's nothing there. But if it really were something, it'd still have to be three-dimensional, as indicated in diagram **3**. It'd have to have width, height, and depth. That's it. Additional dimensions are inconceivable. And two-dimensions, despite being incorrectly justified as merely an analogy, only describe the location of a plane. Without the third dimension of height, existence isn't possible.

Whether you insist that space is actually something or accept the universe as a field of radiant electromagnetic energy that corresponds to all space, in our reality of three actual dimensions, coalescing infalling material, represented by the condensing grey three-dimensional sphere in **3**, would quickly arrive at a common center of mass. There's nowhere else for it to go. It cannot slip over the edge of a nonexistent two-dimensional funnel-shaped space on an endless journey to infinity. That's a make-believe science fiction fantasy that's justified through two-dimensional mathematical gimmickery.

Gravity's exponentially decreasing field density continuously coalesces and condenses matter together in a runaway recycling process where it's increasingly compressed until fusion reactions are triggered that perpetually convert it back to its original state of radiant energy.

Energy can be defined as the capacity for doing work. It expresses in many forms like mechanical, thermal, chemical, atomic, and electrical. But its fundamental essence is always radiant in nature, which is electrical or electromagnetic.[184]

Space is the three-dimensional region of nothingness in between all objects at all scales from subatomic particles to galaxies. So it's everywhere that matter is not. And since radiant energy (the radiation of the electromagnetic spectrum that expresses as a universal field) is everything that is not considered matter, it's also everywhere matter is not. So all radiant energy coincides with all space.

Space, being nothingness, has no property that can vary. But the all-pervasive, universal, electromagnetic field of radiant energy that corresponds to all space does. It innately decreases in intensity, which is the same as density, around any amount of matter. It's this variation in intensity/density that cause gravitation while exhibiting an electrical charge and current.

The notion that space and radiation exactly correspond accounts for the fact that the attractive or repulsive force between any two charges or the attractive force between any two masses is the product of their charges or masses and is inversely proportional to the square of the distance between them.

For electrostatics, this is called Coulomb's law (Charles Augustin Coulomb was a French physicist, 1736-1806). For gravity, this is defined as Newton's law of universal gravitation. Ismaël Bullialdus (French astronomer, 1605–1694) was apparently the first to recognize that gravity obeys the inverse square law.

Because both forces, electrical and gravitational, are rooted in the inverse square law, a fundamental property of a sphere's geometry where its density changes in inverse proportion to the square of the distance from its center, they can now be understood as the same, but whose fields simply dissipate in opposite directions.

This elemental realization "unifies" gravity with electromagnetism and resolves, or undermines depending on your point of view, the quest for a so-called theory of everything. The key to comprehending gravitation though comes with the realization that all matter arises from the radiant energy of the universal field. (See Diagram 7.2 Inverse Sq Fields 8a)

184. "Energy," Encyclopedia Britannica, last modified Oct 18, 2022, https://www.britannica.com/science/energy.

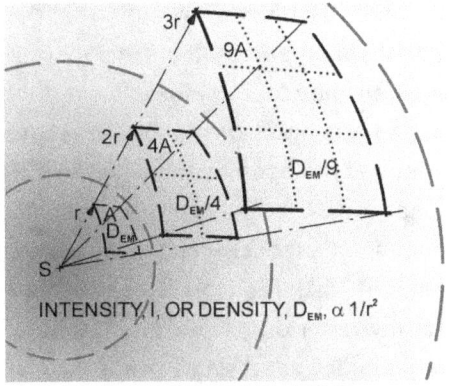

INTENSITY, I, OR DENSITY, D_{EM}, α 1/r^2

DENSITY, D_g, α -1/r^2, OPPOSITE
OF AN ELECTROMAGNETIC FIELD

1. ELECTROMAGNETIC FIELD

ELECTROMAGNETIC FIELD INTENSITY, WHICH IS DENSITY, DECREASES PROPORTIONAL TO THE INVERSE OF THE SQUARE OF THE RADIUS. THIS DIFFUSES ITS OUTWARD ACTING PRESSURE GRADIENT EXPONENTIALLY.

2. GRAVITY FIELD

GRAVITY FIELD INTENSITY, OR DENSITY, ALSO DECREASES PROPORTIONAL TO THE INVERSE OF THE RADIUS SQUARED. BUT ITS DIFFUSION IS INWARD. THIS INCREASES ITS OPPOSITE, NEGATIVE PRESSURE GRADIENT EXPONENTIALLY PER GRAVITY'S FORCE, g.

INVERSE SQUARE LAW, FIELD

An electromagnetic field (EM), depicted in section view as the diffusing background in diagram 1, is subject to the inverse square law that's the product of the three-dimensional geometry of a sphere. So the field's intensity, which is the same as density that produces pressure which is force, twice the distance from its source is diluted by four times the area. This reduces its density to 1/4 the original. At three times the distance, it's spread over nine times the area, which reduces the density to 1/9 the original, and so on where D_{EM} (the density at a given radius) = S (the original density) / $4\pi r^2$ (the area of a sphere).

The big bang's radiant energy, correctly interpreted as a finite universal field, could never express uniformly as we presume. It's subject to the same physical geometry as every other field. It too would have to obey the inverse square law that'd have it diffusing exponentially with its expansion. The homogeneous isotropic expression that we observe is not physically possible for a finite, expanding universe. So in our actual three-dimensional, nontheoretical reality, the big bang is untenable.

The tangible, radiant, EM energy of our real universe that particles condense out of is all-pervasive, continuous, inseparable, and it varies in density. So the remaining ambient radiation that's not been drawn into a congealed particle has to thin inward, diffusing exponentially toward its center. This is what constitutes their, or collectively the bodies they compose, gravity field, portrayed in section view as the diffusing background in diagram 2.

It's the opposite of an EM field. Its lowest density is reciprocal to an EM field's highest. Still bound to a sphere's inverse square law, its density, which is still intensity, which still equates to pressure and force, still has to dissipate exponentially. The gradient remains the same. It just expresses the opposite direction, diffusing inward instead of outward where D_g (the density at a given radius) = -S (the original point source strength or negative density established by a body's mass) / $4\pi r^2$ (the area of a sphere).

So at twice the distance from the center, its original negative density is diffused over four times the area, which is 1/4 the original that reduces the inward acting pressure by the same amount, decreasing gravity's force to 1/4g. At three times the distance, its negative density is spread over nine times the area, which is 1/9 as dense as its original that decreases the inward acting pressure the same, reducing gravity to 1/9g, and so on.

(7.2 Inverse Sq Field 8a)

Our common modeling of Einstein's spacetime is further flawed in that the fabric sheet and the balls incorrectly convey that space and matter are separate discontinuous components of the universe when they're actually just different aspects of the only thing in the universe, radiant electromagnetic energy. What's considered matter is actually just condensed radiation expressing itself as a "particle" that described as possessing a charge.

Just imagine how a subatomic particle might spontaneously condense out of a theoretically uniform field of universal radiant energy. Visualize close up and in slow motion how a burgeoning particle would begin to draw radiation in three-dimensionally, spherically, from all directions, condensing it together toward a single point.

As it continues to grow, the ambient field it's comprised of and drawing from would thin inward, dissipating exponentially toward the particle's center because of the innate geometry of a sphere. The particle and its originating field remain interconnected. They're one and the same. Fields are continuous and uninterruptible. So there can be no separation.

The particle would continue to draw in more radiation until its prescribed mass and density are attained. It can then begin to fully interact with other particles. The influence of their compounded gravity fields first pushes them together until they attain a balanced condition with the repulsive effect of their electromagnetic fields. (See Diagram 27 Fields 12a)

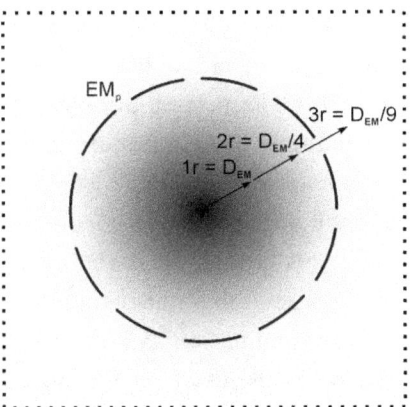

1. ELECTROMAGNETIC FIELD
POSITIVE DENSITY DIMINISHES, $D_{EM} \propto 1/r^2$

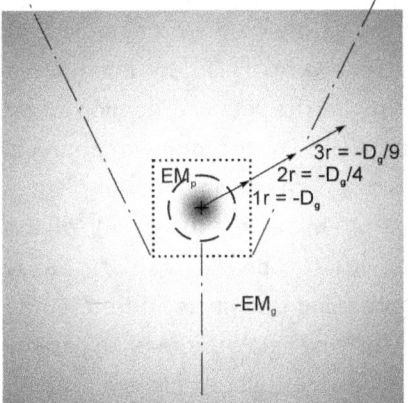

2. GRAVITY FIELD
NEGATIVE DENSITY DIMINISHES, $D_g \propto -1/r^2$

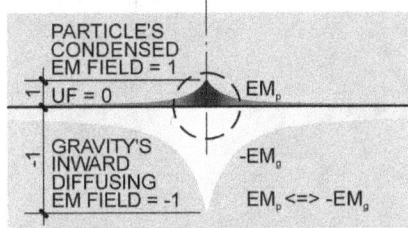

3. RELATIVE FIELD DENSITY & SIZE
A RECIPROCAL EM GRAVITY FIELD (-EM$_g$) WITH NEGATIVE DENSITY & INWARD DIFFUSION IS INNATELY CREATED BY THE RESULTING DEFICIT WHEN A PARTICLE (EM$_p$) CONGEALS OUT OF THE UNIVERSAL FIELD (UF).

(27 Fields 12a)

FIELDS

The fundamental constituent of the universe is radiant electromagnetic energy that manifests as a universal field (UF). It expresses in two ways, as an electromagnetic field (EM) and its inverse, a gravity field (-EM$_g$). They're the same UF but with opposite diffusion and different densities that are the intrinsic byproduct of a particle's inception. They coexist and coincide. But they're reciprocal. And one cannot exist without the other because of their uninterruptable continuity.

When the UF "spontaneously" congeals into a subatomic particle (EM$_p$), it's defined as matter with an assigned amount of mass. But at its essence, it remains radiant EM energy. It's portrayed in an expanded section view by the diffusing background in diagram **1**. The dashed circle represents its theoretical limit or that of any individual body.

The radiation drawn into and composing the particle is the radiation from the UF. That deficit is what constitutes the particle's gravity field (-EM$_g$). They're equivalent because they're the same. They're inverse because of the particle's innate condensing, which is also why it's much smaller, as represented in **2**.

If the UF were assigned the neutral value of zero (0) and the condensed particle (EM$_p$) a value of one (1), the resultant inward diffusion of the UF, the particle's gravity field (-EM$_g$), would have a corresponding negative quantity equivalent to negative one (-1). They naturally reciprocate despite their divergent size and opposite diffusion, as implied in diagram **3**.

It makes no difference whether the EM field's diffusion dissipates inward or outward. The gradient still has to diminish exponentially. Both are subject to the inherent geometry of a sphere that's bound to the inverse square law (Intensity or Density $\propto 1/r^2$).

The exponential outward diffusion of a particle's EM field from higher inner density to lower outer density creates outward acting radial pressure. This should be interpreted as a positive charge, having a male or originative quality.

The exponential inward diffusion of gravity's EM field from higher outer density to lower inner density creates inward acting radial pressure, which should be viewed as a negative charge, having a female or receptive quality.

With the all-pervasive universal field of radiant energy understood as corresponding to all (three-dimensional) space, and because it inherently decreases in density, which is the same as intensity, at particles that originate from it, which creates their gravity fields, it then becomes clear that there is no force of gravity. There's only one force in nature - electromagnetism.

When the universal field is drawn inward as it innately does to create burgeoning particles, and the objects they compose, this creates the decreasing density of their gravity fields. As their gravity fields interact, the particles begin to gravitate toward one another as they naturally seek equilibrium. Constantly pushed by the highest density toward the lowest that always lies directly between them, they begin coalescing and binding to form atoms as they reactively search for balanced states.

Every particle manifests a charge and current as their electromagnetic field repulses outward while their gravity fields repulse inward. This determines their grouping and distance from one another along with the relative location of an electron's orbital, its density cloud, all simply established by the same mechanical search for equilibrium elicited by the varying density of their compounding fields.

This is essentially the impetus behind all chemical reactions. All of the interactions between subatomic particles and their geometry can be easily explained when their electromagnetic field's increasing density is interpreted as a positive charge that repulses outward, and their gravity field's decreasing density is interpreted as a negative charge that repulses inward. This plainly reveals how gravity, along with the apparent strong and weak nuclear forces, are all unified at the subatomic level under one force. (See Diagram 18.1, 18.2, 18.3, 18.4 Atom 8a)

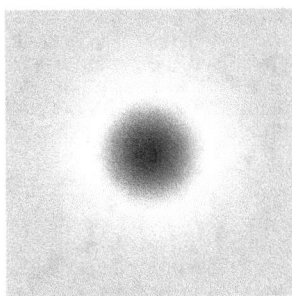

1. PROTON

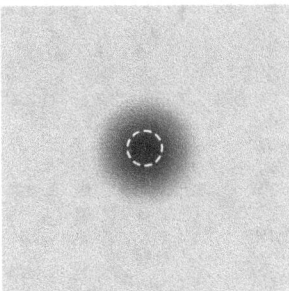

2. ELECTRON

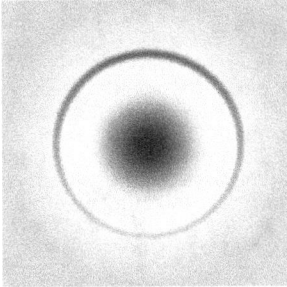

3. HYDROGEN ATOM

ATOMS

Protons should not be considered tiny physical particles within an electromagnetic field but as the field itself. There's no surface where the field stops and matter begins. The field becomes progressively more dense until it peaks at its center, represented in section view by the darker diffused circle in **1** & **3**. But because that proton field has condensed out of the universal field of electromagnetic radiation, the ambient radiation not drawn into the proton has to decrease in density around it diffusing exponentially like any field because of the geometry of a sphere.

But its diffusion disperses inward not outward, which defines its gravity field, depicted in section by the diffusing background in **1** & **3**. Because the decreasing density of a proton's gravity field is larger than the increasing density of its electromagnetic field, the gravity field's compounding with the fields of other particles tends to push them together as they naturally pursue equilibrium, mechanically seeking the lowest density that always lies directly between them. So protons should actually be considered negatively charged.

Convention has protons positive charged and electrons negative. Apparently, this has been mostly an arbitrary designation. But it doesn't correspond to physical reality. It's one of the reasons why gravitation and electromagnetism are not recognized as being the same effect.

Electrons should also be considered as having condensed out of the universal field. Its charge is considered equal to that of a proton. But its mass is 1/1837th as much. So it yields a much smaller gravity field, indicated by the small white dashed circle. For graphic clarity, it's shown proportionally much larger than it would actually be.

Being that the decreasing density of its gravity field is smaller than the increasing density of its electromagnetic field, it has a repulsive effect that when compounded with the fields of other electrons tends to push them away. So in reality it's positively charged. With the electromagnetic field of the electron still smaller than the gravity field of the proton, the compounding of their fields still pushes them toward one another.

An atom's electrons should not be envisioned as small objects that rapidly orbit the nucleus as always portrayed. They're more accurately conceived as having been pressed down and smeared out all over and around the entire nucleus, spherically, three-dimensionally, by the decreasing density of the universal field enveloping it, the atom's gravity field. It's compressed to a level where the repulsive effects of all the fields balance out and find equilibrium, as is implied in the section view through a hydrogen atom that has only one electron and one proton.

(18.1 Atoms 8a)

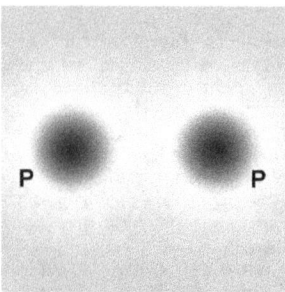

1. TWO PROTONS

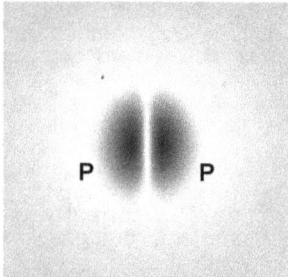

2. PAIRED PROTONS

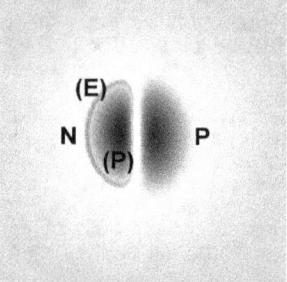

3. NEUTRON & PROTON

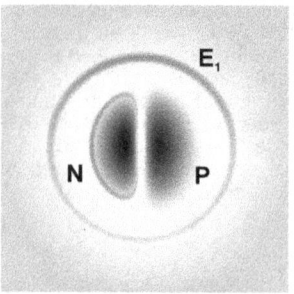

4. HYDROGEN ISOTOPE

(18.2 Atoms 8a)

NEUTRONS & ISOTOPES

A neutron can be considered a merging of a proton and an electron. The compounding of their two electromagnetic fields and their two gravity fields are theoretically balanced to yield no charge, that is if it could stand alone. Its neutral charge suggests that the electromagnetic fields of protons and electrons are half as strong as their combined gravitational fields.

For numerical convenience, if we assume the strength of a proton's gravity field is (-1), negative because of the field's decreasing density, and we know that an electron's is 1/1837th of that (-.00054) then their electromagnetic fields would have to be half of (-1) + (-.00054) or (+.50027), positive because of its increasing density. So a proton's relative charge would be its gravity field (-1) plus its electromagnetic field (+.50027) or (-.49973). And an electron's relative charge would be its gravity field (-.00054) plus its electromagnetic field (+.50027) or (+.49973).

Neutrons usually only exist, though, through the initial pairing of two protons, located at **P**. The compounding of the decreasing density of their fields, (-.49973) + (-.49973) or (-.99946), first draws them together mechanically as they naturally seek equilibrium. Then the even higher decrease in density of their combined fields draws in and tightly holds an electron, which is positively charged (+.49973), located at **E**, to create, or define, a neutron, located at **N**.

It's likely that the electron may move back and forth between protons or at times envelop both at once. But the three together still have a negative charge, or a field of decreasing density of (-.49973), that can draw in another electron (+.49973), located at E_1, to achieve a balanced state, in this case deuterium an isotope of hydrogen.

The actual distance to the electron would be over 60,000 times the radius of the nucleus. At the scale depicted that would put it more than 100yds away. The important principle that's trying to be conveyed here is that it's the sequence in which the particles assemble, which is facilitated by the relative densities or actual charge of their fields, that is responsible for the creation of a neutron. Otherwise, you'd just end up with a hydrogen atom.

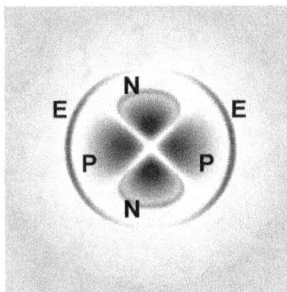

1. HELIUM ATOM

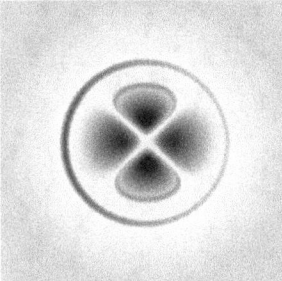

2. NEGATIVE ION (-E)

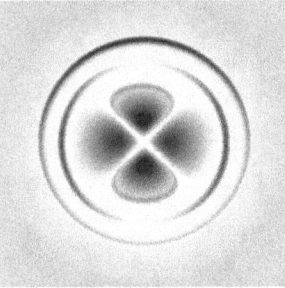

3. POSITIVE ION (+E)

IONS

The actual charge of an ion is also opposite of convention. If we begin with a ground state helium atom, shown theoretically in a section view through its center, the electromagnetic and gravity fields of its two protons, located at **P**, and two electrons, at **E**, balance to neutralize its charge. Its neutrons, at **N**, already a combination of an electron and proton, remain neutral.

If one of the electrons is removed, as depicted in section view in **2**, the density of its combined fields would be decreased where its relative charge, as calculated in the previous diagram, would be (-.49973), where its gravity fields dominate, which would tend to draw in other particles, making its charge negative.

If an electron were added, as represented in section view in **3**, the density of its combined fields would be increased. Its relative charge would be (+.49973), where its electromagnetic fields dominate, which would tend to push away other particles, making its charge positive.

A decreasing density in the universal field, a gravity field, is a negative charge that tends to push inward. The increasing field density of a particle's electromagnetic field is a positive charge that tends to push outward. It's the inherent repulsive nature of a particle's, or any object's, electromagnetic field that mechanically causes them to seek equilibrium in the universal field that innately decreases in density around every particle, or object.

Their reactive search for the lowest density in their combined fields that always lies directly in between them, or toward a common center of mass for multiple objects, causes them to move toward one another in an apparent attraction. It's the same repulsive effect of their interacting fields that pushes or holds them apart when they attain equilibrium.

Protons and neutrons and electrons are not bound together or repelled by imaginary strong and weak nuclear forces that are magically transmitted by unseen massless particles. Gravitation resulting from electromagnetism is simply governing all their interactions.

(18.3 Atoms 8a)

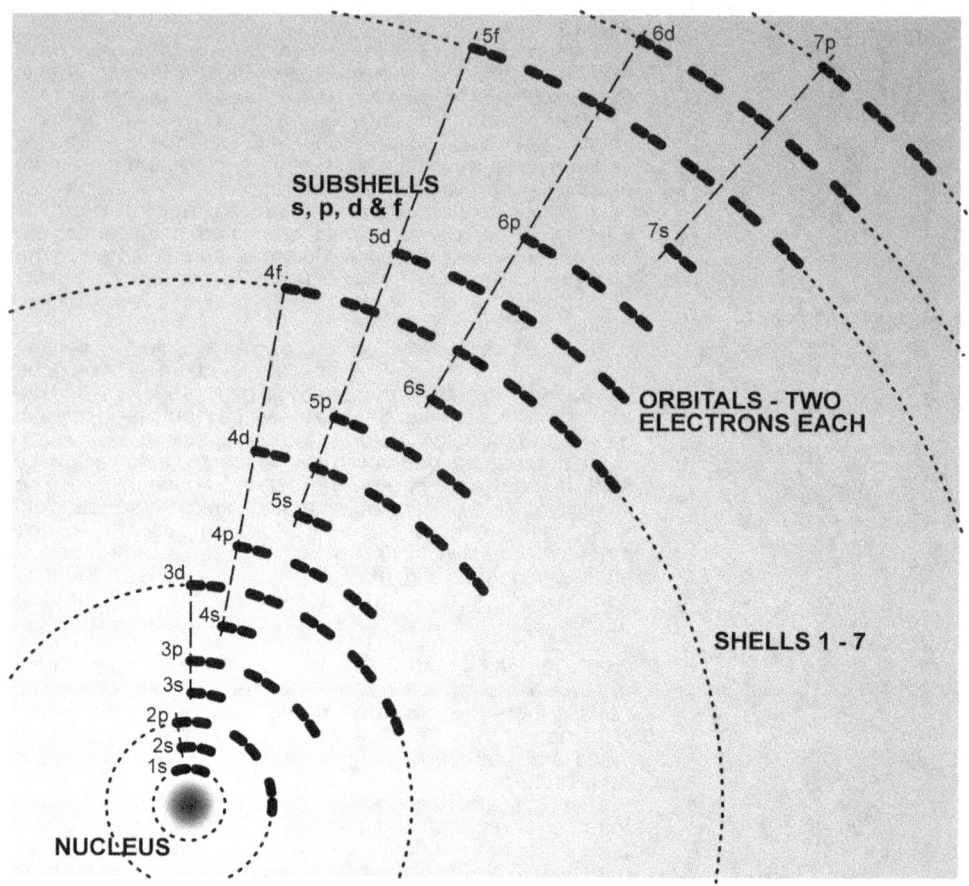

SUBSHELLS
s, p, d & f

ORBITALS - TWO
ELECTRONS EACH

SHELLS 1 - 7

NUCLEUS

AUFBAU OR BUILD-UP PRINCIPLE OF ELECTRONS

This diagram shows how the electrons of all the known elements theoretically distribute themselves around an atom's nucleus according to the build-up principle. It depicts seven shells (1 -7) and their four subshells (s, p, d & f) that contain either 1, 3, 5 or 7 orbitals. Each orbital consists of two electrons.

Electrons tend to fill lower "energy" levels, or gravitate toward the nucleus, pushed inward and smeared out all around it by the increasing force of the decreasing density of the atom's gravity field, portrayed in section by the inward diffusing background, while the repulsive nature of their electromagnetic fields holds them apart. It always keeps them, along with the protons and neutrons at the nucleus, at their maximum distance from one another, which increases from the center out as gravity's force decreases exponentially.

It's important to note the electrons' outward diffusion. They disperse exponentially because of the geometry of a sphere. It's also important to know that they're not paired up side by side in each orbital as the diagram implies. Their repulsive nature ensures their even distribution over the entire nucleus. The same is true in the outward direction. Gravity's ever-decreasing field density pushes them inward causing them to nestle in between one another, naturally pursuing their most balanced and stable distribution that ends up forming shells and subshells that express symmetrically at each consecutive level.

(18.4 Atoms 8a)

The particles', or any object's, mechanical reactive search for equilibrium suggests two things: that the universal field is a physical something and that gravitation acts instantaneously. If mechanical reactions are essentially instantaneous, this means the universal field acts as an instantaneous conveyor of information through gravitational effect. Gravitation cannot be a nonphysical process imparted through waves similar to electromagnetic phenomena propagated at the speed of light as Einstein contradictorily contends and we now consider fact.[185]

Our accepted modeling also does not convey the idea that an object contracts as the density of its gravity field decreases. Since the universal field is everywhere, its density could not just stop at the object's "edge." It also has to continue on through the object in between its molecules and even in between its subatomic particles, causing them to draw inward, naturally reducing the size of the object's molecules and the distance between them, causing their continuing overall condensing.

Since it's not possible to identify any point where the coalescing of a grouping of many small, loosely dense objects transitions to the condensing of a large single object then the actual contraction of any particular object due to self-gravitation can only be interpreted as the same coalescing/condensing process of gravitation. Deciding whether its gravitation or self-gravitation is just a matter of size. Are the Sun and its neighboring stars gravitating toward one another or are they under the influence of the Milky Way's self-gravity? It's the same thing.

A galaxy would of course be correctly viewed as a single object where the density of it constituent material increases from its perimeter to its center, just as any other object because of its self-gravity. But with its material's ongoing gravitational coalescing that causes its ceaseless inward migration, gravity's field for those objects would be ever-thinning commensurately. This means in a very real sense that all the objects that compose a galaxy have to also be continually contracting in size while compacting, becoming more dense as they approach the galactic center.

An object's potential for compaction is clearly indicated by the relative distances between subatomic particles. If the nucleus of a hydrogen atom were expanded to the size of a baseball, its electron would be roughly the size of a grain of sand and located more than 100 miles away. Despite the natural repulsive effect of their interfacing charged fields that's maintaining their distance from one another, this vast distance in addition to the much greater distances between the atoms and molecules suggests that there's ample room for objects to compress significantly given the right circumstances.

It's routinely claimed that the density of a neutron star is so great that a teaspoon full of its matter would weigh more than a ton on Earth. That's a lot of compaction.

185. Einstein, *Relativity*, 53.

And there's still no restriction on the size of the spoon or the matter in it, which is relative based on the density of its enveloping gravity field. It would vary from its least dense at the center of its galaxy that's determined of the galaxy's overall mass to a maximum density located at its greatest distance away from the center of all other surrounding galaxies.

Visualizing any two gravitating objects of any size or material, we could say they're composed of congealed radiant energy, which has made them tangible. They could be envisioned as malleable, sponge-like spheres of matter that'd be enveloped and permeated by the omnipresent universal field they condensed out of, that we could qualify as intangible and corresponding to all space.

With the universal field naturally forming gravity fields that disperse inward around the particles that have spawned from it, they consequently have to diffuse through the objects the particles compose. The density of their gravity fields has to vary from its least dense at the center of every object to a maximum density found at its maximum distance from every other object.

But the density of their compounded gravity fields is always less directly in between them, toward their common center of mass/gravity. So the highest density constantly pushes them toward the lowest, mechanically compelling them to seek equilibrium. As they relentlessly gravitate toward one another in the ever-decreasing density of their ever-combining fields, runaway coalescing/condensing naturally ensues.

This continues until enough material accumulates that the resultant pressure triggers fusion reactions that ultimately transmute every particle back into the electromagnetic radiation from which it arose. (See Diagram 28 Gravitation 12.1a & 11.1, 11.2, 11.3, 11.4 Shape 13a)

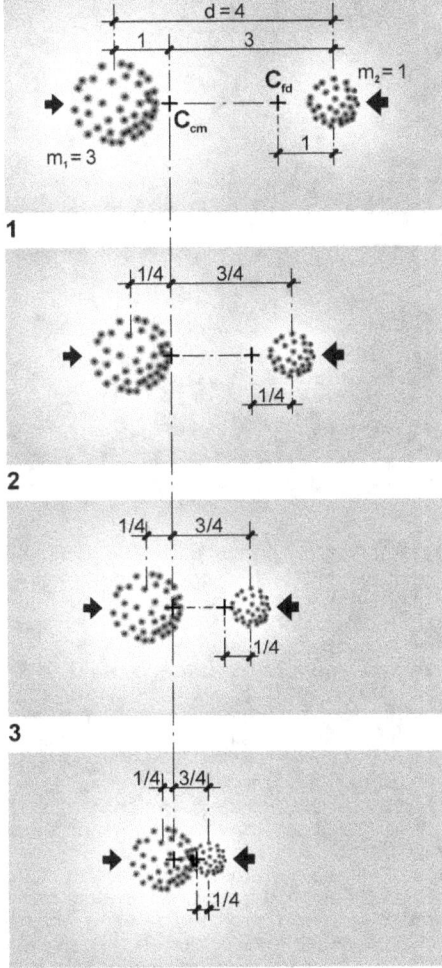

1

2

3

4

COALESCING BODIES
BODIES AREN'T ATTRACTED TO ONE ANOTHER. THEY'RE CONSTANTLY PRESSED TOGETHER BY HIGHER FIELD DENSITY TOWARD LOWER FIELD DENSITY AS THEY MECHANICALLY REACTIVELY SEARCH FOR EQUILIBRIUM IN THE EVER-DECREASINGDENSITY OF THEIR EVER-MERGING GRAVITY FIELDS.

(28 Gravitation 12.1a)

GRAVITATION
A natural consequence of a particle's emergence, gravity fields, depicted in section view by the diffused background, necessarily diffuse inward exponentially because of basic spherical geometry and the uninterruptible continuity of radiant electromagnetic energy.

Gravity fields' innate compounding causes that inward diffusion to always be at its least directly in between the particles and the bodies they surround at their common center of mass, C_{cm}, which is the same as their common center of gravity.

Mechanically pursuing equilibrium in the ever-decreasing density of their ever-compounding gravity fields, all bodies, be it particles or galaxies, are constantly pushed by the highest field density toward the lowest. This inexorably leads to runaway coalescing that ultimately ends with fusion reactions transmuting all matter back into the radiant energy it originated from.

Because gravity fields not only surround but also permeate all bodies, including atoms, depicted as the small spheres comprising the spherical bodies, their compounding simultaneously causes both coalescing and condensing at all scales consistent with Newton's law of gravitation: $F = G(m_1 m_2) / d^2$, where F is the "attractive" force, G is the gravitational constant, m the mass, and d is the distance between their centers.

The distance to their C_{cm} from m_1 is $d_{cm} = m_1 d_1 + m_2 d_2 / m_1 + m_2$, where $d_{cm} = 3(0) + 1(4) / 3+1$ or 1. From m_2, it'd be $1(0) + 3(4) / 3+1$ or 3.

C_{fd} indicates the location in between them where they share a common field density. The distance to their C_{fd} is opposite of or naturally reciprocal to their C_{cm}. Both their C_{cm} and C_{fd} could be interpreted as non-centrifugal Lagrange points where the gravitational influence remains in equilibrium.

Actual Lagrange points incorporate orbital motion's centrifugal force. It's not included in this example for clarity. If it were, their C_{fd} would become the L_1 Lagrange point that'd have to be closer to m_1 to compensate for the outward centrifugal force.

The distance to their C_{cm} and C_{fd}, their relative rate of motion toward each other, and their relative condensing, all remain proportional to their masses as they relentlessly gravitate in the ever-thinning density of their ever-compounding gravity fields, conceptually portrayed in the sequence of diagrams **1-4**.

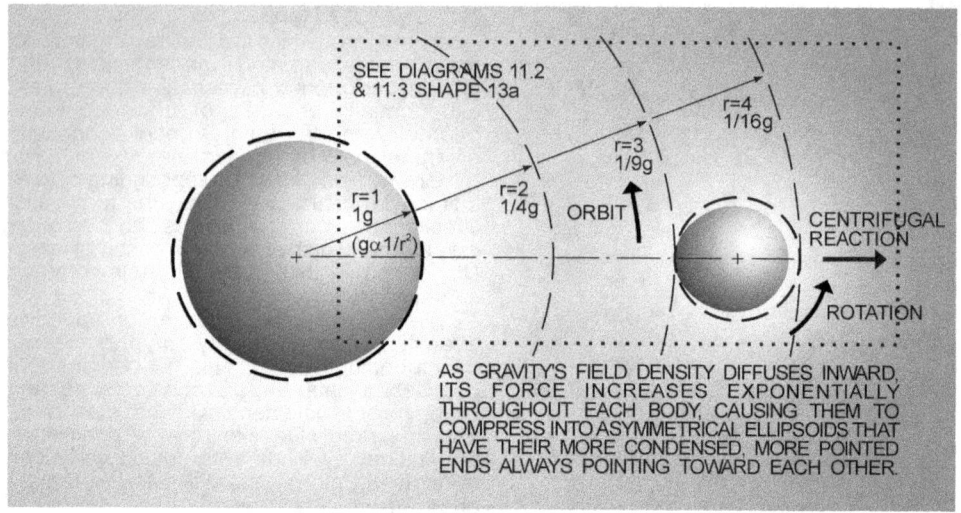

r=4
1/16g

r=3
1/9g

r=2
1/4g ORBIT

r=1
1g

(g∝1/r²)

CENTRIFUGAL
REACTION

ROTATION

AS GRAVITY'S FIELD DENSITY DIFFUSES INWARD,
ITS FORCE INCREASES EXPONENTIALLY
THROUGHOUT EACH BODY, CAUSING THEM TO
COMPRESS INTO ASYMMETRICAL ELLIPSOIDS THAT
HAVE THEIR MORE CONDENSED, MORE POINTED
ENDS ALWAYS POINTING TOWARD EACH OTHER.

GRAVITY'S CONDENSING (TOP-DOWN VIEW)

THE SHAPE OF GRAVITATING BODIES - 1

Bodies don't stretch or "spaghettify" as they gravitate. They continue to condense, contracting omnidirectionally into ellipsoidal shapes that are slightly asymmetrical similar to an egg. This is due to the exponential decrease in density of their compounding gravity fields that permeate each other's bodies, portrayed in section view by the diffusing background.

For simplicity, if we set the smaller body's diameter equal to the larger's radius and locate it three radiuses out then the larger's force of gravity, defined as 1g at its surface, would radially affect the smaller, sweeping across/through its entire body, exponentially decreasing from 1/9g at its closest point to 1/16g at its farthest, causing more condensing/compression at the closer end.

The smaller's gravity field would affect the larger in the same way but much less, creating a slight asymmetry in their condensing that has their more pointed, more compressed ends always pointing toward one another. Or more precisely, they point toward their common center of mass. This applies for any number of objects.

If the smaller body had a decreasing orbit or none at all, the asymmetry of its deformation would increase while it continued to condense/compress until they merged. If it had an increasing orbital and/or rotation rate where its outward centrifugal force began to exceed gravity's inward condensing, its material would begin to loosen, become weightless, and start dispersing.

But that dispersion would begin first from its backside, from its outermost point where the centrifugal force would be the greatest and gravity's compounded force would be at its weakest. This is routinely observed as the fanned dust tails of comets that always diffuse to the outside of their elliptical orbits opposite the Sun.

An obvious example of a body's asymmetrical ellipsoidal deformation is the Moon's, and to a lesser degree the Sun's, effect on the Earth's oceans. Water's pliability causes it to more readily deform than the rocky crust below, making its distortion easily perceivable. Tides are simultaneously at their highest both facing and opposite the Moon where they're slightly lower.

This distortion is not the product of the "pull" of the Moon's gravity as many believe. If it were, there'd be a gravity source on the opposite side pulling those oceans into their high tide. Those tides are often explained as the result of no pull, or sometimes more rationally but still incorrectly, the product of the Earth-Moon system's centrifugal force.

(11.1 Shape 13a)

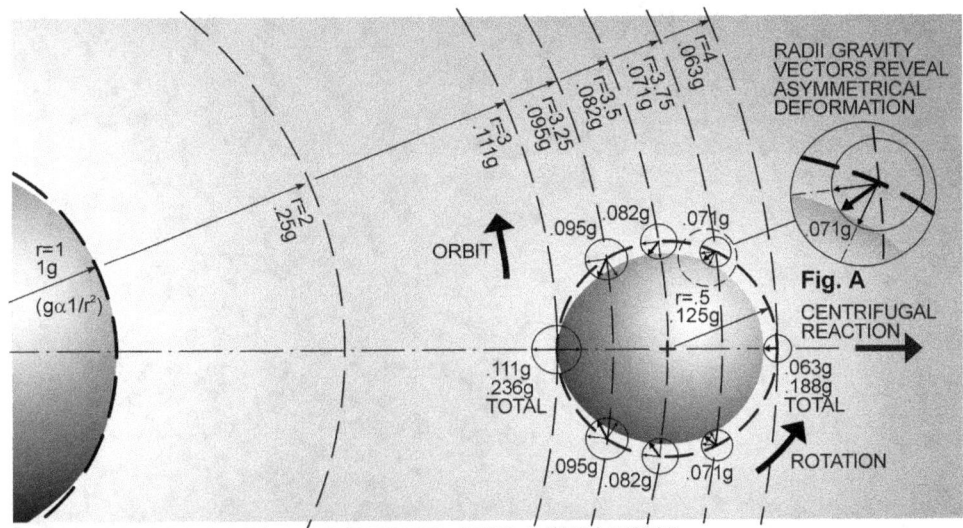

Fig. A

CENTRIFUGAL
REACTION

ROTATION

PLOTTING GRAVITY'S CONDENSING (TOP-DOWN VIEW)

THE SHAPE OF GRAVITATING BODIES - 2

The inward exponential diffusion of compounding gravity fields permeates gravitating bodies causing their ongoing condensing into evermore compact asymmetrical ellipsoids. This effect can be approximated by bisecting the angle established by the larger body's gravity expressed as proportional radii vectors to both bodies, as indicated in **Fig. A**. The bodies' uniform compression from their own self-gravity is assumed, represented by the heavier dashed circle.

Each body's volume ($V = 4/3\pi r^3$) can be used to establish their relative gravity. But any valve could be adopted. Setting the larger's radius at 1.0 (the unit of measure doesn't matter), its volume will be 4.19. For the smaller whose radius is half of the larger's, its volume will be .524. So the smaller body's volume will be about one eighth of the larger's. That's the ratio we'll use for their gravity: the larger, 1g, the smaller, .125g.

At 3.0 radiuses out, the bodies' closest point, the larger's gravity (g α 1/r^2), will be .111g $(1/(3)^2)$. With gravity's radii vectors (exaggerated for clarity) pointing in opposite directions, they counteract. So no distortion is produced. The total gravity here would be .236g (.111g + the smaller's .125g).

At 3.25 radiuses out, the larger's gravity decreases to .095g $(1/(3.25)^2)$. An approximation of the smaller's distortion at this location can be plotted by applying a gravity vector from the smaller body's surface that bisects the angle set by each body's radii.

Using the same method at 3.5 radiuses where the larger's gravity diffuses to .082g $(1/(3.5)^2)$ and at 3.75 radiuses where it's weakened to .071g $(1/(3.75)^2)$, the smaller's deformation at those locations can be charted.

At 4.0 radiuses out, the farthest point from the larger, its gravity diminishes to .063g $(1/(4)^2)$. A radii vector proportional to that gravity defines the outer limit of the smaller's condensing. The total gravity here will be only .188g (.063g + .125g).

With the total gravity at the outermost location always being less than the closest, the material of any body with an increasing rotational and/or orbital velocity will always begin to become weightless, dislodge, and start dispersing from their outermost point first as their increasing centrifugal force's outward dispersal begins to exceed their gravity's inward coalescing/condensing.

This simplified representation reveals how gravitating bodies distort into asymmetrical ellipsoids that continue condensing until they merge unless subject to high enough centrifugal forces that cause them to begin shedding material, which always occurs from their backside first.

(11.2 Shape 13a)

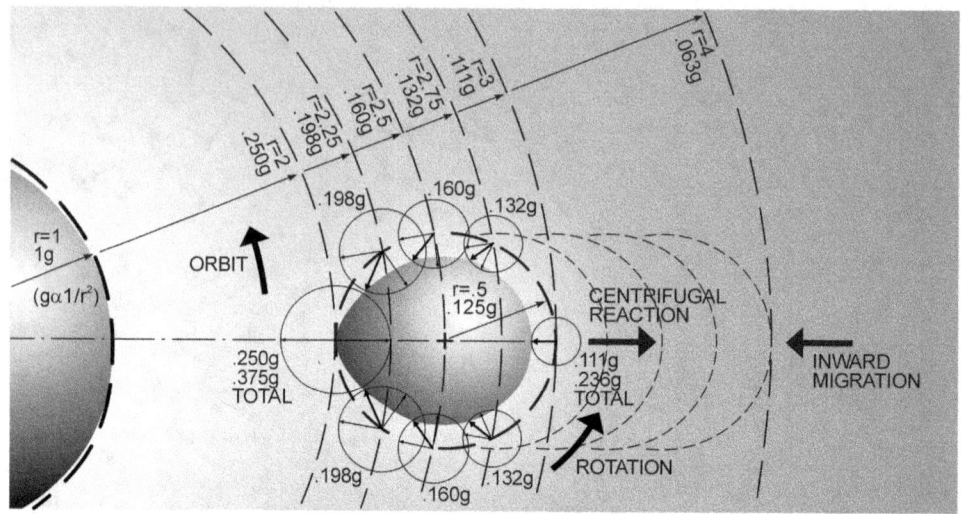

PLOTTING GRAVITY'S INCREASING CONDENSING (TOP-DOWN VIEW)

THE SHAPE OF GRAVITATING BODIES - 3

The ongoing contraction and increasing distortion of gravitating bodies can be conceptually demonstrated by graphically charting gravity's effect at the closer distance of 2.0 radiuses in the same manner that was done at the 3.0 radius distance (also exaggerated for clarity). The conclusion is self-evident: The increasing pressure from the exponentially increasing gradient that's produced by the inward diffusion of gravity's field causes ever-increasing asymmetrical ellipsoidal condensing. No stretching. No spaghettifying.

The relentless coalescing of gravitating bodies, that's a natural byproduct of gravity's inherent runaway nature, has to continue unabated until they merge unless increasing outward centrifugal forces begin to exceed gravity's inward coalescing/condensing. This causes the smaller body's material to begin to loosen, dislodge, and disperse, which is always initiated from the farthest point on its backside.

When the merging gravity fields of coalescing bodies create enough inward pressure, fusion reactions are triggered that begin converting their matter back into the radiant electromagnetic energy it originated from.

(11.3 Shape 13a)

340

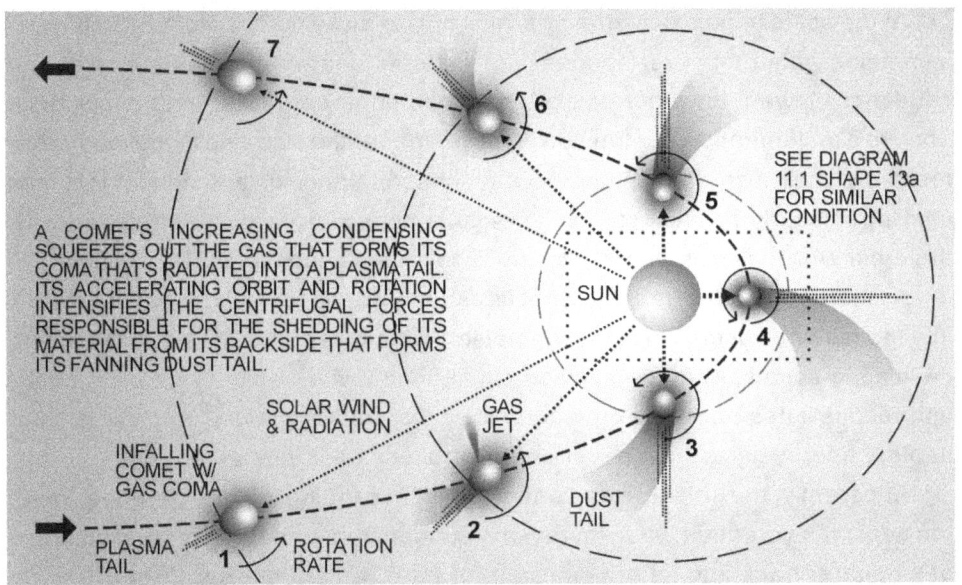

GRAVITY'S EFFECT ON COMETS (TOP-DOWN VIEW)

Within the figure:

7

6

SEE DIAGRAM 11.1 SHAPE 13a FOR SIMILAR CONDITION

5

SUN

4

A COMET'S INCREASING CONDENSING SQUEEZES OUT THE GAS THAT FORMS ITS COMA THAT'S RADIATED INTO A PLASMA TAIL. ITS ACCELERATING ORBIT AND ROTATION INTENSIFIES THE CENTRIFUGAL FORCES RESPONSIBLE FOR THE SHEDDING OF ITS MATERIAL FROM ITS BACKSIDE THAT FORMS ITS FANNING DUST TAIL.

SOLAR WIND & RADIATION

GAS JET

3

INFALLING COMET W/ GAS COMA

DUST TAIL

PLASMA TAIL

ROTATION RATE

1

2

THE SHAPE OF GRAVITATING BODIES - 4

1. A comet that was theoretically uniform and pliable would assume the shape of an asymmetrical ellipsoid that was continually condensing while its smaller, more compressed end always pointed toward the Sun due to the exponential decrease in the density of the Sun's gravity field, portrayed in section view as the diffusing background. That decreasing density in its gravity field at first gently squeezes out the comet's gas, mostly hydrogen, to form its coma, that may or may not have been evaporated/sublimated from internal material by the intensifying pressure and heat from its increasing compression. The Sun's radiant energy then begins to ionize the gas and blow it straight back to form the comet's plasma tail.

2. As its condensing continues, the gas is sometimes seen jetting out at high velocities. This confirms its internal origin that's more likely caused by the pressure originating from the increasing compression than the external heat from increasing sunlight.

3. The comet's increasing condensing also increases its rate of rotation, indicated by the increasing length of the radiused arrows. When its outward acting rotational and orbital centrifugal forces begin to exceed gravity's inward acting condensing, its material begins to dislodge, fall away, and disperse into an arcing fan shape to form its dust tail. This always occurs from the comet's backside opposite the Sun where the combined centrifugal forces are the strongest and gravity's compounded condensing is the weakest.

4. The comet's coma along with its plasma and dust tails continue to increase until it reaches its closest point to the Sun, perihelion, where its condensing and centrifugal forces and the Sun's radiant energy are all at their maximum.

5. As it begins to leave the Sun's vicinity, the now increasing density of the Sun's gravity field begins to reverse the comet's condensing/contraction that in turn slows its rate of rotation. Together with its slowing orbital velocity, its rotational and orbital centrifugal forces weaken, curbing its loss of material, which reduces the size of its dust tail.

6. As it continues to move farther away, the compression responsible for its outgassing also eases while the solar wind and radiation diminish. This reduces the size of its plasma tail as well.

7. The comet's decompression proceeds while its rotation and orbital velocity continue to decrease all the way to its aphelion, its farthest point from the Sun, where the sequence begins again.

(11.4 Shape 13a)

At the scale of galaxies, gravity's inherent runaway nature produces an ever-increasing infall of ever-condensing material that ceaselessly coalesces/condenses inward toward a galaxy's common center of mass, not a black hole. They're a mathematical abstraction that's permanently relegated to the theoretical realm by their two-dimensional, funnel-shaped, nonexistent space that has infalling objects impossibly stretching, "spaghettifying," not condensing spherically, three-dimensionally, as they endlessly accelerate descending toward infinity.

As material nears a galaxy's core, its exponential condensing collapses it back into the radiant/plasma energy it originated from and radiates back out. Or in well-developed spirals, it's often spewed out in huge jets. Eventually it slows, cools, and reconstitutes back into ordinary matter that begins gravitating back to its or another nearby galaxy in a never-ending process of perpetual recycling.

Apparently, the universe has one singular overriding inexorable imperative: the ceaseless creation, ever-increasing condensing, and the eventual collapse of all matter back into its original state of pure radiant energy. (See Diagram 19.1 G Recycling 5a)

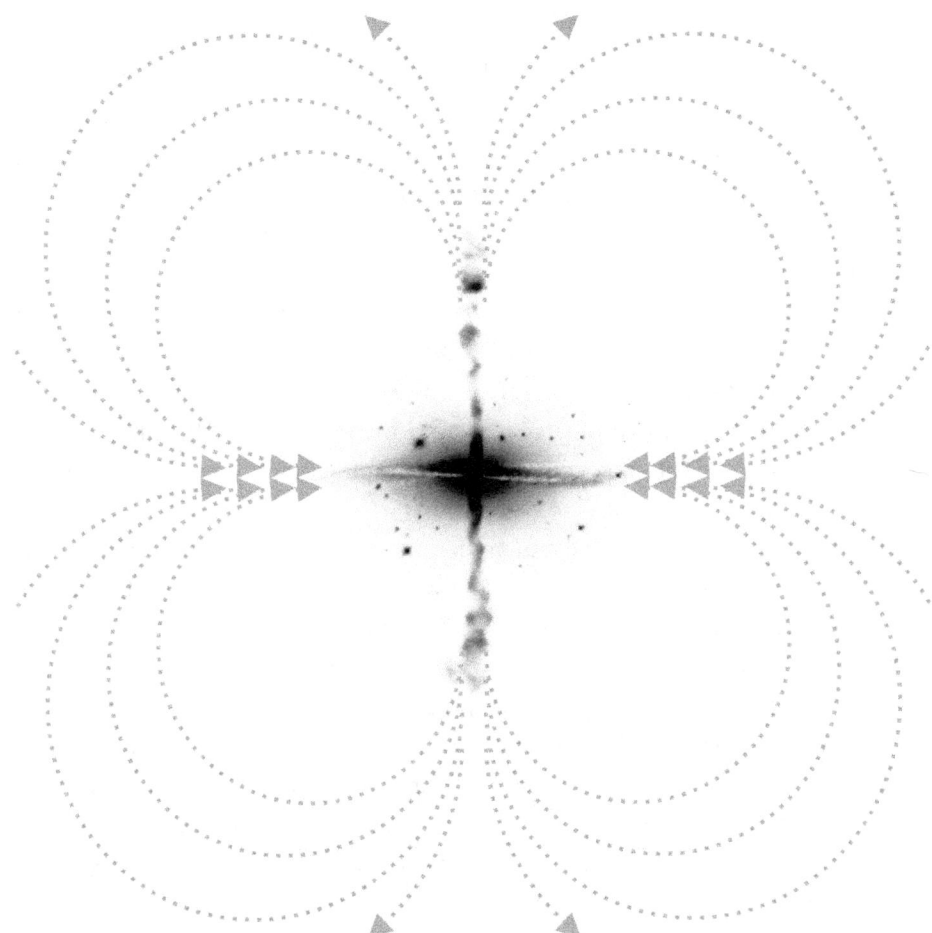

GRAVITY'S PERPETUAL RECYCLING

At the largest scale, protogalactic material begins coalescing spherically, three-dimensionally. Any asymmetry causes it to start rotating, which begins to form an axis. Because the resulting centrifugal force is less at the ensuing poles, it coalesces more readily down the axis toward its center while the more perpendicular material tends to flatten out into the whirlpooling disks that define a fully developed spiral galaxy, shown in side view.

But its coalescing doesn't stop here. Gravitation is a runaway process. It continues unabated facilitated by its exponential condensing that naturally allows its vortex to flatten out. As the material spirals in toward the galactic center, the compression becomes so great that fusion reactions are triggered that begin converting it back into the radiant energy it originated from.

It's then ejected out in huge, high-velocity jets in a bipolar fashion, shown as visible. As the radiation slows and cools, it eventually congeals back into ordinary matter and begins gravitating back toward its or another nearby galaxy, the dotted arrows. The material of every galaxy is subject to this never-ending process of perpetual recycling, which is the real source of cosmological redshift: the Doppler effect from the material's ever-increasing recessional velocity from every other galaxy coupled with our own infall velocity at our own galaxy.

The image is a highly modified black and white negative of galaxy ESO 510-G13 taken by NASA and the Hubble Heritage Team with the HST.

(19.1 G Recycling 5a)

Our solar system is positioned about two-thirds the distance from our galaxy's center. As it free falls in a flat whirlpooling vortex, it's slowly spiraling inward, continuously condensing with the rest of our galaxy's material. Both the Earth's size and orbit along with all the other bodies in our solar system are also slowly condensing, which is increasing their natural frequency. An outside observer might interpret our rate time as increasing.

For us, though, our condensing, the Earth's increasing rotation rate, its smaller and faster orbit, the quickening of time's passage, and our increasing weight would always appear normal, completely the same without change. Any variation in these properties would essentially remain imperceptible. The effects of our contraction and that of our entire solar system would be omnipresent and omnidirectional.

This means that we, along with everything else in our immediate environment, or reference frame, would be contracting in unison nearly equally in all directions, making it virtually impossible to ever directly perceive any of these changes. There's no physical way to maintain a record of our previous condition in a previous time period for comparison in the current time period.

Contrary to Einstein's unworkable, non-Euclidean assertions for his curving, two-dimensional gravitation, our normal Euclidean geometry is always preserved. All measuring devices and techniques are contracting and light's velocity is slowing in unison in gravity's ever-decreasing field density. The most we could ever perceive is an unaccounted for slowing in the orbital and rotational rates of other bodies outside our solar system in a different reference frame. It may be possible that at some point we might hold memories of a slower time or feel like life's pace is quickening. (See Diagram, **33** Vortex 6a & **44** Distance 8a)

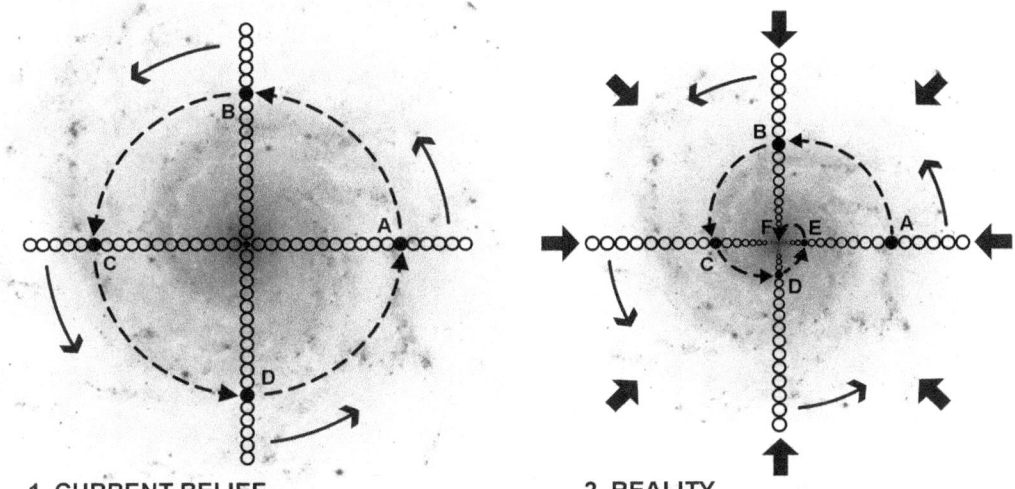

1. CURRENT BELIEF   **2. REALITY**

A SPIRAL'S FLAT WHIRLPOOLING VORTEX

Both diagrams depict a top view of our or any typical spiral galaxy. The small circles arrayed linearly represent a line of stars. Our location would be about 2/3 the distance from the center at **A**. Locations **B**, **C**, **D**, **E** & **F** indicate our position with every 90° of rotation.

Diagram **1** reflects our current belief that stars orbit in a circular fashion around a central supermassive black hole as if in a solid disk of material due to some extra unseen dark matter. Stars don't comprise a galaxy's spiral arms. They pass through them. The arms are fixed created by standing waves of some unknown origin. The centrifugal force resulting from a galaxy's rotation counters any inward migration. Stars only make their way to the central black hole after being displaced from their orbit by another nearby star or galaxy.

Diagram **2** portrays reality. Gravitation is inherently a runaway process. Because of the universal field's exponentially decreasing field density, not the impossible two-dimensional curvature of nonexistent space, all of a spiral's material is continuously coalescing and condensing exponentially as it infalls toward its center of mass, not a supermassive black hole, through the spiral's flat whirlpooling vortex. The vortex couldn't develop if its material wasn't compacting and contracting exponentially, as suggested by the decreasing size of the circles representing stars, and ultimately collapsing back into radiation or plasma that's spewed equally out the top and bottom as high velocity jets that act as the spiral's drain.

Its vortex also couldn't rotate like a solid disk if its material wasn't condensing and compressing exponentially per the geometry of the area of a circle ($A = \pi r^2$). As the area decreases toward the center the amount of material condenses exponentially. This inward exponential distribution of mass simply explains the inference of invisible dark matter.

A galaxy acquires its rotation from the gravitational energy of its coalescing material. So the resultant centrifugal force can never exceed gravity's force and forestall its inward migration. This constant infall of material along with our own infall velocity produces a recessional velocity with a corresponding redshift, which is misinterpreted as every galaxy moving away from every other. The universe is not expanding. Each galaxy is continuously contracting, but remains essentially static through the perpetual recycling of its material that's constantly being replenished.

The actual recessional velocity between infalling stars of different galaxies varies depending on many factors like the galaxy's rate and direction of rotation, a galaxy's mass, the location and infall velocity of each's material, the line of sight angle of the infalling material, and a galaxy's gravitational motion. Compounding all this, is that determining the actual red or blue shifting of a galaxy's light is apparently as much art as it is science.

(33 Vortex 6a)

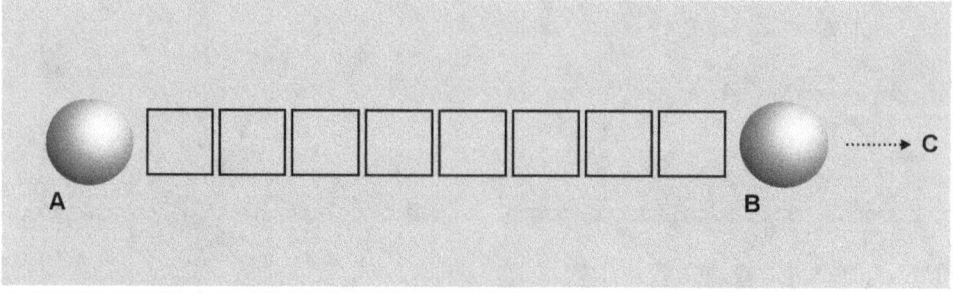

1. PERCEPTION: NO CONDENSING OR DISTORTION IN UNIFORM FIELD

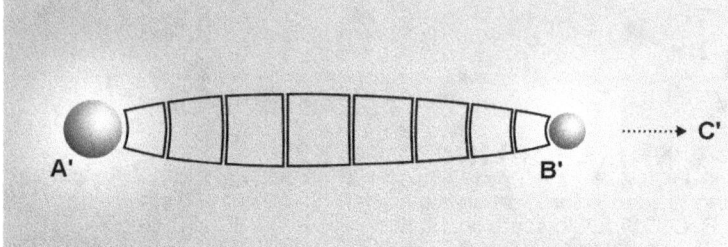

2. REALITY: OBJECTS CONDENSE & DISTORT AS FIELD DENSITY DECREASES

DISTANCE BETWEEN OBJECTS

Our perception is that distance and size are absolute, portrayed by the series of boxes in diagram **1**, that correspond to any consistent unit of measurement that remain the same any distance from a massive body, represented by the spheres, in a space that's uniform, represented by the grey background.

Diagram **2** conceptually portrays reality where the boxes and the spheres condense and distort, three-dimensionally, in gravity's exponentially decreasing field density, depicted in section view by the diffusing background. But to us, they appear the same size, shape, and distance apart as in diagram **1** because distance is a function of size. With measuring devises necessarily distorting the same as the object being measured, and with time's rate also varying with the varying size of the object if we're using the natural frequency of an object in the same vicinity to establish its rate, and with light's velocity changing with the changing density of gravity's field, then there's no way to perceive the distortion, contraction, and the reduced distances between objects, and our Euclidean environment remains intact.

The line of sight distance to stars nearer our galaxy's center, toward **C**, appears to be an absolute distance away, as if we were looking from **A** to **B** in diagram **1**. But from outside our galaxy, they'd measure a much closer distance, like from **A'** to **B'** in **2**. **C'** would be much closer also. If we were to attempt to travel from **A'** to **B'**, gravity's exponentially decreasing field density, which diminishes each consecutive unit of measurement exponentially while its reading remains constant, would make the trip as long as it appears from **A** to **B** in **1**.

This unperceived exponential gravitational condensing of mass towards our or any spiral galaxy's center explains the motion of outer stars that appear to be orbiting faster than their angular momentum should permit and how a spiral's material can be constantly infalling, swirling down an ever-tightening vortex, while at the same time rotating in a disklike manner, all of which being the product of the simple geometry of a circle.

(44 Distance 8a)

Einstein's misinterpretation of gravitation extends to acceleration/braking and rotation. He actually believes they create real gravity.[186] This is in addition to the gravity that's already innately created by an object's mass. He doesn't see that they're not the same or that multiple types of coexisting gravities would inherently conflict. Equivalency's fallacy was covered extensively in Section One, but it warrants a quick review in this context.

He reasons that if we were standing on the floor of an upward accelerating spacecraft, enclosed and theoretically free from any "preexisting" gravitational fields, it would be perceived like the gravity we experience on the Earth's surface. Because we could not tell the difference, all accelerating objects must be producing another gravitational field that acts opposite the direction of their motion.[187]

He reasons something similar for rotation. If we were standing near the edge of a rotating disk like a merry-go-round (or on the surface of any rotating body like the Earth), the outward pull of its centrifugal force could not be distinguished from real gravity. Despite that it only mimics the Earth's gravity in the most rudimentary way, he believes that because we wouldn't immediately be able to perceive the difference, they actually are the same. So all rotating objects must all be producing another gravity field that acts outward in addition to the gravity field created innately by their mass.[188]

This has him first accepting the existence of inertial mass from acceleration and inertial mass from rotation and gravitational mass from natural gravity. Then he infers that these inertial masses must actually be gravitational mass. They're different but they're the same.[189] Nothing contradictory about that either.

This is how he contrives his "principle of equivalence" despite that there's no such thing as either inertial mass or gravitational mass. He invents them. There's only mass (the amount of material an object has) that has inertia when in motion and weight when motionless in a gravity field like when resting on the surface of the Earth.

Our accelerating free fall, produced by gravity's decreasing field density that's resisted by the Earth's surface, only generally appears to be the same as if we were being pushed upward by the floor of the accelerating spacecraft with a one g acceleration or flung outward by rotation with the same centrifugal force. Neither acceleration nor rotation could ever generate a real gravity field.

Real, natural, mass-created gravity always manifests from the center of every mass, even the tiniest, or from the compounded common center of multiple masses. Its strength always dissipates exponentially per the geometry of a sphere,

186. Einstein, relativity, 73-74, 75-78, 79, 88-89, 172.
187. Einstein, relativity, 75-78.
188. Einstein, relativity, 88-89.
189. Einstein, relativity, 73-74, 172.

diffusing outward in an omnidirectional manner, three-dimensionally. The resulting distortion that's imparted to an object is completely different from the distortion that's created by acceleration or rotation. And acceleration and rotation's distortion is completely different from each other.

Natural, mass-created gravity doesn't require acceleration or rotation or any other type of motion. It causes motion. It's continuously coalescing and condensing, pushing objects together as they mechanically seek equilibrium in their compounding fields in a ceaseless runaway process that continues unrestrained until enough mass accumulates to trigger fusion reactions that convert them back into the basal radiant/plasma energy they originated from.

Established dogma maintains that gravitational attraction acts at the speed of light via waves by a force similar to electromagnetism. That attraction is supposedly also somehow simultaneously mitigated by unobservable graviton particles that theoretically exist physically without mass. If they actually were real-world particles, they'd have mass. So according to relativity, they wouldn't be able to act at the speed of light. They'd become infinitely large.

Acceleration's reaction is uniform throughout all locations of a reference frame. It acts mechanically, instantaneously, only in one dimension, opposite the direction of motion. It doesn't emanate from mass. It only requires acceleration. And it doesn't coalesce and condense objects like natural, mass-created gravity.

Rotation's centrifugal force acts outward mechanically. So it also acts essentially instantaneously but in only two dimensions, perpendicular to the rotation's axis. It disperses objects outward and becomes stronger with distance. It also doesn't emanate from mass. It does require rotation and becomes stronger with increasing rotation and distance. It also doesn't coalesce and condense objects.

How can acceleration-created gravity or rotation-created gravity be real when according to Einstein the rate of an object's motion or whether it even has any motion is a subjective choice of each observer? They'd be imparting or removing gravity from accelerating and rotating objects at their whim. Imagine the havoc that'd wreak. And that'd be in conflict with the choices of every other observer.

So any inferred equality between the "inertial mass" of acceleration and rotation and the "gravitational mass" of real gravity, that's redundantly stated in his "principle of equivalence," is imagined. It doesn't and can't exist in reality. Like so much else of relativity, these suppositions are only pertinent to "what if" theoretical discussions that have no relevance whatsoever to our real world. (See Diagram 13.1 Acceleration 8a, 13.2 Centrifugal 8a, 13.3 Distortion 8a)

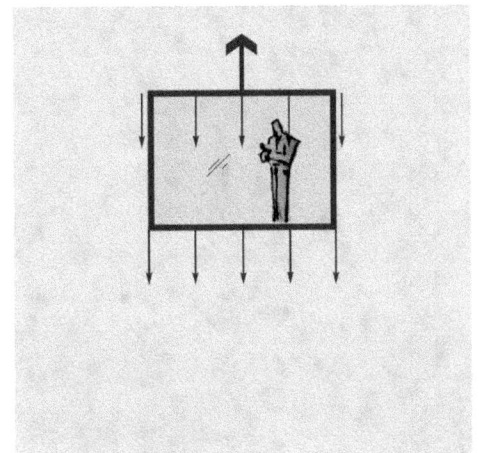

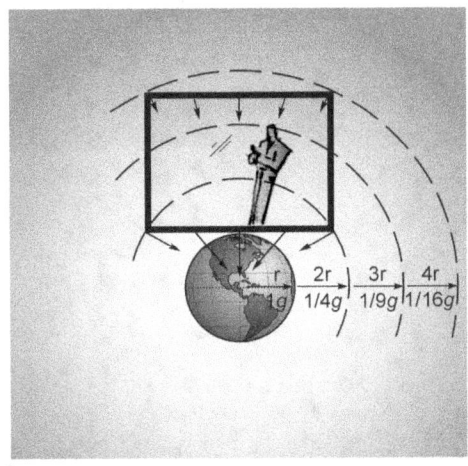

1. ACCELERATION

2. GRAVITY

ACCELERATION & GRAVITY

Einstein asserts that acceleration, which for him also includes braking as if a train were coming to a stop, creates an actual gravity field. But it can easily be shown that acceleration and gravity's reaction are not the same. Using an example similar to his, let's imagine that we've placed someone inside a large crate. But let's make ours transparent. We'll assume it's an independent reference frame.

For diagram **1**, when the crate is being pulled upward with a one g acceleration out in space somewhere theoretically free of gravitational fields, we can see how the reactive force, indicated with the smaller arrows that we'll say corresponds to weights on springs, has to be essentially instantaneous. It's mechanical in nature. This is contrary to Einstein's claim that gravity acts at the speed of light via waves by a force similar to electromagnetic radiation.

It's also equal everywhere throughout the crate and acts in only one dimension, opposite the direction of motion. This might initially appear or feel like gravity to the person experiencing the acceleration until they look outside or acquire the proper equipment to detect the uniform distribution of the reaction.

Now for diagram **2**, let's exaggerate our condition here on Earth and imagine that it's been compressed down to a few feet in diameter but where gravity's force remains the same at the surface. For clarity, let's keep our crate the same size and shape, compensating for any distortion that would also reveal that gravity is entirely different from acceleration.

Because the strength of a gravity field always dissipates exponentially per the inverse square law and because it always radiates spherically, three-dimensionally, from the center of a mass, gravity's force, as indicated by the varying length and direction of the smaller arrows that still corresponds to weights on springs, has to vary in strength and direction at every location within the crate as it rests on the Earth's surface.

Gravity coalesces. Acceleration doesn't. Gravity acts inward three-dimensionally and increases/decreases exponentially. Acceleration acts only one-dimensionally and uniformly. A cursory comparison easily reveals how gravity can never be created by acceleration, which completely undermines Einstein's principle of equivalence.

(13.1 Acceleration 8a)

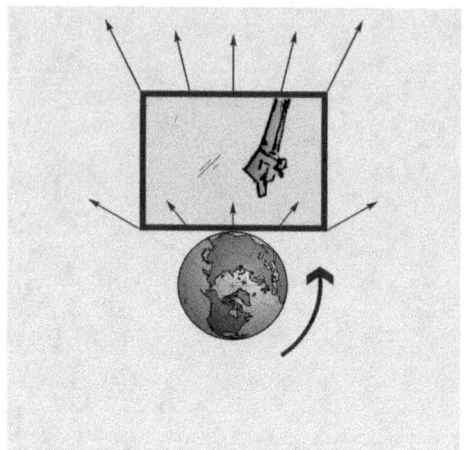

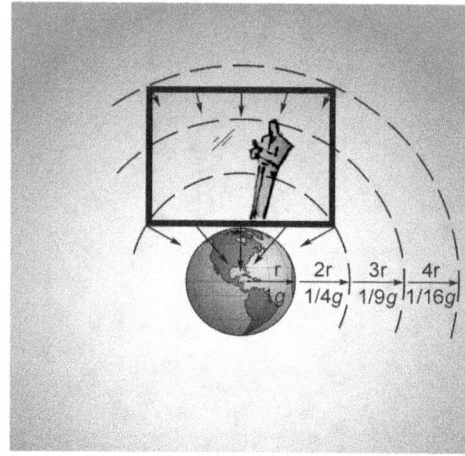

1. CENTRIFUGAL FORCE

2. GRAVITATION

ROTATION & GRAVITY

Einstein's inference of equivalency with gravity isn't limited to acceleration and braking. He actually believes that rotation creates gravity as well. But this too can easily be shown to be false. Rotation's centrifugal force and gravity are not anywhere near the same. Neither are rotation and acceleration. For our discussion, let's continue to use our reference frame that's similar to his and imagine that we've put someone inside a large transparent crate. But let's say that it's attached to the Earth that we've again compressed down to a few feet in diameter.

For diagram **1**, if we were to eliminate the Earth's gravity and set its rotation rate to where the centrifugal reaction approximated gravity's, we can see how that reaction, as indicated with the smaller arrows that could still correspond to weights on springs, would again be mechanical, essentially instantaneous. Which is still contrary to Einstein's assertion that gravity acts at the speed of light via waves.

Its reaction would vary throughout the crate like gravity's but only in two dimensions, perpendicular to the axis defined by its rotation. It would also act outward from the axis where it'd be nonexistent, becoming increasingly stronger farther out. It might appear like gravity to the crated person unless they could see outside or had the equipment to detect the reaction's opposite direction and two-dimensional dispersal.

For diagram **2**, let's again keep gravity's force the same for our shrunken Earth and let's eliminate its rotation for clarity. Let's also keep the crate's same size and shape as it rests on the surface, omitting any distortion for now for simplicity. The difference in the rotating crate's distortion, just like for the accelerating crate, would also reveal that they don't create gravity.

Because the strength of gravity always dissipates exponentially, radially, three-dimensionally, from the center of any mass, gravity's force, as indicated by the varying length and direction of the smaller arrows that could also represent weights on springs, would vary in strength and direction but at every location within the crate, three-dimensionally, not two-dimensionally. And its reaction would be inward, opposite of the centrifugal reaction, and weaken from the center out.

Gravity coalesces and increases inward, three-dimensionally. Centrifugal forces disperse and increase outward, two-dimensionally. Acceleration neither coalesces nor disperses and it acts uniformly in only one dimension. None are the same. This simple analysis clearly shows that rotation and acceleration do not create gravity, which again invalidates Einstein's equivalence principle.

(13.2 Centrifugal 8a)

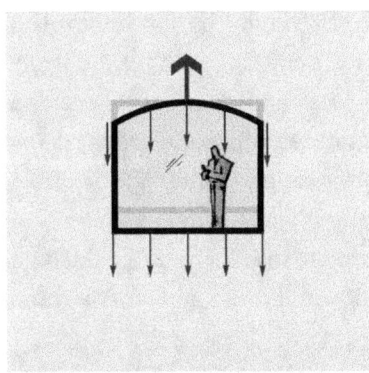

1. ACCELERATION/BRAKING

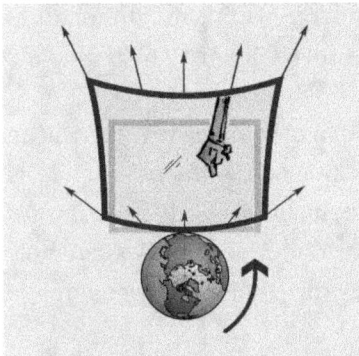

2. CENTRIFUGAL REACTION

3. GRAVITY

(13.3 Distortion 8a)

REFERENCE FRAME DISTORTION

To continue the argument, let's examine our reference frame's distortion conceptually when accelerating and rotating and compare the results to gravity's.

Imagine that our original crate is now theoretically no longer rigid but made out of some pliable rubber-like material that consistently stretches and compresses to some degree when subjected to external forces. The result indicated by the darker distorted crate that's superimposed over the original rectilinear crate that's lighter gray.

Diagram **1** shows acceleration/braking's one-dimensional stretching. If our crate was being pulled uniformly from the top, not from a single point as indicated, it'd remain rectilinear. If it was being pushed evenly from the bottom, it'd also remain rectilinear. But it'd be uniformly compressing.

Diagram **2** portrays the two-dimensional outward expanding diffusion of rotation's centrifugal reaction. Diagram **3** depicts the three-dimensional inward compressing of gravity's innate coalescing. It's impossible for them to produce the same distortion. This indicates that they are not at all experiencing the same force.

Try to imagine the result if you had a rotating reference frame that was subject to linear acceleration. Not only would its distortion from acceleration and rotation impossibly conflict with each other, but they'd also conflict with that of real gravity, which is always present for any quantity of mass, including that of subatomic particles.

They innately have mass and spin. So they have gravity and centrifugal reactions. And they're routinely accelerated. If equivalency were actually real, what would the physical effect be of commingling their compaction, diffusion, and stretching all at the same time?

This again easily establishes that acceleration and rotation do not and cannot create gravity fields. So Einstein's principle of equivalence is a fallacy.

Einstein further reveals his misunderstanding of gravity by his attempt to correlate a curving "ray of light" in a massive body's gravity field with his incorrect interpretation that a perpendicular ray of light curves in an acceleration-created gravity field. Under the pretense of arguing for light's curvature in the gravity fields of massive bodies, he contends that if a perpendicular ray of light arches downward in the acceleration-created gravity field of someone experiencing acceleration then because of general relativity (where the laws of nature must hold true for all reference frames regardless of their motion) it must also curve in the gravity field of a massive body.[190]

The elusive argument he's really making is that acceleration creates gravity to validate his "principle of equivalence." We also covered that in Section One. What we're more interested though here is highlighting his misinterpretation of how gravity works at massive bodies and acceleration. Neither of which produce light's actual curvature.

Einstein concedes that "The velocity of propagation of light varies with position [in gravity fields]."[191] It doesn't curve. Its change in position is the product of its refracted slowing through the field's decreasing density. It is not because it's following the impossible two-dimensional curvature of nonexistent spacetime's geodesic or because its photons are being "pulled" from their otherwise straight path by gravity.

His claim that light curves when its source is accelerating is false. He deceitfully portrays how a perpendicular ray of light would behave under acceleration to support his assertion that acceleration creates a gravity field. He realizes that if it doesn't, there'd be no "law of the equality of inertial mass and gravitational mass" or his redundant "principle of equivalence."

So he suggests that a ray of light projected perpendicular from the path of its accelerating source would arc downward like water streaming out from a hose under the influence of gravity here on Earth. This is supposed to correspond to light's behavior through the acceleration-created gravity field. Increasing velocity would cause increasing downward bending that would correspond with increasing gravity, which would validate acceleration-created gravity and in turn his "principle of equivalence."

But any defined ray of light would always propagate perpendicular in a straight path from its moving source whether its velocity was increasing or constant. It would never arc downward. If we quantify the light into a series of projected photons, each individual photon consecutively connected would be seen as

190. Einstein, *Relativity*, 83-85.
191. Einstein, relativity, 85.

defining an inverted arc with decreasing curvature. But each of their individual paths would actually remain straight relative to their point of origin and perpendicular from the source's path.

Because light, in reality, doesn't curve under acceleration then according to his reasoning this indicates that no gravity field is present. Without acceleration-created gravity fields there can be no law of "the equality of inertial mass and gravitational mass" or its invented duplicate, "the principle of equivalence." (See Diagram 14 Ray 4a)

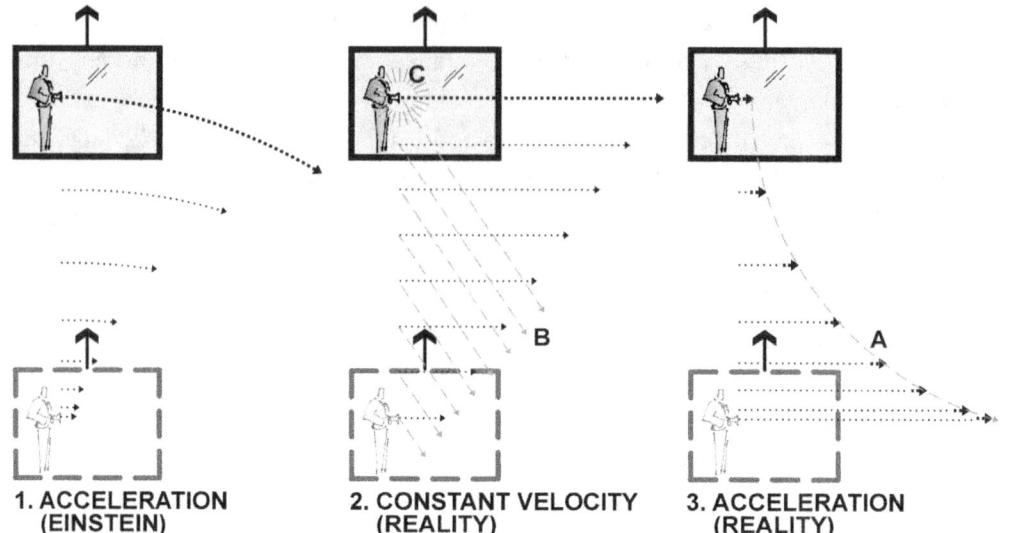

1. ACCELERATION (EINSTEIN)

2. CONSTANT VELOCITY (REALITY)

3. ACCELERATION (REALITY)

LIGHT UNDER PERPENDICULAR ACCELERATION

Using our example that's similar to Einstein's, let's again imagine that our transparent crate is being pulled upwards with a one g acceleration somewhere out in space theoretically free of any gravity fields so that the person standing inside would not be able to perceive the difference between the crate resting on the ground or its upward acceleration.

Einstein contends that because the conditions appear the same they must be the same. He argues that the crate's uniform acceleration must actually produce a real gravitational field. In a backhanded way, he indirectly correlates how a perpendicular light ray transmitted by the upward accelerating person, as depicted in diagram **1**, would curve in a manner no different from the curving light rays passing through the gravity field of a massive body like the Sun.

But to make the argument, he misrepresents how a perpendicular ray of light propagates under acceleration. He implies that it moves upward in unison with the crate as it would under constant velocity, as shown in diagram **2**. And then he wants us to infer that it should be bending downward as if being "pulled" by gravity similar to how water arcs downward as it leaves a garden hose.

In reality, an upward accelerating source would produce a new quantum of light each moment, leaving behind the previous, each propagating parallel in the perpendicular direction with an upward velocity that would remain constant at the rate it was emitted, as indicated in diagram **3**. Once the light separates from its source, there's no force causing its upward velocity to increase.

Inferring an inverted, downward curving path, indicated by the dashed line at **A**, as a ray of light would not be correct either. It would be connecting different quanta of light propagating separately in parallel perpendicular paths. The angled dashed lines at **B** though could be considered a single ray of light.

Light emits radially in all directions from its source, not just in the direction the flashlight is pointing, as implied at **C**. Under constant velocity, it would still be moving in unison with its source. But it's not bent or in any fashion being pulled downward or following the presumed geodesic of nonexistent spacetime's impossible curvature.

Just as Einstein uses light's nonexistent curving under acceleration to infer a gravity field, we have to use its actual perpendicular propagation under acceleration as evidence of the absence of a gravity field. This has us concluding again that acceleration and gravitation are not the same. So his principle of equivalence cannot be valid.

(14 Ray 4a)

Given that space, if it were actually something, would be three-dimensional, applying the two-dimensional property of curvature to it is impossible. When coupled with his entirely unworkable assertions that accelerating/braking objects create another gravitational field that acts opposite their motion and their rotation also creates another gravitational field that acts outward, both in addition to the one innately created by their mass that has him concluding that inertial mass is gravitational mass, all of which comprise much of general relativity, and when integrated with the metaphysical assertions of special relativity that have moving or accelerating objects contracting in the direction of motion while their rate of time slows and their mass increases, all becoming infinite at the speed of light because of its incorrectly assumed and impossible constancy, then all these contradictory ideas truly become ludicrous.

Not only are they inherently flawed by themselves, but they're also at odds with one another. Some may have learned and memorized this dogma as fact, judging and professing that they've conquered relativity. But don't be fooled. Memorization is not the same as understanding. Relativity is and will always remain incomprehensible nonsense.

Einstein argues that because of light's variability at gravitational fields, which outright contradicts relativity's founding premise, its constancy, the phenomena of special relativity can only be considered when disregarding the influence of gravity. But that's physically impossible. Gravitational fields are felt over an infinite distance.

Objects themselves also manifest their own field of self-gravity, even in the subatomic realm. There are no locations or conditions that exclude gravity fields. So whether we're considering the vast distances between galaxy clusters or the interaction between subatomic particles, the effects of special and general relativity are commingled. There's no way around it. So they're fundamentally at conflict, which essentially undermines all of it. (See Diagram 48.1, 48.2 SR&G 6a)

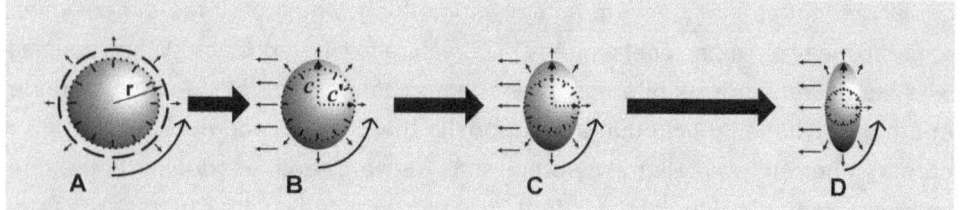

A B C D

1. RELATIVITY

SPECIAL RELATIVITY & GRAVITY

Let's say we could place an object like a heavenly body somewhere out in Einstein's two-dimensional curving but uniform space and let's superimpose a clock face on it that allows us to monitor time's changing rate and then let's give the object some rotation, as implied by the radiused arrows. He asserts that because light's velocity is fixed, the object's perimeter contracts as its time slows while its radius remains the same, depicted in top view at diagram **1.A** where the dashed circle indicates the object's original size.

If the radius of a circle is always proportional to its circumference, just how does that work exactly? Also, the velocity of its rotating perimeter varies from its equator to zero at its poles. This would yield a varying rate of time over the object's entire surface. And what about the velocity of light perpendicular or at any angle to a tangent at its surface? That would also give it an endless number of conflicting rates of time.

Einstein also claims that in addition to its self-gravity, the centrifugal force from the object's rotation creates another gravity field that acts outward, as denoted by the smaller outward pointing arrows. Now let's imagine we could give the object an increasing push that imparts some constant acceleration. He also contends that this would create another gravity field that acts opposite the direction of travel, as indicated by the array of small arrows at **1.B**, **1.C** & **1.D**, despite that its reaction is uniform and gravity's is nonuniform. So we have three types of conflicting gravity fields, one from self-gravity that coalesces three-dimensionally, one from rotation that disperses two-dimensionally, and one from acceleration that acts uniformly in only one dimension that neither coalesces nor disperses.

He also maintains that the moving object's light has to slow in the forward direction to maintain its constancy, as indicated by the dotted arrow at c', that becomes infinitely slow at 186,000mi/s, and the object's time has to slow or dilate concurrently, as portrayed by the shrinking clock-like circle that represents its changing rate, that also becomes infinitely slow at the speed of light, and the object has to contract correspondingly in the direction of motion, as depicted at **1.B**, **1.C** & **1.D**, that becomes infinitely small at 186,000mi/s as well. But how can time be slowing when light's slower forward velocity and the object's contracted size would require a smaller, faster running time to maintain light's constancy? And what about the light perpendicular to the object's linear motion, at c, or in any direction except directly forward? It's not required to slow to maintain light's constancy but is still subject to time's slower rate. This creates a conflicting velocity that exceeds 186,000mi/s.

Also because of light's fixed velocity, Einstein asserts through his mass-energy relation, $E = mc^2$, that an object's mass increases with its acceleration, becoming infinitely large at the speed of light. How can that be if light's fixed velocity also requires its contraction in the direction of motion to become infinitely small at the speed of light?

Beyond that, he actually avows that because of light's variability in gravity fields all of special relativity, light's constancy, time's slowing, an object's length contraction, and its ever-increasing mass, is only valid outside of gravity fields. How can any of it ever be valid when every object generates its own self-gravity field, and every rotating or accelerating object generates another gravity field? And let's not forget the gravity fields from every other object in the universe. They extend indefinitely. So there's no place where light's velocity can be fixed. But his assertion of light's variability itself undermines nearly all of relativity. It's founding premise is light's constancy.

(48.1 SR&G 6a)

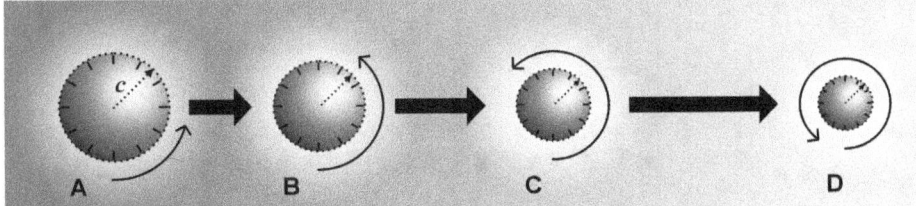

A   B   C   D

SPECIAL RELATIVITY & GRAVITY - CONTINUED

You have to also remember that Einstein asserts that it's a completely subjective choice by each observer as to which object is in motion and their relative rate of motion. That even includes heavenly bodies subject to gravity. So right now I can choose that the Moon is no longer orbiting around the Earth but that the Earth is orbiting the Moon. And this is perfectly acceptable to him despite that my choice conflicts with the laws of gravitation and everyone else's choice, which is also deciding light's velocity, time's rate, an object's contraction in the direction of motion, and the amount of mass it has.

But there is a real-world way that an object in motion can contract and its rate of time can change while light's velocity appears to remain constant. Let's take our same object and instead of placing it in Einstein's nonexistent space that impossibly curves two-dimensionally, let's put it in a real gravity field, the universal field of radiant energy of the electromagnetic spectrum that innately diffuses inward exponentially at each object, as represented in section view by the varying background in diagram **2**. Objects naturally gravitate toward one another as they mechanically seek equilibrium in its decreasing density that's always at its least directly in between them or at their common center of mass for multiple objects.

Pushed by the higher density, the rotating object begins accelerating toward the lessor density while contracting in size in a nearly uniform omnidirectional manner, as portrayed in top view by the sequence **2.A, 2.B, 2.C & 2.D**. Its contraction increases its natural frequency, which increases its rate of rotation, as indicated by the increasing length of the radiused arrows. If its rate of rotation is used to establish its rate of time like the Earth's is for us then it could be said that its rate of time is increasing with motion, as the contracting clock face implies. Light's velocity also slows correspondingly in the gravity field's diffusing density, as suggested by the decreasing length of the dotted arrows at c. This renders its variability along with the object's contraction imperceptible for an experiencer, which maintains a normal Euclidean environment.

(48.2 SR&G 6a)

Size Is All That Matters

It's now widely accepted that supermassive black holes lurk at the center of nearly every galaxy. It's never questioned but just taken for granted that anything that draws too close is pulled over its so-called event horizon and devoured. At which point it becomes permanently disconnected from the rest of the universe, lost forever as it endlessly falls toward the "singularity" at the hole's center. Light too is supposedly trapped forever unable to escape the hole's mass, hence its blackness.

Despite their near universal acceptance, the whole notion of a black hole is fundamentally flawed from the beginning. Since radiant energy is not matter, it doesn't have mass whether it's quantized into a photon or not. So it can't be affected by gravity's alleged pull. Yet, light does slow in a gravity field because of its refraction. As the field thins, light's velocity decreases. The density of a gravity field could theoretically thin indefinitely.

So light's commensurate velocity could never come to a complete stop. It could only do so theoretically in a field of zero density where the field would cease to exist altogether, a place where nothing existed at all - a conceptual impossibility. There is no place where there's nothing. There's always radiant energy. So light's velocity can only approach zero. It can never attain it as the field's density approaches zero. Its slowest velocity occurring at the center of mass of galaxies.

The real reason black holes are never observed is not because light can't escape their tremendous pull, a condition that conveniently makes them inherently unobservable, it's because there's no such thing. They don't exist. They're incorrectly inferred from math. Or maybe it's more accurate to say they're invented with math to wrongly suggest infinite collapse. In reality, everywhere we believe black holes exist, there's actually nothing there but a common center of mass, or a common center of gravity, same thing, that's established by coalescing material, usually a galaxy or a cluster of galaxies.

You'd think it'd be obvious that the whole concept of a black hole, which has infalling material endlessly falling down an impossible two-dimensional funnel of inconceivable four-dimensional, curving spacetime toward the infinite condition of a mathematical singularity, is an untenable idea. But it isn't. It is entirely reasonable, though, to infer that the ongoing coalescing and ultimate collapse of condensing infalling material is transformed back to pure radiant energy as it approaches the galactic center. That is an easy and obvious notion to grasp.

The radiation from increasing nuclear reactions that result from the increasing compaction of infalling stars radiates outward in all directions for elliptical galaxies but is expelled mostly as highly focused, bipolar jets from the active nuclei of mature spiral galaxies. These jets can be huge and extremely powerful, sometimes spanning more than a hundred times the galaxy's diameter and are thought to be exiting at nearly the speed of light. Some have even been measured at speeds many times that of light. But of course, this is explained away as an illusion to preserve the relativity myth.[192]

This seemingly unrestrained energy is envisioned as originating from the heat generated from the collision of material that's thought to be rapidly swirling around the black hole just before it disappears over the event horizon where it's channeled up and out by the galaxy's magnetic fields. Some believe that the gaseous material is somehow ejected at right angles to a quickly rotating accretion disk just before encountering the event horizon. And others even suggest that infalling gas could radiate sufficient energy to account for these gigantic jets.[193]

But all of these speculations avoid confronting the obvious. They fundamentally conflict with the whole basis for black holes. How can a huge constant outflow of radiation and plasma be originating from a source that supposedly endlessly devours and permanently traps all infalling radiation and matter? Taken at face value, our observations suggest that black holes actually expel not consume.

Beyond the fact that none of these solutions adequately explain how the energy is ejected in the first place, its enormous quantity, tremendous force, and extreme velocity does not reasonably correlate with these imagined processes. But it would be entirely consistent with the notion of increasing radiation from increasing nuclear reactions resulting from the increasing pressure that's produced from the ever-decreasing density of the galaxy's gravity field. It's the accumulated mass of all of the galaxy's coalescing material that determines the thinness of the field's density at the galactic center, which sets the rate of matter's collapse and its transmutation back to radiant energy and/or plasma.

But rather than rationally accepting this unremarkable, practical, commonsense conclusion that can hardly be rationally disputed, we're instead constrained by our indoctrinated fantasies and delusions. While at the same time, we're highly incentivized to maintain our entrenched doctrine.

So we concoct mysterious and exciting yet altogether preposterous black hole illusions. And because they're rooted in relativity dogma, we unquestioningly accept them as gospel, fantasizing that infalling objects pass over a threshold of no escape toward a condition that conveniently can never be unobserved where they begin to "spaghettify" while falling endlessly down an inconceivable, two-dimensional funnel of overly-curved, inconceivable four-dimensional nonexistent

192. "To set the record straight: Nothing can break the speed of light," Popular Science, last accessed Apr 2, 2023, https://www.popsci.com/science/whats-faster-than-the-speed-of-light.
193. "What powers a black hole's mighty Jets?" Science, last access Apr 2, 2023, https://www.science.org/content/article/what-powers-black-holes-mighty-jets.

spacetime never to be seen again, banished forever to another universe, or another dimension, or whisked off through a wormhole to pop out at some other location in our universe.

Or, they remain locked away indefinitely inside a tightly compact object that's somehow endlessly condensing to the infinitely small, infinitely dense condition of a singularity, which can only be a mathematical artifact. We prefer absolutely anything to what otherwise would be an unremarkable and mundane reality.

It does seem, though, that our ideas concerning black holes in general are slowly evolving toward something more rational. We may be inching our way toward more correctly realizing that black holes have to be interpreted as a regenerative process found at a galaxy's common center of mass where matter is simply being converted back to radiant energy, similar to interpretations that attempt to explain gamma ray bursts as the product of bipolar jets of hypernova.[194]

Hopefully, the evolution of our understanding will continue in this direction and eventually become the prevailing opinion. But don't bet on it happening any time soon. We enjoy playing make-believe with our black hole fantasies far too much.

Aside from our delusive preference for fantasy over the practical, our misunderstanding of black holes begins with our misinterpretation of galaxy evolution. The consensus appears to be that galaxies originate as spirals in the early evolution of our big bang universe and evolve into ellipticals only after colliding and merging with another galaxy.

The merging of any two spiral galaxies would tear apart their disks and the reassembling stars would clump back together first in an irregular shape that would eventually evolve into an elliptical galaxy that could remain indefinitely.[195] But the question quickly arises, how do spiral galaxies develop in the first place? And what is it that maintains an elliptical's shape, preventing it from collapsing in on itself?

Some believe that whether a galaxy becomes a spiral or not is only a timing issue where gas and dust would evolve into a disk before congealing into stars. If stars form first out of the gas and dust then the galaxy would remain elliptical. But since the same gravitational and centrifugal forces affect both the gas and dust and stars equally in the same manner, why is this even plausible?

It's much more logical that all protogalactic material first condenses together three-dimensionally, spherically, in an elliptical shape that evolves into a spiral with irregulars always being the product of collisions and mergers that eventually reassemble into an elliptical that continues to evolve into a spiral.

194. "Hypernova," Wikipedia, last modified Mar 10, 2023, https://en.wikipedia.org/wki/Hypernova.
195. "Galaxy formation and evolution," Wikipedia, last modified Mar 6, 2023, https://en.wikipedia.org/wiki/Galaxy_formation_and_evolution.

Just try to visualize the realistic coalescing-congealing process of gas, dust, and stars of protogalactic material. Since space really has three dimensions, not two, and the process of coalescing is omnidirectional then protogalactic material has to draw together from all directions, causing it to initially congeal toward a common center of mass in a spherical shape like a globular cluster or an elliptical galaxy. It could never coalesce first into the flat disk of a spiral galaxy. That'd be a two-dimensional coalescing process where the protogalactic material was initially arrayed in a plane.

Each star's, or dust particle's, or gas cloud's asymmetrical relationship with one another produces varying rotational velocities for that material that begins to sort itself around a developing axis with an equator and poles. Faster rotating material naturally migrates out toward an equator forming a disk. The slower material remains closer to forming poles that define the axis.

The lesser centrifugal force at the poles allows that material to coalesce more readily down along the axis, moving inward down from the top, and bottom, before the faster rotating material that's coalescing into the disk. Gravity innately creates a denser environment toward the galactic center that exponentially increases the weight of the inward migrating stars that increases their fusion reactions causing them to burn off their fuel more rapidly.

This radiates more of their energy outward, which is increasingly channeled into an ever-tightening fountain by the disk material. As the polar material's inward coalescing continues, the galaxy takes on a more elliptical shape, flattening and elongating along the equatorial plane, eventually becoming the near flat disk of a fully developed spiral with an active nucleus that will also continue to shrink to nothingness unless it has a source of new material to draw from.

The key to understanding galaxy formation is the centrifugal influence that naturally organizes the material around an axis with the slower rotating material near the pole that coalesces more readily down the axis than the faster rotating material that's collecting into a disk at the equator. This fundamental interaction between gravity and the centrifugal force is responsible for all galaxy types from the initial formation of globular clusters or dwarf elliptical all the way through to a fully developed flat spiral that at times spews huge, highly focused, bipolar jets.

In concept, this simple process that's solely the product of the interaction between gravitation and the centrifugal force is responsible for all bipolar expressions, including the collapsing stars like the Hourglass Nebula, MyCn18, or the gas plumes of Eta Carinae, and others. (See Diagram **34** Galaxy Formation 2a & **20** Bipolar 3a)

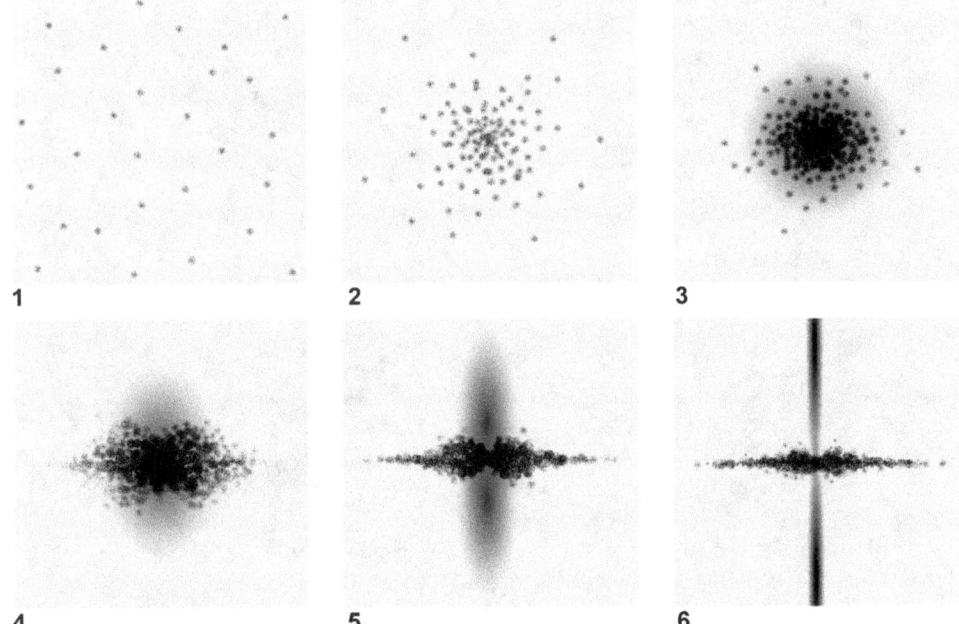

1   2   3
4   5   6

GALAXY FORMATION

Diagram **1** depicts a section view through protogalactic material, essentially gas, dust, and stars that may themselves be in the initial stages of forming. Gravity coalesces the material three-dimensionally in a spherical, omnidirectional manner around a common center of mass as a globular cluster or dwarf elliptical galaxy, portrayed in **2**. Gravity's exponential increase accelerates the infalling stars' condensing and collapse that increases the output of their radiant energy, which initially radiates outward evenly in all directions, indicated by the diffused shaded area in **3**. But because of each star's different mass and distance from one another, they begin to rotate at varying velocities, organizing around a developing axis with faster stars moving outward toward an evolving equator collecting into the beginnings of a disk while the slower remain near the poles as suggested in **4**.

Since the resultant centrifugal force is less at the poles, those slower stars are also coalescing more readily downward along the axis which evolves the galaxy into more of an elliptical shape while their radiation begins to be channeled out toward the poles by the faster rotating material that's coalescing into a flatter disk. As the stars near the poles continue their migration further down the axis, the galaxy compacts to the point that it could be classified as a spiral with a central bulge, implied in **5**. But as the bulge material continues its downward migration, evolving through a torus shape (like a doughnut), the galaxy finally transitions to a pure, full-fledged spiral with almost no bulge with an active nucleus that spews large quantities of radiant energy in highly confined jets, depicted in **6**.

As this radiation and plasma cools, it reconstitutes back into ordinary matter and begins gravitating back to its or another nearby galaxy, perpetuating an endless recycling of matter to radiant energy and back again. A galaxy may remain in a spiral stage indefinitely if it can continually draw in new material. Or it might dwindle down to its core and even go extinct after consuming itself if no new material is available. Or it may wax and wane over varying periods as material becomes available. If it collides with another galaxy, the resultant merger yields an irregular galaxy or galaxies that over time also coalesce around an axis and evolve a disk to rejoin the process.

(34 Galaxy Formation 2a)

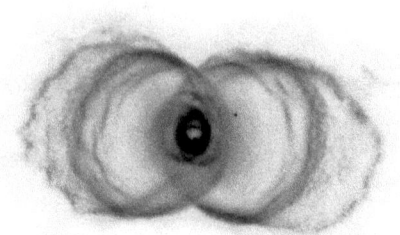

1. HOURGLASS NEBULA

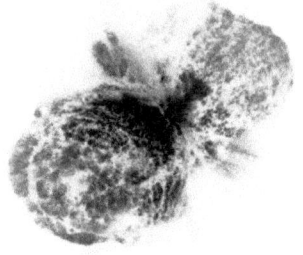

2. ETA CARINAE

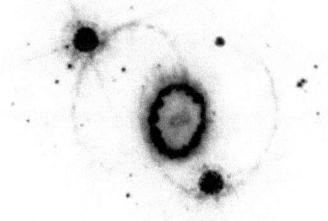

3. SUPERNOVA 1987A

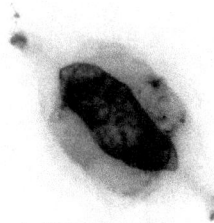

4. NGC 7009

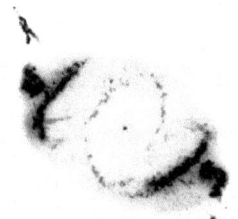

5. NGC 6543B

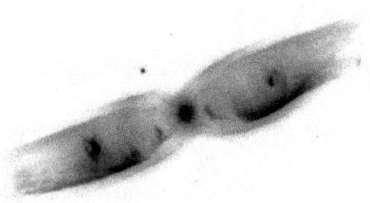

6. NEBULA M2-9

BIPOLAR CATALYST

All of the above photos of "planetary" nebula demonstrate some form of bipolar expression. With the counteracting centrifugal force of any rotating object, or many gravitationally bound objects like a galaxy's stars, always being less at the poles, all collapse manifests in a bipolar manner, beginning first and proceeding faster from the top continuing inward toward the center along the axis.

Each is an altered black and white negative taken by the HST unless noted otherwise. Credits: **1**. Raghvendra Sahai and John Trauger (JPL), the WFPC2 science team, and NASA. **2**. Jon Morse (University of Colorado), and NASA. **3**. X-ray: NASA/CXC/PSU/S.Park & D.Burrows.; Optical: NASA/STScI/CfA/P.Challis. **4**. Bruce Balick (University of Washington), Jason Alexander (University of Washington), Arsen Hajian (U.S. Naval Observatory), Yervant Terzian (Cornell University), Mario Perinotto (University of Florence, Italy), Patrizio Patriarchi (Arcetri Observatory, Italy), and NASA. **5**. J.P. Harrington and K.J. Borkowski (University of Maryland), and NASA. **6**. Bruce Balick (University of Washington), Vincent Icke (Leiden University, The Netherlands), Garrelt Mellema (Stockholm University), and NASA.

(20 Bipolar 3a)

For fully developed spiral galaxies with active nuclei, it's just not that difficult to envision how the constant coalescing and collapse of their infalling material increases exponentially toward the central region producing extremely powerful, high-velocity, bipolar jets and how those jets could be tightly focused by the disk's vortexing material. But instead of expanding forever, the ejected radiation and plasma slows and cools and eventually returns to a normal state of matter where it starts gravitating again, destined to return to the same or other galaxies.

The overall result being a large bipolar, fountain-like effect that's constantly replenishing each galaxy's material that's perpetuated indefinitely by gravity. The universal field's natural decrease in density at galaxies, or around any amount of matter, that's responsible for all gravitation can be interpreted as yielding a positive and negative charge that defines a current as well. (See Diagram 19.1, 19.2 G Recycling 5a)

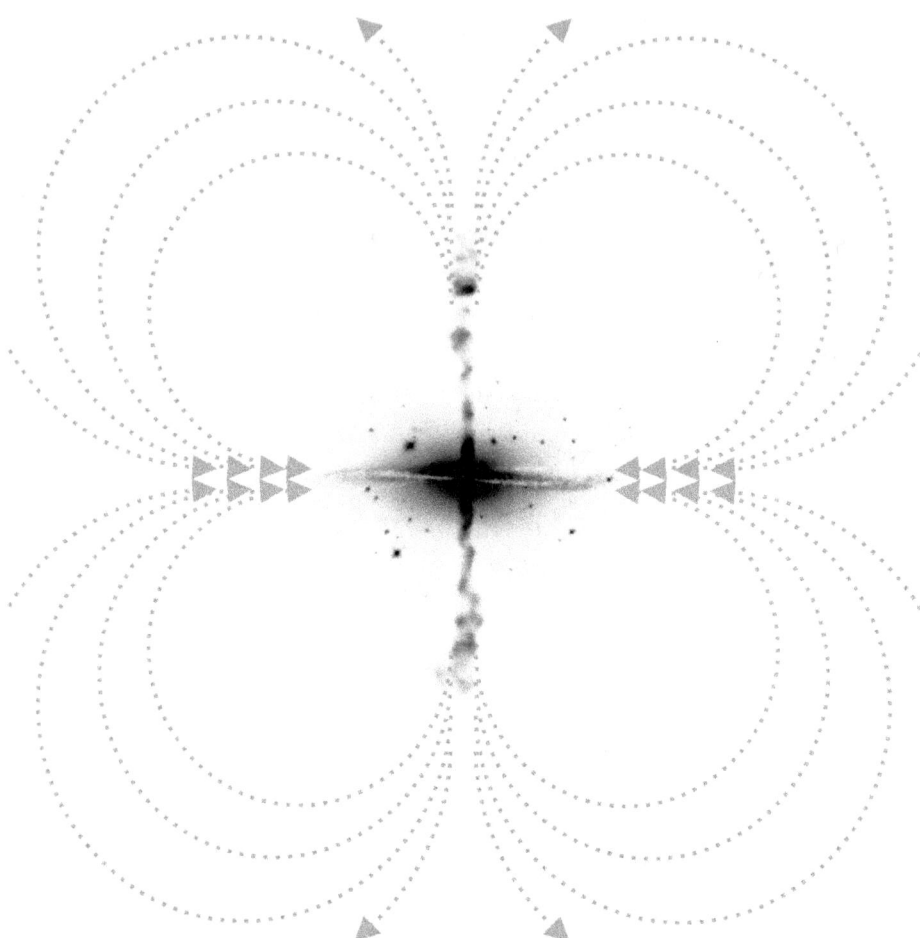

GRAVITY'S PERPETUAL RECYCLING

At the largest scale, protogalactic material begins coalescing spherically, three-dimensionally. Any asymmetry causes it to start rotating, which begins to form an axis. Because the resulting centrifugal force is less at the ensuing poles, it coalesces more readily down the axis toward its center while the more perpendicular material tends to flatten out into the whirlpooling disks that define a fully developed spiral galaxy, shown in side view.

But its coalescing doesn't stop here. Gravitation is a runaway process. It continues unabated facilitated by its exponential condensing that naturally allows its vortex to flatten out. As the material spirals in toward the galactic center, the compression becomes so great that fusion reactions are triggered that begin converting it back into the radiant energy it originated from.

It's then ejected out in huge, high-velocity jets in a bipolar fashion, shown as visible. As the radiation slows and cools, it eventually congeals back into ordinary matter and begins gravitating back toward its or another nearby galaxy, the dotted arrows. The material of every galaxy is subject to this never-ending process of perpetual recycling, which is the real source of cosmological redshift: the Doppler effect from the material's ever-increasing recessional velocity from every other galaxy coupled with our own infall velocity at our own galaxy.

The image is a highly modified black and white negative of galaxy ESO 510-G13 taken by NASA and the Hubble Heritage Team with the HST.

(19.1 G Recycling 5a)

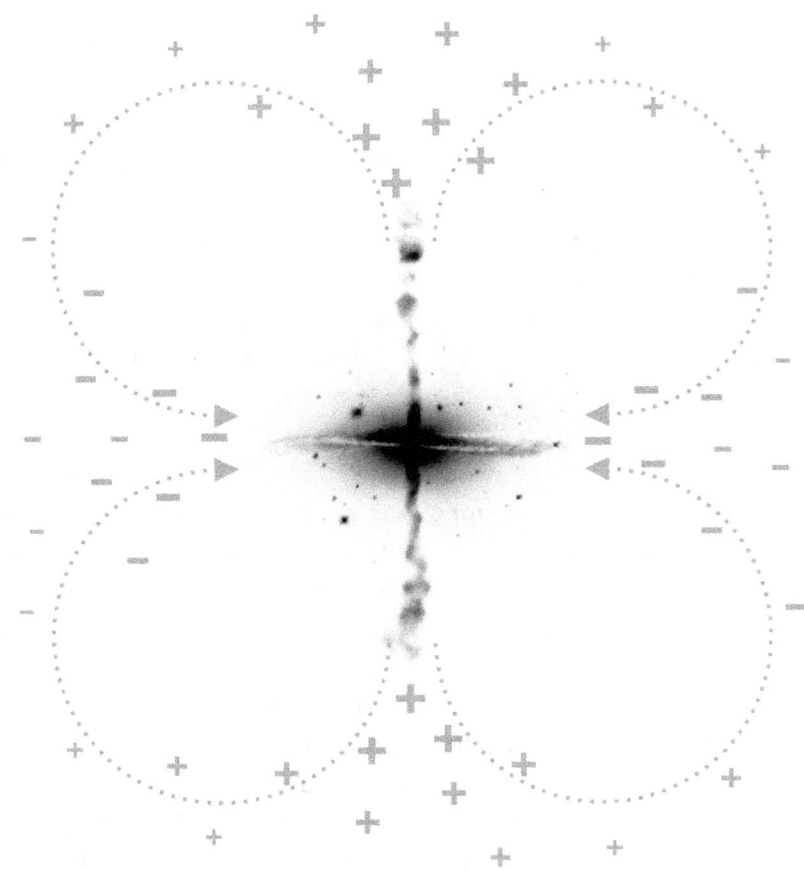

GRAVITY'S BIPOLAR CHARGE
The universal field's exponentially decreasing density that naturally occurs at every object, subatomic particle or galaxy, which is their (electromagnetic) gravity field, manifests a charge and current. This is explicit in the ambient field of mature spiral galaxies.

Their material constantly spirals inward toward their center along their plane in a flattened whirlpooling vortex that's facilitated by their gravity field's ever-decreasing density that's bound to the inverse square law of spherical geometry. That inward, ever-decreasing density corresponds to a negative charge.

The ever-increasing pressure from their gravity field's ever-decreasing density begins to set off fusion reactions near the galaxy's core that eventually convert all matter back into the radiant electromagnetic energy it spawned from. It's then spewed back out perpendicularly in enormous bipolar jets. These jets begin with the highest field density. So they're positively charged by comparison. But they continuously diffuse.

As the radiation slows and cools, it reconstitutes back into ordinary matter. It then naturally starts to gravitate back to its or an adjacent galaxy to repeat the perpetual recycling process. Opposing charges are at their greatest at the galaxy's center as the transformations occur.

The image is a highly modified black and white negative of galaxy ESO 510-G13 taken by NASA and the Hubble Heritage Team with the HST.

(19.2 G Recycling 5a)

All coalescing has to manifest in this basic bipolar fashion. Whether it's congealing protogalactic material or the condensing gas and dust of stars, or even the bonding of subatomic particles, it all has to obey the same underlying gravitational process that will always in some way express in a bipolar manner that will establish a charge and current.

Still, we can be sure that somewhere there exists a computer simulation that mathematically disputes this sensible inference of galaxy formation that somehow has protogalactic material initially coalescing into the flat two-dimensional plane of a spiral galaxy. We shouldn't be surprised, though. We have our entire universe expressing two-dimensionally as the surface of a sphere.

But computer modeling is nothing more than a reflection of biased input. Any simulation is full of parameters that require assumptions. Often they're just guesses. And often they're wrong. And often when simulations produce an undesirable result, rather than revising the theory to match the outcome we revise the values of the parameters, tweaking this constraint or that assumption adjusting until we obtain the desired result. Eventually, finding one way or another that our hypothesis was correct after all, deluded by the belief that it was confirmed by a computer whose high degree of accuracy can't be wrong.

But even if you reject this rational, commonsense modeling of matter's ongoing collapse and perpetual transmutation, you still have to concede that the fusion reaction of infalling stars has to be increasing exponentially toward the center of every galaxy. Gravity's exponentially increasing force, a product of its exponentially decreasing field density, causes an exponential increase in their rate of collapse, which exponentially increases their energy output. This dramatic increase in radiant energy is not even recognized, much less accounted for. Their answer is that stars are not constantly infalling.

This exponential increase in radiant energy at the galactic center explains why ellipticals tend to be brighter than spirals. When viewed from the side, a spiral's visible radiant energy is limited to what's emanating from the edge of its equatorial plane. So it'd be less bright. When viewed face-on, it's less bright than an elliptical because the disk has fewer stars collapsing at a lesser rate because of its two-dimensional, planar coalescing. So it inherently radiates less energy.

The three-dimensional spherical nature of an elliptical galaxy has its stars and gas condensing from all directions, not just two-dimensionally along a disk. So they're considerably more numerous toward the galactic center, which further increases their rate of collapse, which substantially increases the elliptical's brightness from all directions. It's the exponential increase for a three-dimensional spherical volume versus the smaller exponential increase for the two-dimensional area of a circle.

Because of our incorrect conflicted understanding of gravitation, but mostly because of our obsession with preserving our entrenched big bang-relativity ideology, we refuse to accept the obvious. Stars are constantly migrating inward, continually coalescing and collapsing toward a galaxy's center. Instead, we have them orbiting, circling endlessly around a presumed central supermassive black hole in a near fixed trajectory.

Despite the blatant whirlpooling vortex of spiral galaxies, it's believed that the centrifugal force produced by its rotation exactly balances the gravitational attraction produced by its coalescing material. It's argued that the orbital angular momentum (the product of the angular velocity of a rotating body and its moment of inertia) of any individual star increases as its infall velocity increases because its angular momentum must be conserved.

This supposedly continues until its orbital velocity becomes so fast that it begins to counteract gravity's tendency to attract, preventing any further migration to the galaxy's center. It's theorized that the resultant loss of angular momentum brought about by chance collisions between stars, which is believed to occur once every 10,000 years or so in our galaxy, is responsible for what would be a trickle of infalling material that comprises the assumed near static mass of its supermassive black hole.

The inference that stars endlessly revolve around a spiral galaxy's supermassive black hole in a circular orbit conflicts with the runaway nature of gravity. Both can't be true. Since it's the asymmetric gravitational coalescing of a galaxy's material that establishes its rotation around a common center of mass, it could never create the additional centrifugal force necessary to forestall its eventual merging. Extra energy would be required from an exterior source that increased its rate of rotation such as the effect a nearby galaxy might have. But even then this influence would only be temporary. The galaxies themselves are being pushed together and will also eventually merge.

What's also being missed is gravity's exponential condensing. It's the exponential compaction of material toward a galaxy's center due to gravity's exponentially decreasing field density that's not being perceived. This results in more mass at a closer distance, producing a constantly increasing infall of coalescing material that creates a two-dimensional whirlpooling vortex toward the galactic center that condenses and compresses it until it collapses where it's transformed back to radiant energy and ejected out the galactic center in a bipolar fashion, which acts as the whirlpool's drain.

For galaxy clusters, though, we do correctly interpret their motion as spiraling inward toward one another, but around a common center of mass, not a fantastically large black hole. Yet, we refuse to apply this same commonsense reasoning to the material of galaxies even though it is, in every way, exactly the same process of gravitational coalescing. We also assume that the material of the theorized accretion disk that's believed to surround supermassive black holes is constantly infalling, and the same with gas in the intergalactic medium.

But not a galaxy's stars despite that all are driven by the same exact coalescing process of runaway gravitation. We could accept that the ongoing coalescing of a galaxy's material might slow for periods of time as its asymmetrical distribution interacts with itself, either hastening or forestalling its impending infall. But gravitation's runaway nature will always win out, perpetuating the never-ending recycling of its material that might produce surges of collapse that correspond with increases in ejecting energy.

What about a galaxy's eventual merging with another galaxy or any assimilation of new material? Wouldn't that increase and redistribute its mass and disrupt the presumed balance between its centrifugal force and gravity, which could either increase or decrease its infall velocity that would either promote or inhibit its collapse?

For elliptical galaxies, which are thought to exist indefinitely, without any rotation to generate a counteracting centrifugal force, what is it that prevents their collapse from self-gravity? The same has to be asked of the central bulge region of spiral galaxies. Einstein even recognized this problem. That's the whole reason why he had to make his finite universe two-dimensional and invent a cosmological constant that would theoretically resist its uniform self-gravity and prevent it from collapsing in on itself.

And what about a star's continual loss of mass? All the stars across the entire galaxy are constantly burning off their mass as it's converted to radiant energy. With the mass of their radiation continually leaving the galaxy while the mass of the stars decreases, wouldn't their constant angular momentum and centrifugal force cause the galaxy to disburse? It's either that or their angular momentum and/or centrifugal force are continuously decreasing, exactly compensating for the loss of mass to maintain their circular orbit. This is highly unlikely, which indicates a condition that would always be out of balance.

If galaxies aren't disbursing but their stars are continuously losing mass, contracting/condensing, while apparently holding the same distribution pattern, doesn't this suggest ever-increasing infall velocity? It has to. Their increasing compaction and density toward a galaxy's center also suggest increasing coalescing, regardless of centrifugal effects.

And what about the effect expanding space would have? Wouldn't it upset any presumed centrifugal-gravity balance, causing a galaxy's outer material to begin disbursing? The conclusion has to be that many conditions exist that would disrupt the presumed balance between a spiral's centrifugal force and its self-gravity.

The spiral arms that compose the whirlpooling vortex of a spiral galaxy that extend all the way to its very center plainly evidence the continual infall of its material. But we refuse to recognize or accept the possibility that this conspicuous vortex could be an actual whirlpool of infalling material. Instead, it's thought that the spiral arms are the product of standing density waves from some unknown source that the orbiting stars pass through.

Gravity holds and delays them for some period of time while they're still continuing to move through to the other side. After they escape, they proceed on to the next arm. Still, we have no idea how or why these imagined density waves originate in the first place. It's reasoned that if the spiral arms were composed of a string of infalling stars and not the product of a wave, their faster orbital rate near the center would cause the arms to quickly disintegrate.

It's also believed that a spiral galaxy's material not just rotates around the galactic center in fixed circular orbits, but that it also maintains its relative position. It behaves as if it comprises a solid disk like a spinning phonograph record. With the centrifugal force increasing from the center out while gravity's decreases exponentially, how can the two ever synchronize to attain a counteracting balance? Conceptually, they'd always be mismatched, especially when gravity is innately ever-increasing.

The observation of stars near our galaxy's edge "orbiting" faster than the inferred mass of our galaxy should permit also suggests that this balanced condition doesn't exist. Something else must be occurring. The obvious inference is the constant infall of ever-condensing material that naturally occurs from gravitation's runaway coalescing is exponentially increasing the galaxy's mass toward its center.

A simple way to simulate the infalling material of spirals is to imagine that we're at a sink that's full of water. We've unplugged the drain and have water running from the tap at nearly the same rate that it's draining. A flushed toilet produces the same effect.

After the whirlpool motion has begun, we see the surface of the water forming a warping plane that curves downward as it falls through the drain. It'd be similar to the familiar two-dimensional, funnel-shaped grid that's commonly misused to represent spacetime's two-dimensional curvature around black holes.

If we could now somehow instantly stop the water's motion and suspend it in place and then compress its curved surface flat horizontally, we could then visualize how the molecules located along the curved surface would have to compress and contract to accommodate the flattening. The most compaction would correspond to where the surface curved the most, at the drain, and the least would correspond to the edge of the sink.

If we could put this flattened whirlpool back into motion, it'd exactly simulate the necessary exponential contraction and condensing of a spiral's material. It's eventual collapse back into radiation that's ejected out as high-velocity, bipolar jets becomes its drain. All of this is simply accomplished through gravity's exponentially decreasing field density. It begins with three-dimensional spherical coalescing that naturally evolves into a disk-shaped vortex with bipolar ejections. (See Diagram **35** Whirlpool 5a)

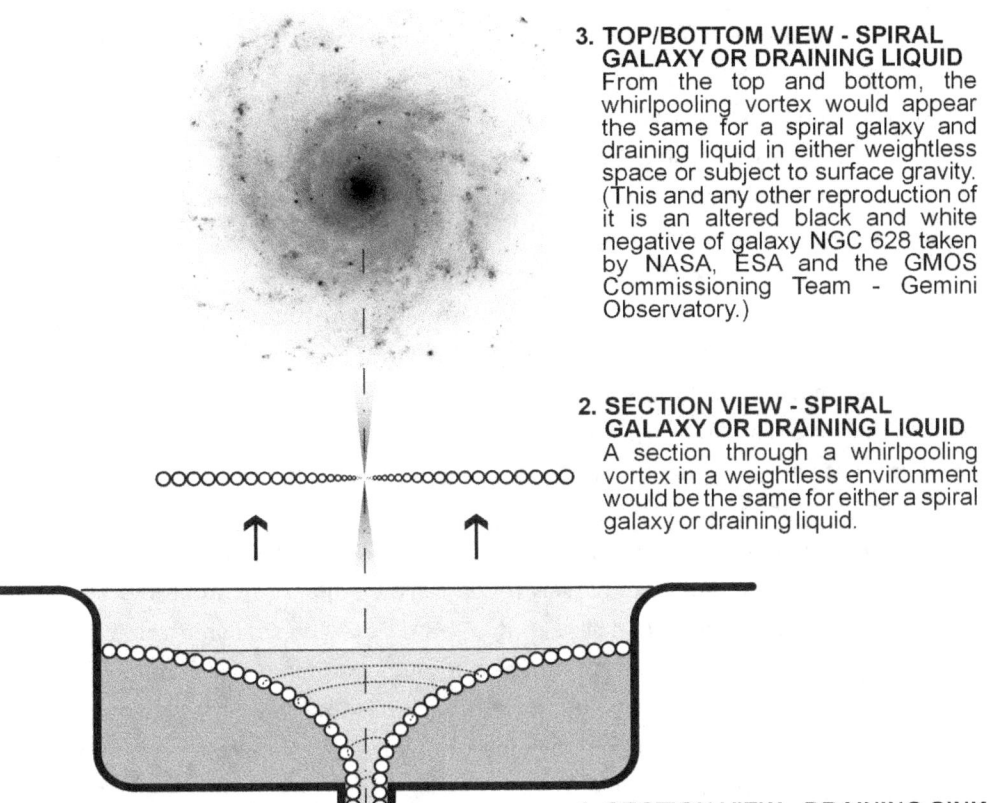

3. TOP/BOTTOM VIEW - SPIRAL GALAXY OR DRAINING LIQUID
From the top and bottom, the whirlpooling vortex would appear the same for a spiral galaxy and draining liquid in either weightless space or subject to surface gravity. (This and any other reproduction of it is an altered black and white negative of galaxy NGC 628 taken by NASA, ESA and the GMOS Commissioning Team - Gemini Observatory.)

2. SECTION VIEW - SPIRAL GALAXY OR DRAINING LIQUID
A section through a whirlpooling vortex in a weightless environment would be the same for either a spiral galaxy or draining liquid.

1. SECTION VIEW - DRAINING SINK

WHIRLPOOL VORTEX

The cross-section view of a sink in diagram **1** conceptually models how the molecules on the water's curving surface, depicted as small circles, remain a constant size as they drain out in a whirlpooling vortex. But if you could project those same molecules straight up into the weightless environment of space, portrayed in diagram **2**, where they'd be subject to their combined self-gravity three-dimensionally just like the stars of a spiral galaxy, they'd have to physically compact exponentially to be able to spiral in toward their center of mass before they could "drain" out of both the top and bottom equally in a bipolar manner.

This simple illustration shows how the flat whirlpooling vortex of a spiral galaxy, like that at **3**, could only develop if its material were continually contracting and condensing as it spiraled in toward the galaxy's center and had somewhere to drain. The exponential thinning of a galaxy's gravity field exponentially increases the pressure on its infalling stars exponentially increasing their fusion reactions that are transforming their material back to energy/plasma that first radiates out as a galactic halo but ultimately is ejected, or drained, out as high velocity, bipolar jets that spew from active nuclei. There are no unseen black holes lurking at the center of galaxies endlessly devouring matter. There's only a common center of mass where matter is being converted back to radiant energy.

It's the inward diffusion of its gravity field that's condensing and collapsing all of its infalling material exponentially that facilitates its whirlpooling vortex while keeping the rotation rate of its material nearly equal like a solid spinning disk. This simple realization of basic geometry completely resolves the mystery of any so-called "dark matter."

(35 Whirlpool 5a)

Do not be deceived by the simplicity of this illustration. The important idea to be gleaned, in addition to matter's perpetual recycling, is the fact that the flat whirlpool vortex of spiral galaxies could not physically develop if its coalescing, infalling stars were not continually condensing. It's the ever-increasing pressure of the ever-decreasing density of their galaxy's gravity field that's transmuting their material back into radiant energy that's radiated and/or ejection back out, which acts as the vortex's bipolar drain.

It's just not possible for a galaxy to ever find an equalized state between its infalling coalescing material and its outward centrifugal force. Only the most conditioned mind could look at a photo of any spiral galaxy with its coiling spiral arms of material plainly arrayed in an ever-tightening whirlpool and not infer the obvious: that its matter is constantly infalling to the galactic center through the motion of a continuous vortex.

Instead, we decide, even contrary to the basic principle of runaway gravitation, that a galaxy's material must endlessly revolve in a circular orbit around its center in a static, balanced condition like that of a flat disk of solid material, never coalescing or collapsing, establish that as a premise, and then invent an invisible dark matter to explain away the incongruities. Most laypeople would be astonished to learn that this is actually the orthodoxy of mainstream astronomy. (See Diagram **36** Spiral Galaxies 3a, **37** Galaxy Density 4a, 31 Condensing 7a, 33 Vortex 6a, **38** Orbital Velocity 3a, **39** Spiral's Mass 5a, **40** Elliptical's Mass 5a & 19.1 G Recycling 5a)

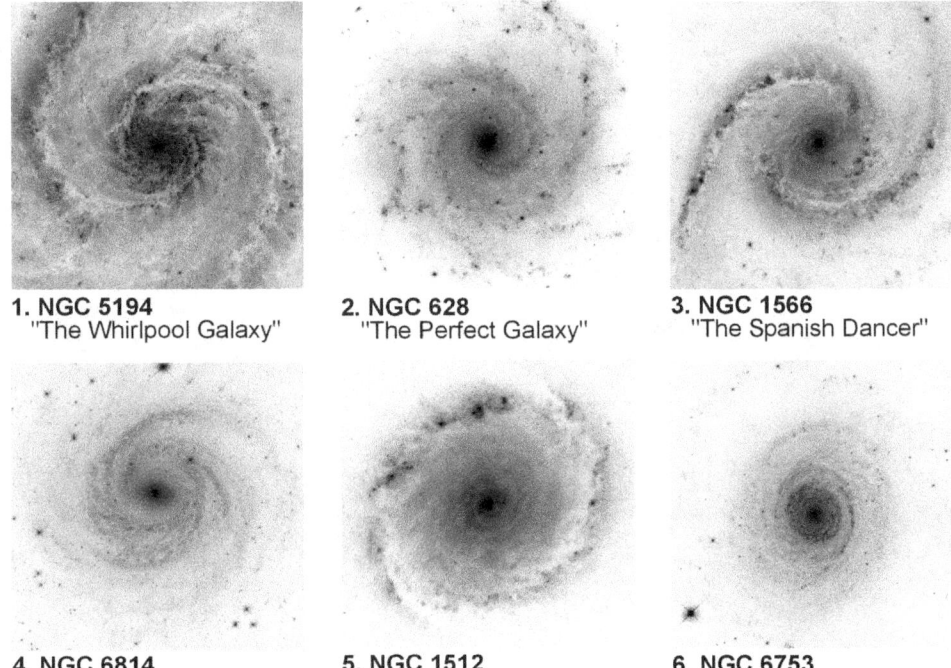

1. NGC 5194
 "The Whirlpool Galaxy"

2. NGC 628
 "The Perfect Galaxy"

3. NGC 1566
 "The Spanish Dancer"

4. NGC 6814

5. NGC 1512

6. NGC 6753

SPIRAL GALAXIES

It's inconceivable how astronomers can look at any spiral galaxy and not see their obvious whirlpool vortex. Gravity's inherent runaway nature would naturally cause all of its material to continuously migrate toward its center of mass, ceaselessly coalescing, exponentially compacting per the inverse square law in an ever-tightening spiral trajectory. But instead, they reason that its centrifugal force is exactly counterbalanced. So all of a spiral galaxy's material is endlessly orbiting around a central supermassive black hole in a circular manner as if it uniformly comprised a solid rotating disk. Stars only fall toward the black hole after being accidentally dislodged from their orbit by another nearby star or galaxy.

All photos are altered black and white negatives. **1**. NGC 5194: NASA and The Hubble Heritage Team taken with the Hubble Space Telescope. **2**. NGC 628: NASA, ESA and the GMOS Commissioning Team (Gemini Observatory). **3**. NGC 1566: ESA/Hubble & NASA, Acknowledgment: Flickr user Det58. **4**. NGC 6814: ESA/Hubble & NASA; Acknowledgment: Judy Schmidt. **5**. NGC 1512: NASA, ESA, and D. Maoz (Tel-Aviv University and Columbia University) taken with the Hubble Space Telescope. **6**. NGC 6753: ESA/Hubble & NASA, Acknowledgment: Judy Schmidt.

(36 Spiral Galaxies 3a)

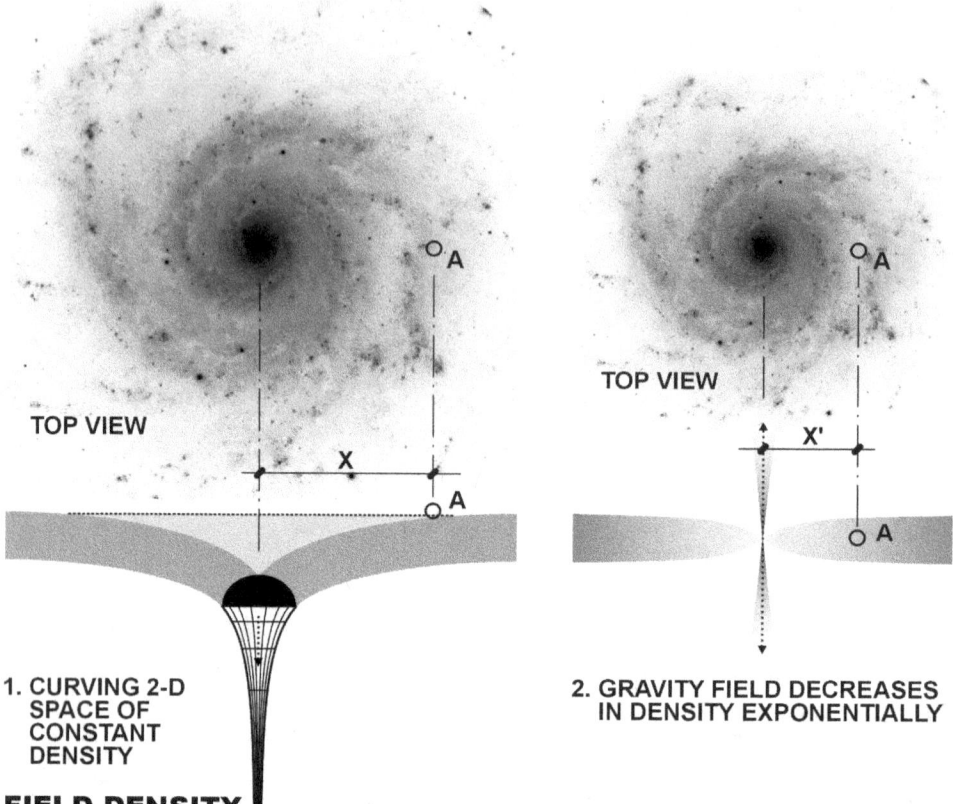

1. CURVING 2-D SPACE OF CONSTANT DENSITY

2. GRAVITY FIELD DECREASES IN DENSITY EXPONENTIALLY

FIELD DENSITY ▌

Diagram **1** portrays a cross-section view through our galaxy that represents our current belief that we reside at location **A** about 2/3 the distance from the galactic center in a space of constant density that curves toward our galaxy's four million solar mass central black hole. Our galaxy's stars orbit in a circular fashion with a velocity that's constant from the center out because of some mysterious extra unseen mass. They only plummet toward the black hole after being accidentally knocked from their fixed path. After disappearing beyond its event horizon, they begin to "spaghettify" as they endlessly fall with ever-increasing velocity toward the infinite mathematical condition of a singularity.

Diagram **2** portrays our or any spiral galaxy's actual condition where its gravity field exponentially decreases in density toward its center of mass, not a black hole, that increases the coalescing, condensing, and eventual collapse that transforms and recycles all of its infalling material back to the radiation/plasma that comprises a galaxy's halo and the energetic jets that spew from its core. This constant infall, compaction, and "draining" of a spiral's material is how its whirlpooling vortex can be maintained while the apparent orbital velocity of its material remains constant like a spinning disk.

Because distance is only a function of size and time, and those vary as a function of a gravity's varying field density and light's synchronized velocity, then from our position at **A** this has the distance at **X** in diagram **1**, where space's density is assumed constant, appear equal to the distance at **X'** in **2**, where gravity's field density decreases exponentially. This has us misperceiving the distribution of our and every other galaxy's mass, which simply resolves the "dark matter" issue where galaxies appear to not have enough mass to explain the orbital velocity of their outer most stars.

(37 Galaxy Density 4a)

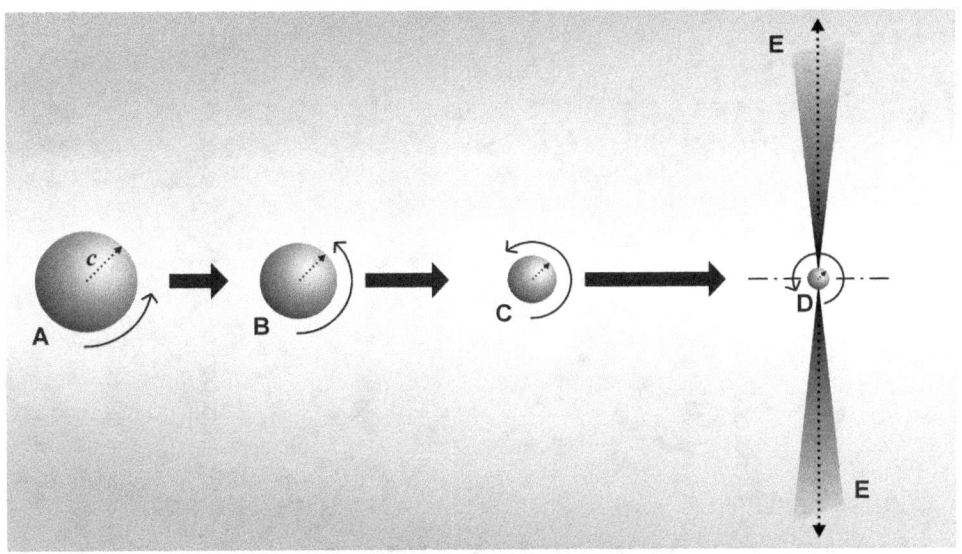

SECTION VIEW THROUGH A MATURE SPIRAL GALAXY

CONDENSING BODIES

The sphere depicted as moving in the sequence from **A** to **D** represents the condensing and ultimate collapse of an infalling star or any heavenly body as it whirlpools in toward a spiral galaxy's center. It is accelerating and contracting while its rotation rate increases, implied by the increasing length of the radiused arrows, as it mechanically pursues equilibrium in the decreasing density of the spiral's gravity field. Light's velocity decreases correspondingly, as indicated by the dotted arrows of decreasing length at c.

Galaxies naturally coalesce from ellipticals into the flat rotating plane of mature spirals because the slower rotation rate nearer their developing poles resists gravity less. That's where the downward migration of its material occurs the most, which produces the disk's overall flattening. So the impetus of its coalescing transitions from the geometry of a full sphere more into the two-dimensional geometry of a circle. With its area, $A = \pi r^2$, it's now condensing exponentially more toward the disk's center within its planar vortex. The diffused background portrays the spiral's flattening, disk-shaped gravity field.

Because the diffusing density of a gravity field causes the omnidirectional contraction of all infalling bodies and a corresponding decrease in light's velocity, all measurements are shrinking while time's rate quickens with the body's increasing rotation, assuming time's rate is established by that rotation. This causes any measurement to always read the same for anyone at the body, making it virtually impossible for them to ever directly perceive their contraction or increasing rate of time.

As infalling bodies approach the galactic center, instead of disappearing over the presumed two-dimensional event horizon of a supermassive black hole that's believed to lurk at the center of almost all galaxies devouring anything that strays to close, the increasing pressure created by the galaxy's exponentially decreasing field density rapidly increases the infalling body's fusion reactions near the spiral's center, at **D**, hastening its collapse. This essentially squeezes its material back into radiant/plasma energy that's then spewed out in a bipolar manner as high velocity jets, as represented at **E**, which act as the vortex's drain. As the expelled radiation slows and cools, it reconstitutes back into ordinary matter. It can then begin gravitating back to its or another nearby galaxy to be perpetually recycled in a ceaseless process of degeneration, transmutation, and rejuvenation.

(31 Condensing 7a)

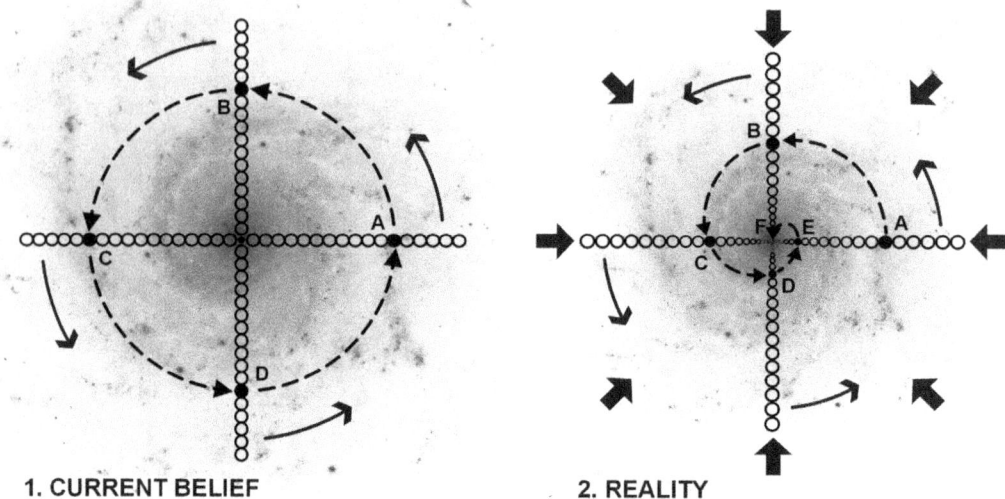

1. CURRENT BELIEF　　　　　　　　　**2. REALITY**

A SPIRAL'S FLAT WHIRLPOOLING VORTEX

Both diagrams depict a top view of our or any typical spiral galaxy. The small circles arrayed linearly represent a line of stars. Our location would be about 2/3 the distance from the center at **A**. Locations **B**, **C**, **D**, **E** & **F** indicate our position with every 90° of rotation.

Diagram **1** reflects our current belief that stars orbit in a circular fashion around a central supermassive black hole as if in a solid disk of material due to some extra unseen dark matter. Stars don't comprise a galaxy's spiral arms. They pass through them. The arms are fixed created by standing waves of some unknown origin. The centrifugal force resulting from a galaxy's rotation counters any inward migration. Stars only make their way to the central black hole after being displaced from their orbit by another nearby star or galaxy.

Diagram **2** portrays reality. Gravitation is inherently a runaway process. Because of the universal field's exponentially decreasing field density, not the impossible two-dimensional curvature of nonexistent space, all of a spiral's material is continuously coalescing and condensing exponentially as it infalls toward its center of mass, not a supermassive black hole, through the spiral's flat whirlpooling vortex. The vortex couldn't develop if its material wasn't compacting and contracting exponentially, as suggested by the decreasing size of the circles representing stars, and ultimately collapsing back into radiation or plasma that's spewed equally out the top and bottom as high velocity jets that act as the spiral's drain.

Its vortex also couldn't rotate like a solid disk if its material wasn't condensing and compressing exponentially per the geometry of the area of a circle ($A = \pi r^2$). As the area decreases toward the center the amount of material condenses exponentially. This inward exponential distribution of mass simply explains the inference of invisible dark matter.

A galaxy acquires its rotation from the gravitational energy of its coalescing material. So the resultant centrifugal force can never exceed gravity's force and forestall its inward migration. This constant infall of material along with our own infall velocity produces a recessional velocity with a corresponding redshift, which is misinterpreted as every galaxy moving away from every other. The universe is not expanding. Each galaxy is continuously contracting, but remains essentially static through the perpetual recycling of its material that's constantly being replenished.

The actual recessional velocity between infalling stars of different galaxies varies depending on many factors like the galaxy's rate and direction of rotation, a galaxy's mass, the location and infall velocity of each's material, the line of sight angle of the infalling material, and a galaxy's gravitational motion. Compounding all this, is that determining the actual red or blue shifting of a galaxy's light is apparently as much art as it is science.

(33 Vortex 6a)

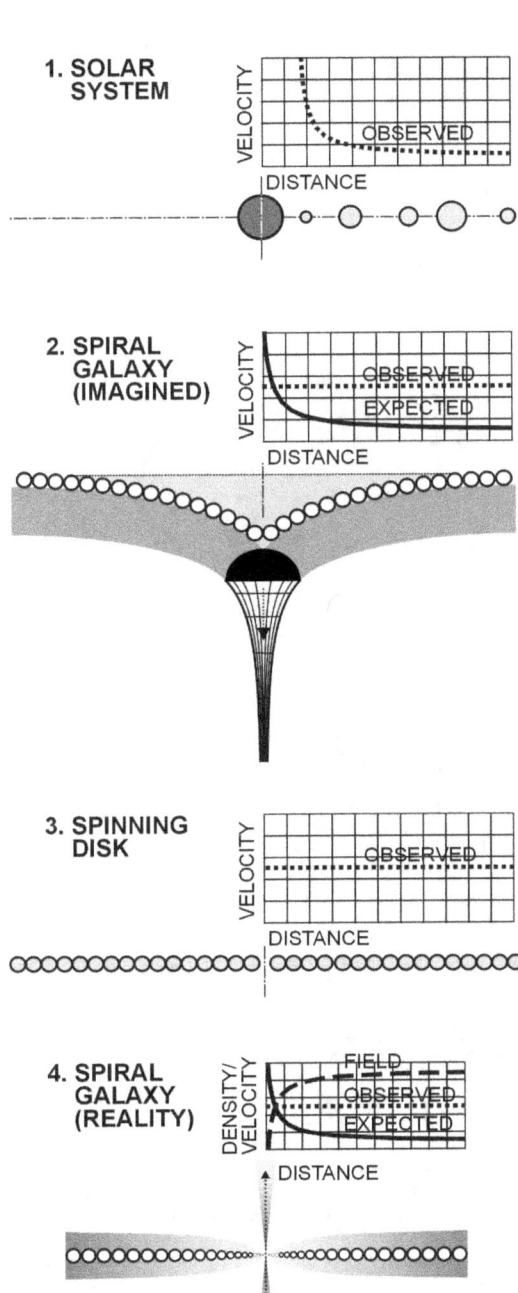

1. SOLAR SYSTEM

VELOCITY

OBSERVED

DISTANCE

2. SPIRAL GALAXY (IMAGINED)

VELOCITY

OBSERVED

EXPECTED

DISTANCE

3. SPINNING DISK

VELOCITY

OBSERVED

DISTANCE

4. SPIRAL GALAXY (REALITY)

DENSITY/ VELOCITY

FIELD

OBSERVED

EXPECTED

DISTANCE

(38 Orbital Velocity 3a)

ORBITAL VELOCITY

Diagram **1** conceptually portrays a cross section view through any hypothetical solar system. The graph charts how the orbital velocity of the planets generally increases exponentially the closer they are to their star. Diagram **2** conceptually portrays a cross section view of how cosmologists envision our or any spiral galaxy. They expect the stars to orbit around a central black hole in the same manner as planets orbit a star. But what they observe is stars orbiting at a constant velocity from the center out, as shown in the graph. It could be argued that gravity interlocks the stars effectively forming a solid disk as pictured in diagram **3**, causing their orbital velocities to remain constant, as indicated in their graph. But the only way for a spiral's vortex to form and continue, pictured in **5**, while the apparent orbital velocity of its infalling stars remains constant is for gravity's decreasing field density to coalesce and collapse all of its infalling matter exponentially, as depicted and charted in **4**, as it's transmuted back to radiation and jetted, or drained, out into space from the top and bottom where it eventually cools and congeals back into ordinary matter that begins to be drawn back to start the process all over again.

Gravity's decreasing field density compacts a spiral's mass exponentially towards its center per the geometry of a circle creating a constant velocity for all of its stars from the center out. Whether it's galaxies, galaxy clusters, or any gravitationally bound objects, it's this same principle of exponentially decreasing field density, not curving space, that accounts for the inferred increase in unseen mass, which again easily explains the dark matter mystery.

5. SPIRAL GALAXY'S WHIRLPOOLING VORTEX

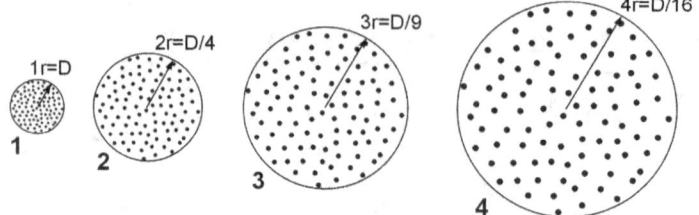

1. THE FLUX DENSITY OF AN EXPANDING DISK DIFFUSES EXPONENTIALLY, D ∝ 1/r²

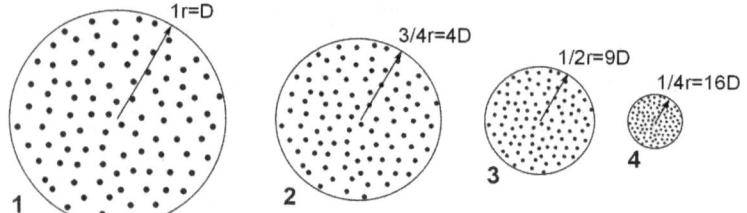

2. FOR A CONTRACTING DISK IT COMPRESSES EXPONENTIALLY, D ∝ r²

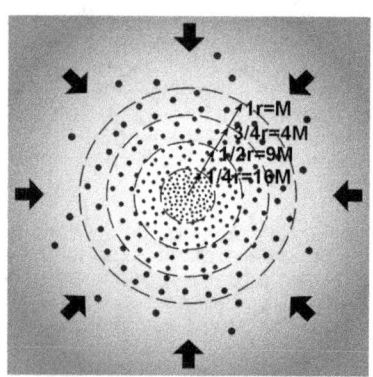

3. THE MASS OF A SPIRAL'S DISK INCREASES EXPONENTIALLY

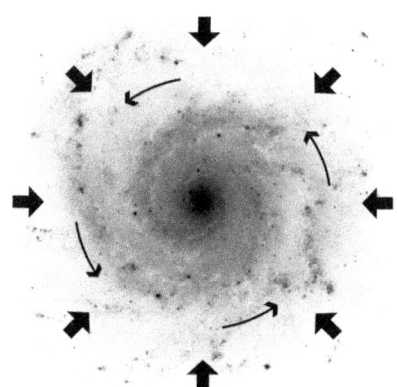

4. SPIRAL'S UNENDING VORTEX OF RECYCLED INFALLING MATERIAL

MASS DISTRIBUTION OF SPIRAL GALAXIES

Gravity's force dissipates per the inverse square law because of the three-dimensional geometry of a sphere. All galaxies begin spherically as a type of elliptical, but evolve into the flat disk of a spiral. At the largest scale, gravity's field for spirals behaves according to the two-dimensional geometry of a disk, which is a circle where the area, $A = \pi r^2$, and a flux density, shown as dots representing stars, that decreases or increases exponentially within its plane as depicted in diagrams **1** & **2**. Diagrams **3** & **4** depict the exponential distribution of mass toward the center of a spiral's disk, the dots still representing stars, that has to occur with the ceaseless vortex of infalling material that's endlessly coalescing and compacting in gravity's ever-decreasing planar field density, portrayed by the diffused background, and ultimately transmuted and "drained" out as high velocity jets of radiant energy/plasma that eventually reconstitutes as ordinary matter to be perpetually recycled.

This unrecognized exponential increase in mass toward a spiral's core simply explains the faster rotation of perimeter stars, eliminating any need for an exotic unseen dark matter.

(39 Spiral's Mass 5a)

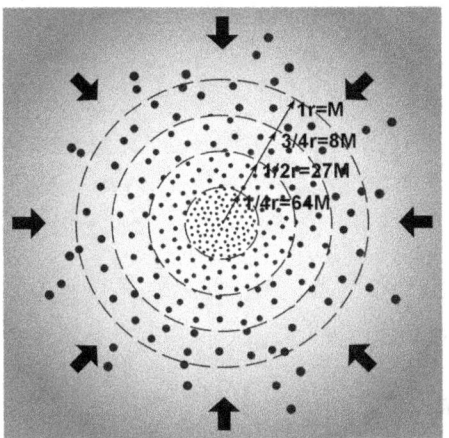

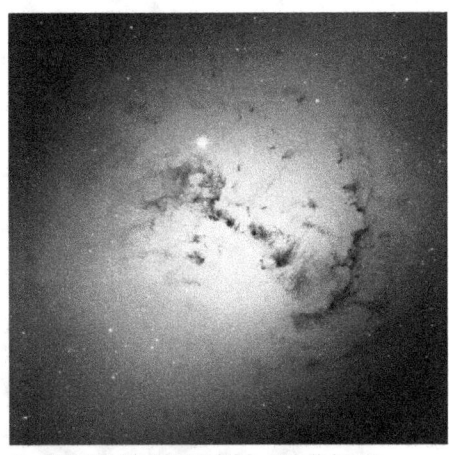

1. TYPICAL ELLIPTICAL GALAXY -
SECTION VIEW DIAGRAM

2. GIANT ELLIPTICAL NGC 1316
NASA, ESA, and The Hubble Heritage Team
(STScI/AURA)

MASS DISTRIBUTION OF ELLIPTICAL GALAXIES

Because of gravity, an elliptical's mass increases exponentially toward its center the same as a spiral's. But being spherical, it's bound to a sphere's geometry where its volume, $V = 4/3\pi r^3$, and its density, shown as dots representing stars in **1**, increases exponentially with D (volume density at a given radius) α $1/r^3$, which would be the same for its mass.

For ellipticals, gravity has yet to fully organize its infalling stars around a developing axis and poles. But their perpetual recycling is still taking place, just not as high velocity jets spewing from its center, but as a spherical halo of radiated energy that's also eventually reconstituted and gravitationally drawn back in to begin the coalescing-collapsing-transmuting process all over again.

Just like for spirals, this gives the infalling material of an elliptical a recessional velocity from every other galaxy, but it occurs almost evenly around its 360° three-dimensional spheroidal shape. Despite gravity's inherent runaway nature, cosmologists maintain that an elliptical's material is also not constantly infalling. But they don't offer an explanation for what prevents it or the associated redshift that its infall would produce that they misinterpret as indicating the recessional velocity of the entire galaxy receding uniformly from every other galaxy.

(40 Elliptical's Mass 5a)

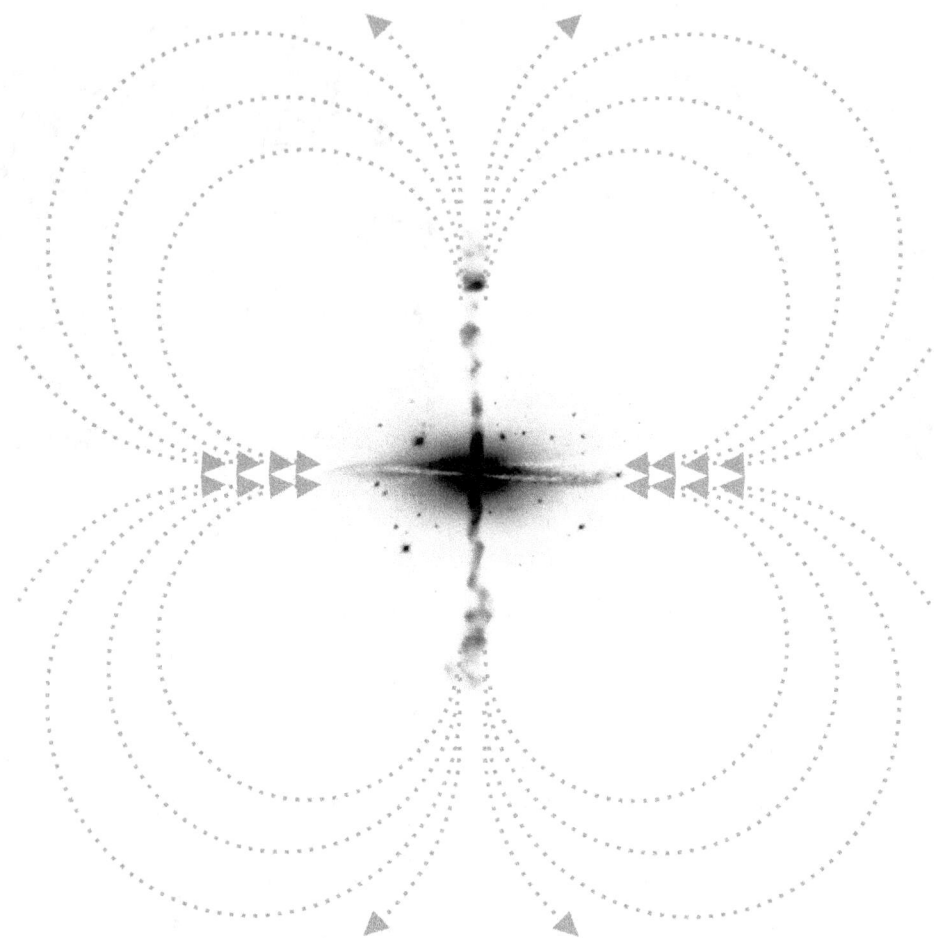

GRAVITY'S PERPETUAL RECYCLING

At the largest scale, protogalactic material begins coalescing spherically, three-dimensionally. Any asymmetry causes it to start rotating, which begins to form an axis. Because the resulting centrifugal force is less at the ensuing poles, it coalesces more readily down the axis toward its center while the more perpendicular material tends to flatten out into the whirlpooling disks that define a fully developed spiral galaxy, shown in side view.

But its coalescing doesn't stop here. Gravitation is a runaway process. It continues unabated facilitated by its exponential condensing that naturally allows its vortex to flatten out. As the material spirals in toward the galactic center, the compression becomes so great that fusion reactions are triggered that begin converting it back into the radiant energy it originated from.

It's then ejected out in huge, high-velocity jets in a bipolar fashion, shown as visible. As the radiation slows and cools, it eventually congeals back into ordinary matter and begins gravitating back toward its or another nearby galaxy, the dotted arrows. The material of every galaxy is subject to this never-ending process of perpetual recycling, which is the real source of cosmological redshift: the Doppler effect from the material's ever-increasing recessional velocity from every other galaxy coupled with our own infall velocity at our own galaxy.

The image is a highly modified black and white negative of galaxy ESO 510-G13 taken by NASA and the Hubble Heritage Team with the HST.

(19.1 G Recycling 5a)

We're highly motivated to maintain our solid disk interpretation. If the material of a spiral galaxy, or any type of galaxy, was continuously coalescing, its constant inward migration would produce a recessional velocity with an associated redshift. With galactic redshifts originating from the recessional velocity of its own constant infall of material, this would directly conflict with their assumption that it originates from space's expansion, or its stretching, which would throw our big bang-relativity cosmology on its head.

What's just as disruptive is our inability to recognize that if any other galactic redshift source exists in addition to space's stretching, it'd completely invalidate the big bang as well. We could never know how much of the total redshift is attributable to it and how much to other sources. So the universe's rate of expansion, or whether it was expanding at all, could never be determined. At least eight other sources of redshift have legitimacy:

1. Tired light results when a galaxy's photons lose energy when they interact with deep space particles as they travel toward us, which shifts their light toward the red end of the spectrum. The theory does not account for galaxies whose light is blueshifted.

2. A charge is induced to atoms by a galaxy's motion and its rotation through the universe's omnipresent electromagnetic field that slows the atom's natural frequency, shifting their light toward the red. Blueshifted galaxies are not addressed.

3. Relativistic time dilation slows an atom's rate of time due to a galaxy's linear and rotating motion that shifts their light to the red. Galactic blueshifting is not explained.

4. Light's velocity slows innately as it traverses the decreasing density of a galaxy's gravity field, shifting it toward the red end of the spectrum. There's no explanation for how the velocity of a galaxy's light increases to shift it toward the blue.

5. Gravitational redshift is where light leaving a galaxy or any massive object's gravity loses energy, shifting it toward the red. Blueshifted galaxies aren't addressed.

6. Einstein's gravitational redshift is where relativistic time dilation slows an atom's rate of time due to its rapid rotation around a massive body like a galaxy. When the atom is regarded as a clock, its slower rate of time slows its frequency, shifting its light toward the red. This corresponds to the outward acting gravity produced by its rotation's centrifugal force. And because of his "principle of equivalence," centrifugal created gravity is equal to or the same as mass created gravity. So all galactic redshifts indicate their galaxy's gravitational potential. The blueshifting of galaxies is not accounted for.

7. The big bang's uniform but increasing expansion causes the recessional velocity of each entire galaxy as they move away from us and every other galaxy that produces a Doppler effect that shifts their light toward the red.

It's been replaced by space's stretching. But its effect wouldn't go away. Galaxies would still have a recessional velocity. Blueshifting occurs with a limited number of galaxies that have a closing velocity when they move toward us against expansion due to an overriding gravitational attraction with our or other galaxies.

8. Each galaxy's continuously coalescing infalling material, the product of gravity's innate runaway nature coupled with our own continuous infall in our own galaxy, produces a recessional velocity with a corresponding Doppler effect that redshifts the light of a galaxy's material. A blueshift in a galaxy's light occurs in rarer instances when their material has a closing velocity caused by a combination of a variety of conditions like their gravitational interplay with our or other galaxies, the type of galaxy, its orientation, its rotation's rate and direction, etc. None of these redshift sources are vigorously disputed despite that their coexistence with space's expansion/stretching would be catastrophic.

But astronomers already accept that other redshift sources exist and interfere with the redshift from universal expansion. That's why they only use the redshift from elliptical galaxies when refining their measurements of the Hubble constant, the universe's presumed rate of expansion. All others are too inconsistent.

A spiral's rotation and the chaotic nature of irregulars are problematic for universal expansion. Ellipticals on the other hand are more predictable. The coalescing and condensing of their material is more uniform. Its infall/recessional velocity increases as the size of the elliptical decreases, which is the real source of cosmological redshift. What they don't realize is that groupthink's confirmation bias has them unconsciously culling data and selectively shaping solutions to conform to established big bang/relativity dogma to insure the status quo is maintained.

But space's stretching is a fundamentally flawed concept. There's no such thing as space. As repeatedly stated, it's the nothingness between objects. So there's nothing there to be stretching, or expanding, or redshifting light from stretching or expanding. It's also interesting that no one seems to notice that space's stretching has no viable explanation for those galaxies with blueshifted light. Are we actually expected to believe that (nonexistent) space is contracting in the direction of blueshifted galaxies? How would that be explained?

Given that universal expansion is wholly untenable on its own, (nonexistent) space's stretching is summarily disqualified as a source of galactic redshifts anyway. The same goes for redshifting from the Doppler effect from an entire galaxy's recessional velocity. They're both out of the question.

Relativistic time dilation isn't plausible. Light's velocity isn't fixed. So it doesn't work. Einstein's version of gravitational redshift is also unworkable. First, it's easily invalidated by its incoherent logic. But it's also founded on light's fixed velocity. So it's not a possibility either.

Ordinary gravitational redshift, tired light redshift, redshift from charged atoms, and redshifting from light's slowing in gravity fields are all reasonable options that may or may not be true but cannot be outright excluded. But the possibility of light's variability, its slowing, is always preemptively rejected. It would undermine all of relativity. Can't have that, even though Einstein himself has repeatedly asserted its variability beginning as early as 1907.

With Doppler redshifting an observable fact while all other optional redshift sources remain speculative theory, that can't account for blueshifted galaxies, and with universal expansion a conceptual impossibility, which excludes the redshifting from an entire galaxy's recessional velocity and space's universal stretching, the only remaining viable source of galactic redshifts is the Doppler effect from the recessional velocity of each galaxy's constant infall of ever-coalescing material. If recessional velocity does in fact redshift an object's light then this source cannot be ignored and has to be accommodated in any cosmological redshift theory. Compounding it with any other redshift source that isn't derived from universal expansion doesn't present a conflict.

Our inability to entertain this alternative, even as a possibility, stems from our extensive conditioning and an unconscious realization of the wide-ranging ramifications of admitting our stupendous error. (See Diagram **41** BB-DSS 8a & **46** G Redshift 3a)

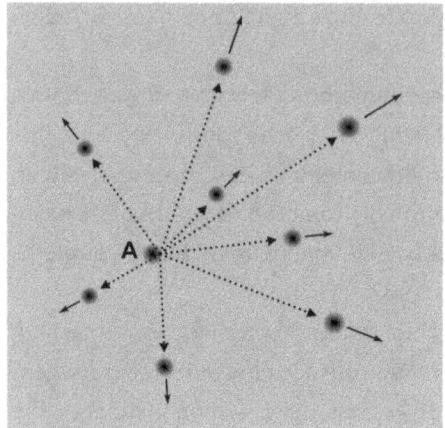

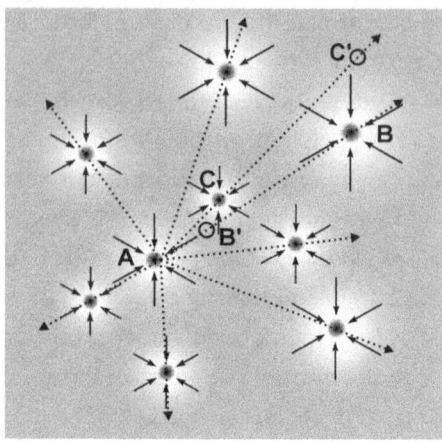

1. BIG BANG　　　　　　　　　　**2. DYNAMIC STEADY-STATE**

REDSHIFT - EXPANDING SPACE / CEASELESS COALESCING

The smooth gray background in diagram **1** portrays the big bang's physically impossible uniform expansion of space. The dots represent galaxies that are assumed to be moving away equally from every other galaxy because of the universe's two-dimensional curvature like the surface of a sphere. The solid arrows indicate each galaxy's direction of motion from our position at **A**, which would be radial in three dimensions as it appears. Their length corresponds to the galaxy's inferred recessional velocity and redshift that's believed to increase with distance due to space's stretching.

Diagram **2** portrays an infinitely vast, eternal universe in a dynamic but steady-state condition where each galaxy is perpetually recycling its material. Their motion is limited to their gravitational interplay while their material continuously migrates inward, constantly infalling, ceaselessly coalescing and condensing because of gravity's inherent runaway nature until it's ultimately collapsed back into radiant energy/plasma and radiated or ejected out where it eventually slows and cools becoming ordinary matter again that begins to be gravitationally drawn back to its or another nearby galaxy to be recycled.

The dots in diagram **2** also represent galaxies with our location at **A**. The diffusing background at each galaxy depicts the universal field's exponentially decreasing density in section view. The inward pointing arrows at each galaxy suggest the ceaseless omnidirectional coalescing of its infalling material that's condensing three-dimensionally from every direction. The arrows' length corresponds to its infall velocity. That infall velocity coupled with our own infall velocity at our own galaxy produces a recessional velocity, which is the true source of each galaxy's redshift, not universal expansion. Depending on the type, orientation, and gravitational interplay, the material of a small number of galaxies has a closing velocity that produces a corresponding blueshift.

With redshift no longer an indication of distance, every galaxy could literally be any distance away along our line of sight. A small compact massive galaxy with a high redshift that appears to be hundreds of millions of light-years away, like at **B**, could literally be right next door at **B'** and we'd never know it. Or a galaxy that appears very close with lower redshift, like at **C**, may actually be a great distance away at **C'**. There's no way to tell using galactic redshifts.

Also, because of an object's decreasing size in the universal field's decreasing density where light's velocity slows correspondingly, the galaxy at **B'** that appears hundreds of millions of light-years away may actually be that far if we were to attempt to travel there. It's entirely possible that it could take less time to travel to its backside by circumventing it along its perimeter than to travel directly to its core.

(41 BB-DSS 8a)

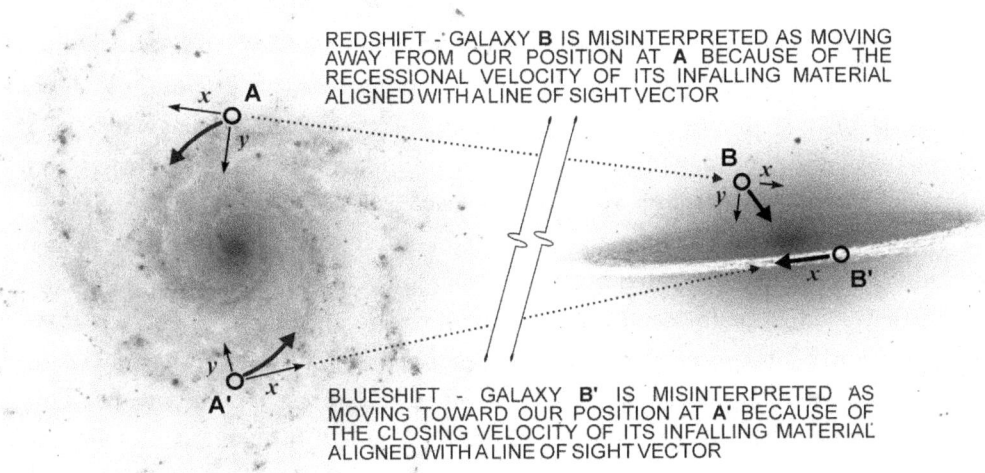

REDSHIFT - GALAXY **B** IS MISINTERPRETED AS MOVING AWAY FROM OUR POSITION AT **A** BECAUSE OF THE RECESSIONAL VELOCITY OF ITS INFALLING MATERIAL ALIGNED WITH A LINE OF SIGHT VECTOR

BLUESHIFT - GALAXY **B'** IS MISINTERPRETED AS MOVING TOWARD OUR POSITION AT **A'** BECAUSE OF THE CLOSING VELOCITY OF ITS INFALLING MATERIAL ALIGNED WITH A LINE OF SIGHT VECTOR

REDSHIFT/BLUESHIFT OF A GALAXY'S INFALLING MATERIAL

Without a viable explanation for blueshifted galaxies or realizing that any redshift theory based on universal expansion is invalidated by the existence of any other redshift source, like the Doppler effect from the big bang's recessional velocity (the previous source that wouldn't just disappear) or gravitational redshifting from a galaxy's mass that's also widely accepted, light's displacement toward longer wavelengths due to space's stretching is now the most accepted source of cosmological redshifts.

In our real nonexpanding infinite universe, the most likely source of a galaxy's redshifted light is the Doppler effect from the recessional velocity of each galaxy's continuous infall of coalescing material. Many factors influence its recessional velocity: the type and mass of the galaxy, the location on the galaxy where the redshift is recorded, the velocity and angle of its infalling material, the direction and rate of its rotation, its gravitational interplay with our or other galaxies, etc. Ultimately, all we can know for certain is the line of sight recessional velocity of the infalling material coupled with our own line of sight infall velocity.

If we were located at **A**, we'd be swirling around our galaxy's vortex of infalling material in the direction of the heavier arrow. If we were looking opposite our direction of infall at a galaxy's stars at location **B**, our rotational velocity would yield line of sight vectors for all three dimensions, x, y & z of our infall velocity. The z vector in the third dimension is not shown for clarity. For the material at **B** that may be located in an elliptical or the central bulge halo of a spiral as pictured, it too would have three infall velocity vectors keyed to our line of sight. But we can only perceive the rate of the line of sight x vector. And even then, we couldn't know how much to attribute to us or the other galaxy's infalling material. All we can be certain of is that x at **A** plus x at **B** gives us a line of sight recessional velocity and a corresponding redshift. As an example of blueshifting, if we were located at **A'** looking in the direction of our infall at the oncoming material at the edge of another spiral at **B'**, there might be a closing rate where -x at **A'** plus -x at **B'** would give us a line of sight closing velocity and a corresponding blueshift in the spectrum of that galaxy's light.

These simple illustrations conceptually represent just two of the countless possibilities for the true source of red or blueshifted galaxies, the constant infall of coalescing material at each galaxy that's perpetually recycling because of gravity's inherent runaway nature in an infinitely vast, ageless universe.

The whirlpool image also used in other diagrams is a modified black and white negative of galaxy NGC 628 taken by the Gemini Observatory - GMOS Team with the Gemini North Telescope on Hawaii's Mauna Kea. The "Sombrero" image is a modified black & white negative of NGC 4594 taken by NASA/Hubble Heritage Team.

(46 G Redshift 3a)

A universe that was ceaselessly coalescing and transmuting its material through the center of each galaxy could never produce a "big crunch," its inward contraction and ultimate collapse due to gravity. This could only occur with a finite, spherical, and three-dimensional universe. In an infinite, ageless universe that extends indefinitely, it's conceptually impossible.

A finite universe that was spewing radiant energy from active galaxies would tend to be promoting expansion. It'd be filling the void regions between galaxies with an ever-increasing field density that would be pushing them apart, increasing its overall size. The unceasing conversion of matter to radiant energy that's taking place at the core of every galaxy that's constantly being replenished with recycled material might seem like an eternally balanced, miniature, continuous big bang that's coupled to a miniature, continuous big crunch where the process might wax or wane at times but only with variations in the availability of material both in and outside each galaxy.

With our acceptance of gravitation as fundamentally a runaway process that's evident at the largest scale as the constant infall of material of galaxies, we have to infer that there's only one underlying imperative for the universe, the unrelenting coalescing of matter. It seemingly produces ever-higher states of order that eventually leads to its collapse back into pure radiant energy that's then violently ejected great distances out of the core of fully evolved spirals.

As it slows, cools, and dissipates, seemingly producing ever-higher states of disorder, it eventually reconstitutes back into ordinary matter that naturally starts gravitating again, drawing back to its or a neighboring galaxy to complete the perfectly balanced, eternal recycling process that harmoniously rejuvenates through an endless ethereal transfiguration that's solely perpetuated by gravitation.

Unfortunately, we've been blindly evolving the big bang into ever-higher orders of nonsense and don't realize it. What initially began only as Einstein's whimsical musings of how the universe, which for him consisted only of stars, the existence of galaxies was not yet known, might be finite while somehow remaining boundless is becoming even more absurd almost daily.

Through the use of contrived, non-Euclidean geometry that impossibly curves in three dimensions and Riemann's delusional two-dimensional space that magically melds our three-dimensional existence with the isotropic properties of a sphere's two-dimensional surface and an invented cosmological constant that mathematically resists gravitation's now uniform coalescing that counteracts the collapse of his entire universe, Einstein formulated his illusive cosmology.

He later substituted Friedmann's notion of expanding space for his cosmological constant, which had to be uniform so that in the same two-dimensional manner it also mathematically prevented gravitational coalescing. Because space's expansion naturally implies a smaller universe at an earlier time and eventually a beginning, we were obliged to imagine a big bang that started it all.[196]

But because of the ever-increasing incongruities and contradictions, we now find ourselves concocting evermore delusive, ad hoc theories like inflation, dark matter, dark energy, an unworkable galaxy evolution, the circular orbit of a galaxy's material, and so on. We're not aware that we're frantically trying to reconcile all the invalidating discrepancies when we should be conceding the obvious. Einstein's finite yet somehow unbounded, expanding, curving, two-dimensional universe is nothing more than surreal make-believe fantasy.

Our fabricated, imaginary explanation for galaxy formation and evolution is a perfect example. It runs contrary to observation and common sense. And it fundamentally conflicts with the runaway nature of gravitation that has a galaxy's centrifugal force counteracting the unrelenting coalescing of its material while ignoring the undermining effect of space's presumed expansion.

Without ever having the possibility of constant coalescing/collapsing material at galaxies, despite our acceptance of the runaway nature of gravity that would ensure the endless infall of all galactic material and a corresponding redshift, this keeps us incorrectly reasoning that galactic redshifts are the product of the recessional velocity of the entire galaxy being whisked away by expanding or stretching space, falsely suggesting a finite universe with a beginning, that we all instinctively know is actually limitless and everlasting.

Gravitation's unceasing coalescing and perpetual recycling of galactic material that's responsible for its continual inward migration that produces a recessional velocity with a corresponding redshift easily and more rationally explains any number of cosmological quandaries. A significant one being how it is that two or more galaxies, or quasars, could be associated or even physically interacting yet differ vastly in their redshifts, brightness, or size that suggests a difference in line of sight distance.[197]

Apparently, this is a common predicament. But it's rarely addressed. But by simply accepting the obvious notion that a galaxy's material has to be constantly infalling, this clears up the matter entirely and implies that all associated galaxies or quasars with these differences should not only be acceptable but actually the norm.

Because of gravity's exponentially decreasing field density, distance cannot remain constant but becomes a function of an object's decreasing size, time's synchronized increase, and light's slowing velocity. This suggests that the true

196. Einstein, *Relativity*, 122-127.
197. Arp, *Seeing Red*, mostly on pages 169-193, but there are many more.

distance to the center of any galaxy could be located almost any distance away along our line of sight, outside of our own galaxy of course, which again allows for the entirely unorthodox possibly that even the smallest, faintest galaxy with the highest redshift could actually be located right in our own backyard without us even knowing it.

So it's not in any way possible to infer a relationship between a galaxy's redshift and their size and brightness and its actual distance. And accurately mapping the true relative location of any galaxy or quasar is virtually impossible. All of our redshift-based maps of the universe only reveal the line of sight recessional velocity of its infalling material coupled with our own recessional/infall velocity, along with other potential sources of redshift like mass and charge. They cannot reveal a galaxy's true distance.

The only possible way to interpret a redshift-based map of galaxies is to make the conceptual leap that all objects, including galaxies, contract in an omnidirectional manner in the exponentially decreasing density of a gravity field. This restricts each galaxy's distance to what it would be based on its contracted size along its line of sight as if we were to attempt to travel there.

Its actual distance from a location somewhere in the universe perpendicular to our line of sight would be completely different. It could literally be any distance away. And someone in a different galaxy would map completely different distances that wouldn't correspond to ours at all. (See Diagram **42.1**, **42.2**, **42.3** Associated G 2a)

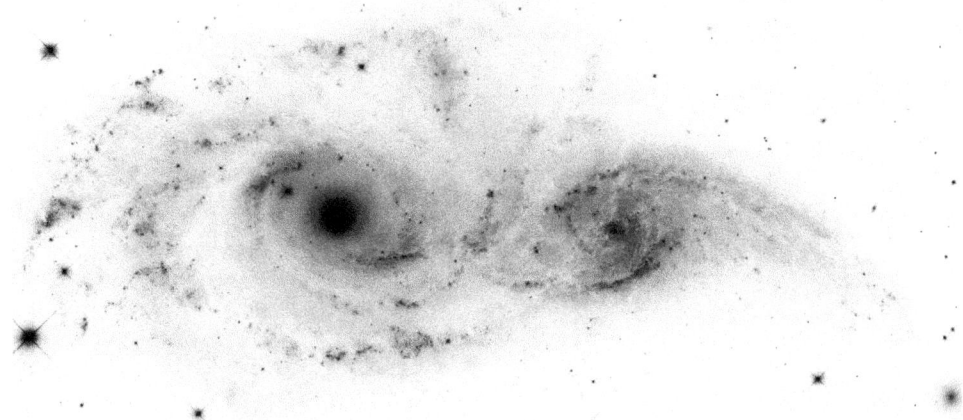

ASSOCIATED GALAXIES

This photo of associated galaxies clearly depicts the apparent conflict where two or more galaxies appear to be physically interacting, as these do, but where one appears more distant than the other like the smaller galaxy on the right. Since farther distances supposedly correspond to higher recessional velocities and higher redshifts, according to our big bang-relativity cosmology, we would expect that the smaller, more distant appearing galaxy has a higher redshift. But when the source of a galaxy's redshift is the recessional velocity of its own perpetually coalescing, infalling material coupled with our own infall velocity, not the recessional velocity of the entire galaxy, either galaxy could have any redshift while occupying the same region of space or even be in the process of merging.

This modified black and white negative of galaxies NGC 2207 and IC 2163 was taken by NASA and The Hubble Heritage Team.

(42.1 Associated G 2a)

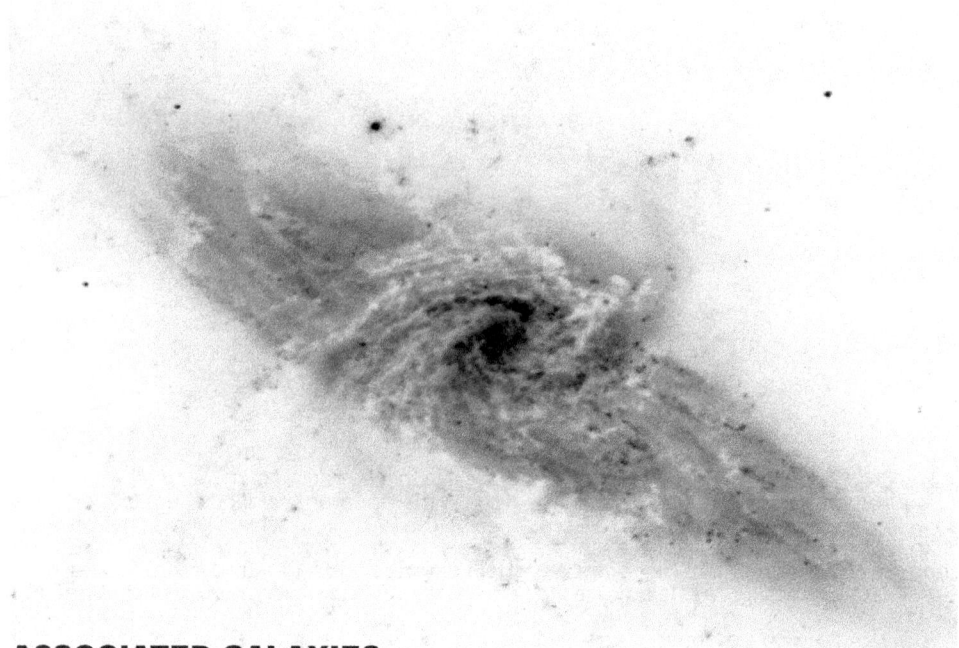

ASSOCIATED GALAXIES

If you look closely at this photo of two galaxies that appear associated, you'll see a much smaller one superimposed in front of a larger one. Their relationship reveals how a smaller, less bright, more distant appearing galaxy can actually be closer than a larger one that appears nearby. If its redshift were higher than the larger one, this too would be at conflict with our big bang-relativity cosmology where higher redshifts also supposedly correspond to farther distances.

But for an infinite, steady-state universe where each galaxy's redshift is mostly the product of its own constantly infalling material that's dynamically recycled through its ceaseless gravitational coalescing, not the entire galaxy's recessional velocity, we would expect the smaller galaxy to have a higher redshift because its higher density indicates higher mass, which would produce faster infall velocities for its condensing material. Or it has consumed more of its infalling material without it being replenished. So it would have contracted in size, leaving more of its core exposed where the infall velocity of its coalescing material is higher. Both of which would produce a higher recessional velocity when coupled to our own infall velocity that corresponded to higher redshifts. But in reality, either galaxy could have any redshift. It's simply a function of the line of sight recessional velocity of the infalling material where the measurement is being taken, not to exclude the other possible sources of redshift like mass, charge, and gravity's field density.

Because of the exponentially decreasing density of a galaxy's gravity field that has objects continually contracting as they migrate toward the galactic center while light's velocity continually slows, we might find that the center of the smaller galaxy is actually farther away than the larger one. It may actually be a shorter distance and take less time to travel to the edge of larger one by completely circumventing the smaller, remaining in the denser region of its gravity field that's farther out. For the same reason of contraction in decreasing field density, both galactic centers would be much farther away than their overall appearance suggests.

This modified black and white negative of galaxies NGC 3314a & b was taken by NASA and The Hubble Heritage Team.

(42.2 Associated G 2a)

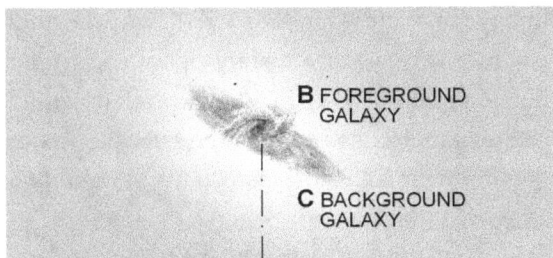

B FOREGROUND GALAXY

C BACKGROUND GALAXY

1. VIEW OF GALAXIES B & C FROM A

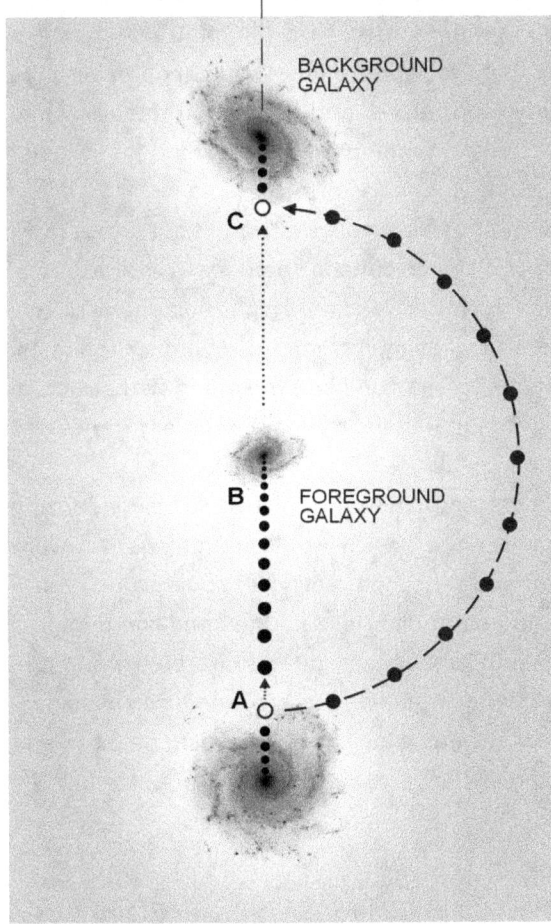

BACKGROUND GALAXY

C

B FOREGROUND GALAXY

A

2. TOP VIEW OF GALAXIES A, B & C

(42.3 Associated G 2a)

ASSOCIATED GALAXIES

If we had a line of sight view of galaxies **B** and **C**, pictured in diagram **1**, that appear to be associated from our vantage point at **A** in diagram **2**, it may very well be that the smaller foreground galaxy, which in reality could be located any distance between **A** and **C**, could also be less bright and have a higher redshift than the larger background galaxy, which according to the big bang would make it farther away despite that it's in front of the larger.

Because gravity fields decrease in density exponentially at all objects, shown in section view as the diffused background around each galaxy, it's entirely possible that the foreground galaxy **B** has more mass, which has condensed its material more tightly, making it smaller and less bright while imparting faster infall velocities, which yields higher recessional velocities and higher redshifts.

Because distance is a function of an object's contracting size, time's smaller synchronized rate, and light's decreasing velocity in a gravity field's exponentially decreasing density, the distance between each dot remains the same for someone traversing them despite their decreasing measurement from an outside perspective.

It's also entirely possible that it would be a shorter distance and take less time to travel from **A** around galaxy **B** to get to **C** than to travel directly to **B**'s center. It would also be a greater distance and take more time than expected to travel from **A** or **C** to the center of their respective galaxies.

The modified photos of galaxies NGC 3314a & b was taken by NASA and The Hubble Heritage Team. The modified photo of galaxy NGC 628 was taken by the Gemini Observatory - GMOS Team with the Gemini North Telescope.

Because of a galaxy's continuous coalescing/condensing, it may be that smaller galaxies or quasars that are inferred as more distant because of their higher redshift just have a denser, more concentrated distribution to their material. This would increase the rate of its gravitational coalescing, causing its faster inward collapse, which would result in higher infall velocities. So it'd be receding more rapidly from us and potentially every other galaxy.

This may be true for those smaller galaxies that are just forming and lack enough merging material to become fully developed as could be the case with quasars. We may simply be looking at their exposed cores where the infall velocities are greatest, producing the highest redshifts. Or it could be that smaller galaxies have just used up their source of new/recycled material and have contracted down to a point where they've become more compact and the view of their faster collapsing material less obstructed. Both scenarios are plausible.

In either case, any galaxy that was considered smaller would not necessarily be more distant. It may just be denser. With its concentrated mass more visible, we would have to see its material coalescing faster, producing higher infalling velocities, which would be receding more rapidly producing corresponding higher redshifts. This is consistent with the observation that we don't see the elongated structure of barred spirals for fainter galaxies, and that these bars are smaller for galaxies with higher inferred mass.

Any galaxy or quasar that was not drawing in new material would be condensing at a rate consistent with the inverse square law. It would appear as if it were accelerating away because its size would be contracting exponentially, rapidly becoming smaller as it approached the end of its life. And the condition of higher infall velocities for galaxies that are smaller because of their more concentrated mass would mistakenly appear as higher redshifts for more inferred distance, incorrectly intimating accelerating recessional velocities for receding galaxies and the ever-accelerating expansion of the universe. (See Diagram **43** G Size 3a)

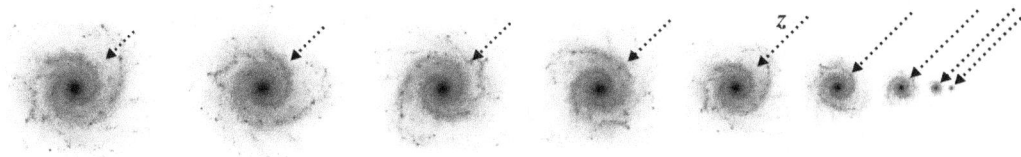

A GALAXY'S CONDENSING

We estimate a galaxy's distance by its size, brightness, the brightness of certain types of its stars and supernovas, but mostly by its redshift. From these, we've inferred that galaxies have an accelerating recessional velocity from us, and every other galaxy. Presumably, they're being swept away by a uniformly expanding space whose rate increases with distance. This is a misinterpretation that's become our conceptually impossible, big bang orthodoxy.

Imagine that we could speed up time while observing a galaxy's continuously infalling coalescing material. For a spiral galaxy, we'd see it swirling down its whirlpooling vortex as it condenses. Let's also imagine that the continuous recycling of its material is waning, or that its replenishment has decreased down to near nothing.

What we'd get is the continuous contraction of the entire galaxy as its infalling material condenses. The resulting pressure from gravity's runaway coalescing continues transforming its material, collapsing it back into the radiant energy it originated from by triggering fusion reactions that's then radiated back out. Or for most spirals, it's ejected out as high velocity, bipolar jets. Eventually, as the galaxy's material is depleted, it shrinks down to nothing as it reaches the end of its existence.

Without any new or recycled material to maintain its size, it contracts exponentially, as indicated by the sequence of shrinking galaxies. The rate of the recessional velocity of its infalling material would also be increasing exponentially along with its corresponding redshift, as suggested by the increasing length of the arrowed lines at z. Its brightness would also be decreasing exponentially. All governed by the inverse square law, an innate property of a sphere's three-dimensional geometry.

In concept, gravitation's runaway exponential condensing would have to apply across the board to all types of galaxies, ellipticals, irregulars, and peculiars, not just spirals. It would have to include quasars as well.

Their much smaller size, reduced brightness, and higher redshifts have us inferring them to be mysterious objects that reside at much greater distances. But it's more likely that they're just normal galaxies whose collapsing material has been devoured down to a point that their dense cores have been exposed where the material's ever-increasing infall/recessional velocity is exponentially faster. Regardless of the disposition of each galaxy or quasar's material, we misinterpret the redshift from its infalling recessional velocity as originating from the recessional velocity of each entire galaxy, or more precisely from space's stretching between galaxies.

This altered photo of NGC 628 was taken by the Gemini Observatory -GMOS Team with the Gemini North Telescope on Hawaii's Mauna Kea.

(43 G Size 3a)

Quasars are explained as extremely bright celestial objects. That's it. Or they may possibly be the bright centers of galaxies. But in either case, they're inferred as very distant because of their very high redshifts. More progressive theories speculate that they might be ejected core material of active spiral galaxies.[198] With a galaxy's constant infall of material that's continuous condensing and collapsing from runaway gravitation, this is perfectly consistent and highly plausible.

The core material's ejection is easily explained as the product of a convulsion. The amount of coalescing, infalling material lessens momentarily, reducing the pressure at the galactic center, disrupting its ongoing conversion to radiant/ plasma energy, which allows for the escape or the ejection of a quantity of collapsing core material. We then label this tightly compact, rapidly condensing material a quasar.

It naturally follows that an ejected quasar should have a higher redshift than that of its originating galaxy as the originating galaxy's material could only be visible farther from its center, which would have a slower infall velocity and lower redshift. For the same reason, an ejected quasar's redshift would lessen over time if its mass were high enough to continue attracting additional material. The slower accumulating material would begin to obscure the faster receding, higher redshifted material nearer its center.

This accumulating material may eventually cause the quasar to transition to a full-fledged galaxy that might initially express as a globular cluster that would evolve into a small elliptical galaxy. Only later, after coalescing much more material would it further condense, evolving into a well-developed spiral galaxy.

With the revved-up, collapsing process of spiral galaxies possibly being the source of multiple quasars that have been ejected in approximately the same direction, where each of which may develop into globular cluster and dwarf galaxies that soon evolve into ellipticals, and with mature spirals being much older and more massive having coalesced all nearby material, this could offer a rational explanation for why elliptical galaxies tend to be more numerous and more tightly clustered than spiral galaxies.

In addition to redshift being the product of recessional velocity, whether it's each entire galaxy or the recessional velocity of each galaxy's own infalling material, it's reasonable that a relationship might exist between redshift and an atom's natural frequency. A quasar, or any object with a high redshift, might simply have a higher rotation rate.

198. Arp, *Seeing Red*, pages 5, 35,42-43, 54... and many others.

A higher rate of rotation in the ambient electromagnetic energy of the universal field might induce them with a charge that would increase the size and mass of their atoms. This would cause the atoms' natural frequency to slow, producing a shift in their light toward the red end of the spectrum. Or it could be that high redshifted quasars are just objects that are passing through a denser region of the universal field, which would also induce a charge that would increase the size and mass of their atoms that would slow their natural frequency and redshift their light.

A third possibility is that the velocity of a quasar's light is slowed in the decreased density of the gravity field of a nearby or line of sight higher mass object. This could just as well impart a redshift to its observed light. None of these potential sources of astronomical redshift can be excluded.

X-ray observations of ours and other galaxies confirm higher temperatures, many more x-ray sources, and far more star forming regions nearer the center than throughout the balance of the galaxy. This is exactly the result you'd expect for a constant, exponentially increasing infall of coalescing/condensing material.

This also simply explains the association between star formation and (presumed) supermassive black holes. They are always located at a galaxy's center of mass where its exponentially decreasing field density is at its least and its gravitational coalescing/condensing is at its most.

Yet, even when directly confronted with the obvious affect gravitation is having on the material of galaxies, it's still not realized that it has to be constantly infalling, continuously coalescing and collapsing while endlessly migrating inward because of gravity's inherent runaway nature, which has to produce a recessional velocity with a correlating redshift. Instead, to preserve our big bang-relativity ideology, we invent some imaginary, unseeable dark matter, decide that it must also somehow have repulsive properties, and declare that it concentrates around the center of galaxies, causing the higher rotational velocities of the material nearer the galaxy's edge.

Since distance is wholly dependent on size, which sets time's rate being that it's usually established by an object's natural frequency, which changes as a function of size, which varies in the varying density of gravity fields, which also alters light's velocity commensurately, the coalescing infalling stars that are condensing/contracting exponentially as they approach our galactic center would also seem farther away than they actually are.

This would cause our galaxy's central mass to appear more diffused than it really is, throwing off any calculation of angular momentum. The exponential condensing of mass with a shorter distance to our galaxy's perimeter easily reconciles the problem of material near its edge orbiting too fast and eliminates any need for exotic dark matter like WIMPs (Weakly Interacting Massive Particles like axions, massive neutrinos, and photinos), MACHOs (Massive Compact Halo Objects like brown dwarfs, white dwarfs, neutron stars, and black holes), or other kinds of supposedly hidden mass. (See Diagram 37 Galaxy Density 4a, 44 Distance 8a & 31 Condensing 7a)

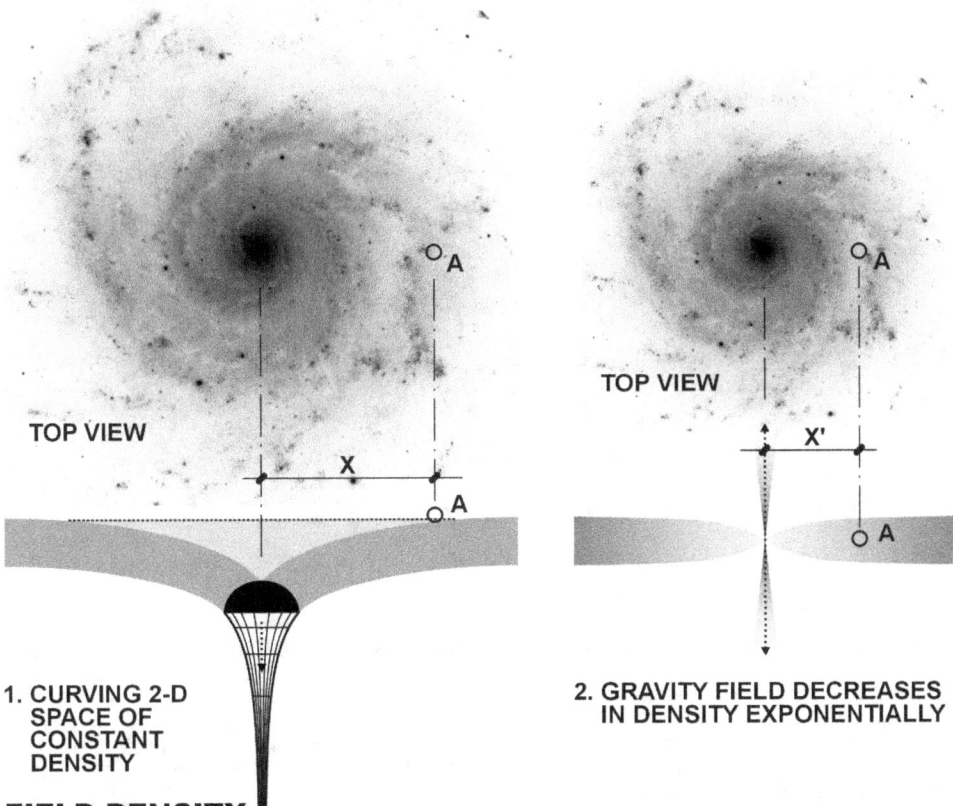

1. CURVING 2-D SPACE OF CONSTANT DENSITY

2. GRAVITY FIELD DECREASES IN DENSITY EXPONENTIALLY

FIELD DENSITY

Diagram **1** portrays a cross-section view through our galaxy that represents our current belief that we reside at location **A** about 2/3 the distance from the galactic center in a space of constant density that curves toward our galaxy's four million solar mass central black hole. Our galaxy's stars orbit in a circular fashion with a velocity that's constant from the center out because of some mysterious extra unseen mass. They only plummet toward the black hole after being accidentally knocked from their fixed path. After disappearing beyond its event horizon, they begin to "spaghettify" as they endlessly fall with ever-increasing velocity toward the infinite mathematical condition of a singularity.

Diagram **2** portrays our or any spiral galaxy's actual condition where its gravity field exponentially decreases in density toward its center of mass, not a black hole, that increases the coalescing, condensing, and eventual collapse that transforms and recycles all of its infalling material back to the radiation/plasma that comprises a galaxy's halo and the energetic jets that spew from its core. This constant infall, compaction, and "draining" of a spiral's material is how its whirlpooling vortex can be maintained while the apparent orbital velocity of its material remains constant like a spinning disk.

Because distance is only a function of size and time, and those vary as a function of a gravity's varying field density and light's synchronized velocity, then from our position at **A** this has the distance at **X** in diagram **1**, where space's density is assumed constant, appear equal to the distance at **X'** in **2**, where gravity's field density decreases exponentially. This has us misperceiving the distribution of our and every other galaxy's mass, which simply resolves the "dark matter" issue where galaxies appear to not have enough mass to explain the orbital velocity of their outer most stars.

(37 Galaxy Density 4a)

1. PERCEPTION: NO CONDENSING OR DISTORTION IN UNIFORM FIELD

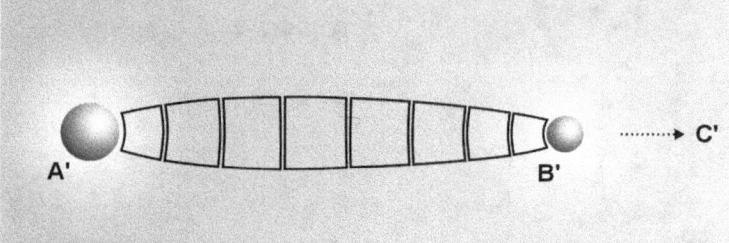

2. REALITY: OBJECTS CONDENSE & DISTORT AS FIELD DENSITY DECREASES

DISTANCE BETWEEN OBJECTS

Our perception is that distance and size are absolute, portrayed by the series of boxes in diagram **1**, that correspond to any consistent unit of measurement that remain the same any distance from a massive body, represented by the spheres, in a space that's uniform, represented by the grey background.

Diagram **2** conceptually portrays reality where the boxes and the spheres condense and distort, three-dimensionally, in gravity's exponentially decreasing field density, depicted in section view by the diffusing background. But to us, they appear the same size, shape, and distance apart as in diagram **1** because distance is a function of size. With measuring devises necessarily distorting the same as the object being measured, and with time's rate also varying with the varying size of the object if we're using the natural frequency of an object in the same vicinity to establish its rate, and with light's velocity changing with the changing density of gravity's field, then there's no way to perceive the distortion, contraction, and the reduced distances between objects, and our Euclidean environment remains intact.

The line of sight distance to stars nearer our galaxy's center, toward **C**, appears to be an absolute distance away, as if we were looking from **A** to **B** in diagram **1**. But from outside our galaxy, they'd measure a much closer distance, like from **A'** to **B'** in **2**. **C'** would be much closer also. If we were to attempt to travel from **A'** to **B'**, gravity's exponentially decreasing field density, which diminishes each consecutive unit of measurement exponentially while its reading remains constant, would make the trip as long as it appears from **A** to **B** in **1**.

This unperceived exponential gravitational condensing of mass towards our or any spiral galaxy's center explains the motion of outer stars that appear to be orbiting faster than their angular momentum should permit and how a spiral's material can be constantly infalling, swirling down an ever-tightening vortex, while at the same time rotating in a disklike manner, all of which being the product of the simple geometry of a circle.

(44 Distance 8a)

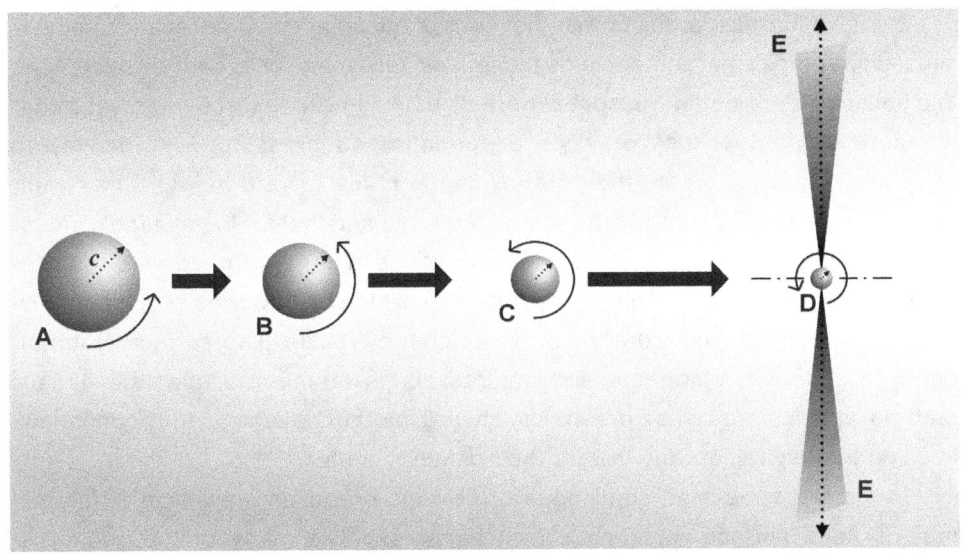

SECTION VIEW THROUGH A MATURE SPIRAL GALAXY

CONDENSING BODIES

The sphere depicted as moving in the sequence from **A** to **D** represents the condensing and ultimate collapse of an infalling star or any heavenly body as it whirlpools in toward a spiral galaxy's center. It is accelerating and contracting while its rotation rate increases, implied by the increasing length of the radiused arrows, as it mechanically pursues equilibrium in the decreasing density of the spiral's gravity field. Light's velocity decreases correspondingly, as indicated by the dotted arrows of decreasing length at c.

Galaxies naturally coalesce from ellipticals into the flat rotating plane of mature spirals because the slower rotation rate nearer their developing poles resists gravity less. That's where the downward migration of its material occurs the most, which produces the disk's overall flattening. So the impetus of its coalescing transitions from the geometry of a full sphere more into the two-dimensional geometry of a circle. With its area, $A = \pi r^2$, it's now condensing exponentially more toward the disk's center within its planar vortex. The diffused background portrays the spiral's flattening, disk-shaped gravity field.

Because the diffusing density of a gravity field causes the omnidirectional contraction of all infalling bodies and a corresponding decrease in light's velocity, all measurements are shrinking while time's rate quickens with the body's increasing rotation, assuming time's rate is established by that rotation. This causes any measurement to always read the same for anyone at the body, making it virtually impossible for them to ever directly perceive their contraction or increasing rate of time.

As infalling bodies approach the galactic center, instead of disappearing over the presumed two-dimensional event horizon of a supermassive black hole that's believed to lurk at the center of almost all galaxies devouring anything that strays to close, the increasing pressure created by the galaxy's exponentially decreasing field density rapidly increases the infalling body's fusion reactions near the spiral's center, at **D**, hastening its collapse. This essentially squeezes its material back into radiant/plasma energy that's then spewed out in a bipolar manner as high velocity jets, as represented at **E**, which act as the vortex's drain. As the expelled radiation slows and cools, it reconstitutes back into ordinary matter. It can then begin gravitating back to its or another nearby galaxy to be perpetually recycled in a ceaseless process of degeneration, transmutation, and rejuvenation.

(31 Condensing 7a)

We have a similar problem with the mass of our solar system. Recent computer simulations predict that another large planet was necessary for its early development. We believe it's still out there, somewhere. It just hasn't been discovered yet. What we don't realize is that gravity's exponentially decreasing field density is exponentially condensing the solar system's mass toward its center as well, which is making the planets' orbits seem too large and fast for the available mass.

Just like galaxies, all celestial bodies increase in density from their surface to their centers because of their own self-gravity, which compounds with the gravity of other bodies, including the galaxy in which they reside. The surface material, being less dense, compacts and contracts at a faster rate than the denser material at their cores when gravitating toward another body or as they continually infall on their spiral journey toward their galaxy's center.

This could appear as continuous or periodic eruptions, upwelling of internal material, and surface rupturing, which better explains many of the geological processes and surface features found throughout our solar system, including here on Earth. Because a body's minute but persistent contraction occurs in an omnidirectional manner, which makes it virtually imperceptible, the resultant geologic activity and formations may very well be the most tangible evidence that we can observe that confirms the ongoing condensing of our solar system and the bodies in it, along with our continuous infalling migration toward our galaxy's center.

It's thought that a comet's coma (the cloud of mostly gas and a small amount of dust that surrounds it) is the product of evaporation or sublimation (the vaporization of a solid directly into a gaseous state) of the frozen material that composes its nucleus. Sunlight supposedly warms the comet enough to transform some of it into gas, mostly hydrogen, that makes its way out past the surface to form the coma.

The gas has also been observed sometimes jetting out at high velocities, fast enough to cause the comet to tumble or even slightly alter its trajectory. So it has to originate from the comet's interior. Otherwise, no jetting could occur. The problem is, how can the internal heat get high enough to produce these energetic outbursts when its source is external? The exterior temperature would rise before the interior and be much higher, evaporating/sublimating the frozen material nearer the surface way before any in the interior, if at all, that the sunlight would then gently blow back. There'd be no violent outgassing.

It's just as likely that hydrogen is a constituent element of a comet's makeup and not necessarily a product of evaporation or sublimation. But either way, the obvious source of a comet's outgassing is not external heat from sunlight but its ongoing condensing as it approaches the Sun, and not just for any jetting but for its coma as well.

The only difference between an asteroid and a comet is their orbit. An asteroid with an elliptical orbit is a comet. It's their elliptical path that's responsible for their increasing condensing. As they plunge in from the outer reaches of the solar system or farther, the exponentially decreasing density of the Sun's gravity field begins to increase the pressure from the outside in.

This creates enough heat to evaporate/sublimate its internal material into hydrogen if that's actually its source. Whether the gas originates from evaporation/sublimation or not, it's still at first softly squeezed out to form the comet's coma. As it draws closer to the Sun, the external pressure increases from its condensing, pushing out more gas that increases the size of its coma, and potentially causing high-velocity jetting. The pressure from sunlight and solar wind (the Sun's continuous discharge of particles) also increases, ionizing the coma's gas while blowing it straight back to form its plasma tail.

The source of a comet's second tail is believed to be small particles and dust. They are first loosened by the escaping gas. Solar radiation and wind then push them away but less effectively. So they dissipate into curving dust tails.

But the more likely source is the increasing centrifugal forces from both the comet's increasing orbital velocity of its elliptical path and its increasing rotation from its ongoing condensing. As it nears the Sun, the outward acting centrifugal forces continue to increase and at some point begin to exceed gravity's inward condensing. This occurs on its backside first opposite the Sun where centrifugal forces will always be greatest and gravity's compounded force will always be at its least, causing its material to begin to dislodge as it becomes weightless then start to diffuse outward in an arcing fan shape. (See Diagram **45** Self-Gravity 9a & 11.1, 11.2, 11.3, 11.4 Shape 13a)

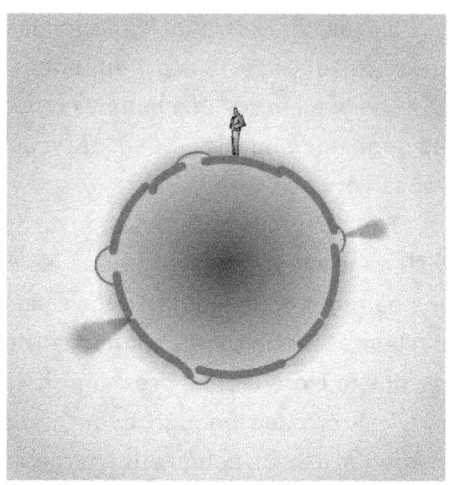

1. CONSTANT CIRCULAR ORBIT

2. CONTINUOUS INFALL OR HIGHLY ELLIPTICAL ORBIT

SELF-GRAVITY'S CONDENSING

The only bodies that could be said to have "self-gravity" are subatomic particles. Even then they're still subject to every other particle's self-gravity. The self-gravity of all others from atoms to galaxies, or even the entire universe if it's finite, is the product of the compounding of the self-gravity fields of all the individual particles they're composed of that condense toward their center of mass, depicted in section view. Any body of fixed mass theoretically in a circular orbit would experience no further condensing, as suggested in diagram **1**.

If it's gravitating, infalling in a decaying or elliptical orbit or even continuously migrating down a spiral galaxy's vortex, the ever-decreasing density of its gravity's field causes the cooler, less dense, crustal material at its surface to condense faster than the already denser, hotter material near its core. Its faster contracting surface shrinks, splits, and ruptures around the denser material below it that works and pushes its way up through expanding fissures and thinner crustal regions, as portrayed in diagram **2**.

This can express in any number of ways: as the water vapor eruptions, the tenuous atmosphere, the cracked surface, and the newly frozen over regions seen on Jupiter's moon Europa, or as the ongoing volcanic eruptions of Io, or as the ice plumes of Saturn's moon Enceladus, or as water vapor exuding from the asteroid Ceres, or as the large canyons that split the surface of Mars, or as the extreme geologic activity theorized to have resurfaced Venus, or here at home as periodic volcanism, earthquakes, and upwelling of internal material that oozes out between separating tectonic plates at mid-oceanic ridges.

Gravity's condensing affects comets the same way. As they plunge closer to the Sun, their increasing condensing intensifies their internal pressure and temperature, vaporizing/sublimating more of their internal material into gas that's more readily squeezed out through the surface. The process reverses as they leave their orbit's perihelion, the closest point to the Sun, where gravity's field density begins to increase, which eases the condensing, lessening the internal pressure and heat, that in turn reduces the outgassing.

The process holds for stars and our Sun as well. Its energy output oscillates as it porpoises in and out of the galactic plane, compressing and relaxing, while its overall condensing continues as it spirals in toward the Milky Way's center.

Because of the omnidirectional nature of gravity's condensing, a person would be unable to perceive their planet's contraction. It'd appear instead as if it was growing from the inside, becoming increasingly larger as the eruptions and the upwelling of internal material seemed to be adding new surface area.

(45 Self-Gravity 9a)

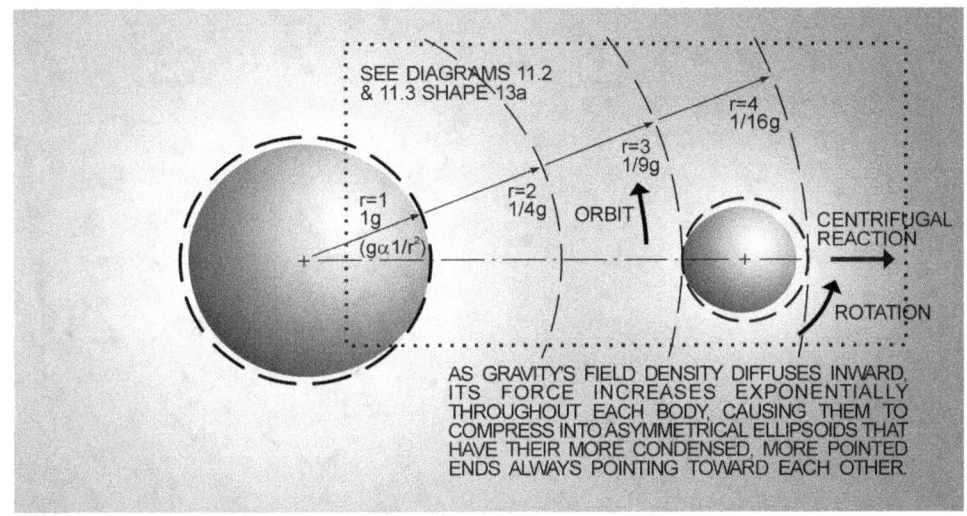

AS GRAVITY'S FIELD DENSITY DIFFUSES INWARD, ITS FORCE INCREASES EXPONENTIALLY THROUGHOUT EACH BODY, CAUSING THEM TO COMPRESS INTO ASYMMETRICAL ELLIPSOIDS THAT HAVE THEIR MORE CONDENSED, MORE POINTED ENDS ALWAYS POINTING TOWARD EACH OTHER.

GRAVITY'S CONDENSING (TOP-DOWN VIEW)

THE SHAPE OF GRAVITATING BODIES - 1

Bodies don't stretch or "spaghettify" as they gravitate. They continue to condense, contracting omnidirectionally into ellipsoidal shapes that are slightly asymmetrical similar to an egg. This is due to the exponential decrease in density of their compounding gravity fields that permeate each other's bodies, portrayed in section view by the diffusing background.

For simplicity, if we set the smaller body's diameter equal to the larger's radius and locate it three radiuses out then the larger's force of gravity, defined as 1g at its surface, would radially affect the smaller, sweeping across/through its entire body, exponentially decreasing from 1/9g at its closest point to 1/16g at its farthest, causing more condensing/compression at the closer end.

The smaller's gravity field would affect the larger in the same way but much less, creating a slight asymmetry in their condensing that has their more pointed, more compressed ends always pointing toward one another. Or more precisely, they point toward their common center of mass. This applies for any number of objects.

If the smaller body had a decreasing orbit or none at all, the asymmetry of its deformation would increase while it continued to condense/compress until they merged. If it had an increasing orbital and/or rotation rate where its outward centrifugal force began to exceed gravity's inward condensing, its material would begin to loosen, become weightless, and start dispersing.

But that dispersion would begin first from its backside, from its outermost point where the centrifugal force would be the greatest and gravity's compounded force would be at its weakest. This is routinely observed as the fanned dust tails of comets that always diffuse to the outside of their elliptical orbits opposite the Sun.

An obvious example of a body's asymmetrical ellipsoidal deformation is the Moon's, and to a lesser degree the Sun's, effect on the Earth's oceans. Water's pliability causes it to more readily deform than the rocky crust below, making its distortion easily perceivable. Tides are simultaneously at their highest both facing and opposite the Moon where they're slightly lower.

This distortion is not the product of the "pull" of the Moon's gravity as many believe. If it were, there'd be a gravity source on the opposite side pulling those oceans into their high tide. Those tides are often explained as the result of no pull, or sometimes more rationally but still incorrectly, the product of the Earth-Moon system's centrifugal force.

(11.1 Shape 13a)

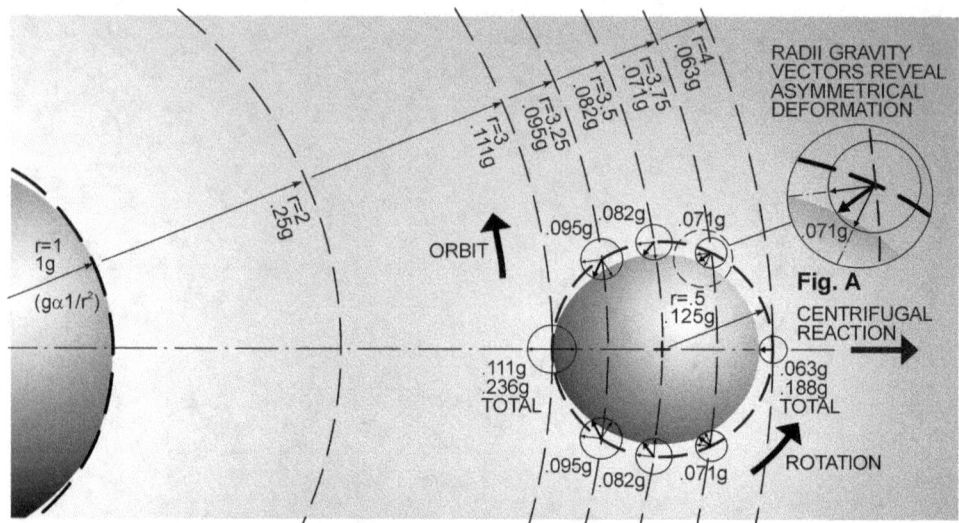

PLOTTING GRAVITY'S CONDENSING (TOP-DOWN VIEW)

THE SHAPE OF GRAVITATING BODIES - 2

The inward exponential diffusion of compounding gravity fields permeates gravitating bodies causing their ongoing condensing into evermore compact asymmetrical ellipsoids. This effect can be approximated by bisecting the angle established by the larger body's gravity expressed as proportional radii vectors to both bodies, as indicated in **Fig. A**. The bodies' uniform compression from their own self-gravity is assumed, represented by the heavier dashed circle.

Each body's volume ($V = 4/3\pi r^3$) can be used to establish their relative gravity. But any valve could be adopted. Setting the larger's radius at 1.0 (the unit of measure doesn't matter), its volume will be 4.19. For the smaller whose radius is half of the larger's, its volume will be .524. So the smaller body's volume will be about one eighth of the larger's. That's the ratio we'll use for their gravity: the larger, 1g, the smaller, .125g.

At 3.0 radiuses out, the bodies' closest point, the larger's gravity ($g \propto 1/r^2$), will be .111g ($1/(3)^2$). With gravity's radii vectors (exaggerated for clarity) pointing in opposite directions, they counteract. So no distortion is produced. The total gravity here would be .236g (.111g + the smaller's .125g).

At 3.25 radiuses out, the larger's gravity decreases to .095g ($1/(3.25)^2$). An approximation of the smaller's distortion at this location can be plotted by applying a gravity vector from the smaller body's surface that bisects the angle set by each body's radii.

Using the same method at 3.5 radiuses where the larger's gravity diffuses to .082g ($1/(3.5)^2$) and at 3.75 radiuses where it's weakened to .071g ($1/(3.75)^2$), the smaller's deformation at those locations can be charted.

At 4.0 radiuses out, the farthest point from the larger, its gravity diminishes to .063g ($1/(4)^2$). A radii vector proportional to that gravity defines the outer limit of the smaller's condensing. The total gravity here will be only .188g (.063g + .125g).

With the total gravity at the outermost location always being less than the closest, the material of any body with an increasing rotational and/or orbital velocity will always begin to become weightless, dislodge, and start dispersing from their outermost point first as their increasing centrifugal force's outward dispersal begins to exceed their gravity's inward coalescing/condensing.

This simplified representation reveals how gravitating bodies distort into asymmetrical ellipsoids that continue condensing until they merge unless subject to high enough centrifugal forces that cause them to begin shedding material, which always occurs from their backside first.

(11.2 Shape 13a)

406

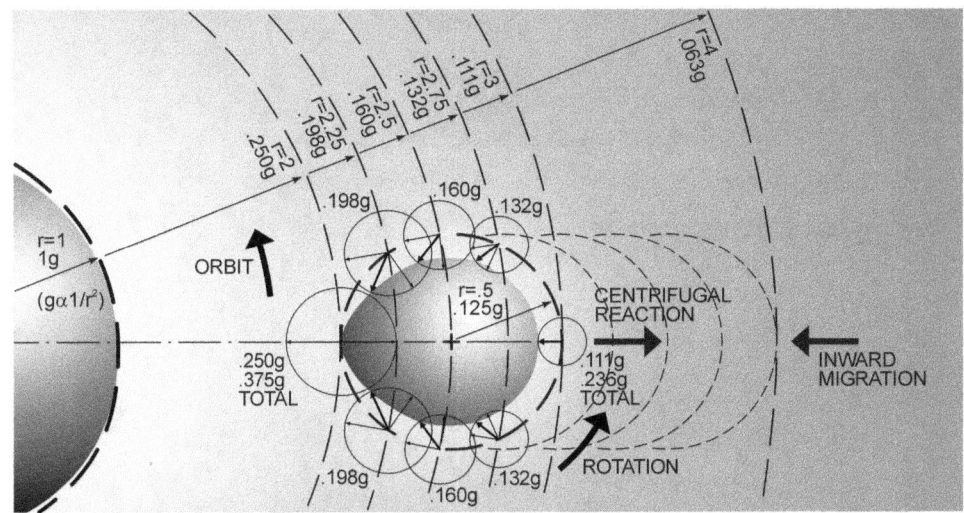

PLOTTING GRAVITY'S INCREASING CONDENSING (TOP-DOWN VIEW)

THE SHAPE OF GRAVITATING BODIES - 3

The ongoing contraction and increasing distortion of gravitating bodies can be conceptually demonstrated by graphically charting gravity's effect at the closer distance of 2.0 radiuses in the same manner that was done at the 3.0 radius distance (also exaggerated for clarity). The conclusion is self-evident: The increasing pressure from the exponentially increasing gradient that's produced by the inward diffusion of gravity's field causes ever-increasing asymmetrical ellipsoidal condensing. No stretching. No spaghettifying.

The relentless coalescing of gravitating bodies, that's a natural byproduct of gravity's inherent runaway nature, has to continue unabated until they merge unless increasing outward centrifugal forces begin to exceed gravity's inward coalescing/condensing. This causes the smaller body's material to begin to loosen, dislodge, and disperse, which is always initiated from the farthest point on its backside.

When the merging gravity fields of coalescing bodies create enough inward pressure, fusion reactions are triggered that begin converting their matter back into the radiant electromagnetic energy it originated from.

(11.3 Shape 13a)

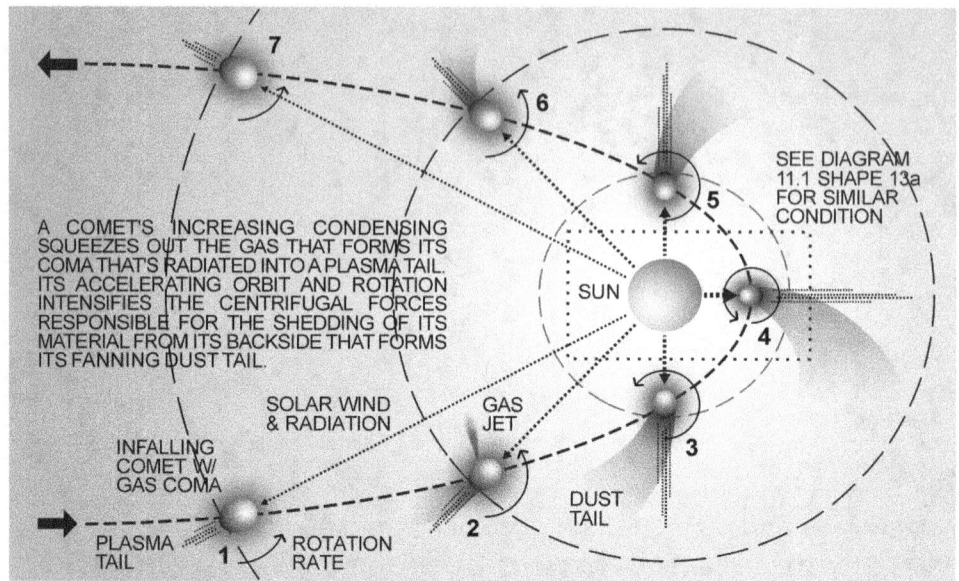

Labels within diagram:

A COMET'S INCREASING CONDENSING SQUEEZES OUT THE GAS THAT FORMS ITS COMA THAT'S RADIATED INTO A PLASMA TAIL. ITS ACCELERATING ORBIT AND ROTATION INTENSIFIES THE CENTRIFUGAL FORCES RESPONSIBLE FOR THE SHEDDING OF ITS MATERIAL FROM ITS BACKSIDE THAT FORMS ITS FANNING DUST TAIL.

SEE DIAGRAM 11.1 SHAPE 13a FOR SIMILAR CONDITION

SUN

SOLAR WIND & RADIATION

GAS JET

INFALLING COMET W/ GAS COMA

DUST TAIL

PLASMA TAIL

ROTATION RATE

GRAVITY'S EFFECT ON COMETS (TOP-DOWN VIEW)

THE SHAPE OF GRAVITATING BODIES - 4

1. A comet that was theoretically uniform and pliable would assume the shape of an asymmetrical ellipsoid that was continually condensing while its smaller, more compressed end always pointed toward the Sun due to the exponential decrease in the density of the Sun's gravity field, portrayed in section view as the diffusing background. That decreasing density in its gravity field at first gently squeezes out the comet's gas, mostly hydrogen, to form its coma, that may or may not have been evaporated/sublimated from internal material by the intensifying pressure and heat from its increasing compression. The Sun's radiant energy then begins to ionize the gas and blow it straight back to form the comet's plasma tail.

2. As its condensing continues, the gas is sometimes seen jetting out at high velocities. This confirms its internal origin that's more likely caused by the pressure originating from the increasing compression than the external heat from increasing sunlight.

3. The comet's increasing condensing also increases its rate of rotation, indicated by the increasing length of the radiused arrows. When its outward acting rotational and orbital centrifugal forces begin to exceed gravity's inward acting condensing, its material begins to dislodge, fall away, and disperse into an arcing fan shape to form its dust tail. This always occurs from the comet's backside opposite the Sun where the combined centrifugal forces are the strongest and gravity's compounded condensing is the weakest.

4. The comet's coma along with its plasma and dust tails continue to increase until it reaches its closest point to the Sun, perihelion, where its condensing and centrifugal forces and the Sun's radiant energy are all at their maximum.

5. As it begins to leave the Sun's vicinity, the now increasing density of the Sun's gravity field begins to reverse the comet's condensing/contraction that in turn slows its rate of rotation. Together with its slowing orbital velocity, its rotational and orbital centrifugal forces weaken, curbing its loss of material, which reduces the size of its dust tail.

6. As it continues to move farther away, the compression responsible for its outgassing also eases while the solar wind and radiation diminish. This reduces the size of its plasma tail as well.

7. The comet's decompression proceeds while its rotation and orbital velocity continue to decrease all the way to its aphelion, its farthest point from the Sun, where the sequence begins again.

(11.4 Shape 13a)

It's believed that the Earth's magnetic field is generated by the motion of its molten iron outer core. Because this process is not understood, it's naturally assumed that it must be complex. But since space is filled with a sea of electromagnetic energy, it may be that it's actually the Earth's rotation in and orbital motion through this radiant energy that has induced it with a charge.

This is the real source of its magnetic field, similar to how a charge is induced by a generator with a conductor moving through any electromagnetic field. And since magnetic fields tend to align themselves, it's not unreasonable that the varying location of the Earth's magnetic field, including its periodic reversal over many thousands of years, originates from its relative position to our solar system's and our galaxy's magnetic field as it porpoises in and out of the galactic plane over millennia.

With newer spectral measurements of the light from quasars passing through gas clouds, we believe we've discovered a discrepancy in the value of what is called "alpha" (the strength of the electromagnetic force that determines how subatomic particles interact), which was thought to be a universal constant. After analyzing many different gas cloud spectra, we speculate that the value for alpha was for some reason slightly smaller in the early universe. But for a universe that's composed of nothing but radiant energy which varies in density at every object, including galaxy and galaxy clusters, a varying value for alpha is exactly what we should expect.

The apparent force between subatomic particles, which is only the product of the interaction between the increasing density of their charged fields and the decreasing density of their gravity fields, would have to change as the density of these fields compounded with the density of the gravity field they inhabited. With the density greater at the weaker gravity field of the gas clouds that are less massive than celestial objects, the distance between subatomic particles would naturally be larger, making the apparent force between them seem smaller by comparison.

Conventional wisdom would like us to accept that the cosmic microwave background radiation measurements obtained by the COBE (COSMIC Background Explorer) satellite and more recently the WMAP (Wilkinson Microwave Anisotropy Probe) satellite finally clenches the argument for the big bang. COBE and WMAP maps of minor temperature fluctuations are interpreted to correspond to galaxy groups and clusters. This presumably echoes the intrinsic anisotropy (variations) of the primeval energy of the expanding universe that supposedly seeded the original development and groupings of galaxies.

Confirmation bias has us misconstruing the source of this background radiation, which could just as well originate from the galaxy clusters, the galaxies themselves, and/or their stars. If the big bang is conceptually impossible, it would of course have to be originating from some other source.

Besides, how could this radiation reflect the universe's condition 14 billion years ago, yet correspond to the location of galaxies and clusters of galaxies as they exist now? It wouldn't be possible in an expanding, three-dimensional universe. It is conceded that gamma-ray and x-ray radiation innately originates from galaxies. So why not microwave radiation? It's all the same radiation, just different wavelengths. This is a much more rational explanation. But it's never even considered an option. It doesn't fit our big bang-relativity narrative.

But more fundamentally, denser regions of an expanding energy that are supposedly associated with these measurements could not lead to coalescing and eventually to galaxy formation. It could only express as a faster expanding region that pushed outward with greater force as it sought equilibrium.

Also, apparently recent investigations of the COBE and WMAP data have led to the discovery of significant discrepancies in large-scale temperature fluctuations that if true completely undermine the big bang. This has some frantically trying to find a way to interpret the disparity that will somehow still allow the big bang to work. But we can't even consider the possibility that the data may suggest that the big bang itself is inherently untenable.

Now equipped with an understanding of how the varying field density of electromagnetism is the source of gravitation, space travel through a sea of radiation might be no more difficult than reducing the field density in front of or more likely increasing it behind a spacecraft to produce a continuous free fall effect in the desired direction of motion. Or even just inducing the entire spacecraft with either a positive or negative charge relative to the ambient field would result in its free fall acceleration either to or from a gravity source.

And because it would be in free fall, its inhabitants would never feel the effects of acceleration. Experiments done with electrodynamic tethers in low Earth orbit where their altitude is either raised and lowered depending on their charge confirm this notion. Yet, it seems that the full implications of these tests remain unrecognized.

When we abandon the impossible, two-dimensional curvature of a nonexistent, inconceivable, four-dimensional spacetime and arrive at the conceptual realization that the density of the radiant energy of the universal field decreases exponentially immediately around every object, from subatomic particles to galaxies, all cosmological difficulties are easily reconciled.

Observations

By simply accepting that Einstein's curving non-Euclidean two-dimensional, finite yet somehow unbounded, expanding universe that somehow expresses two-dimensionally like the surface of a sphere is physically impossible and that the homogeneous uniform distribution of galaxies and their redshifts in an expanding three-dimensional finite universe is just as physically impossible, we have to conclude that our universe has to be something other than the big bang.

If we could then accept that gravity is inherently a runaway process at all conditions including galaxies, this would allow us to view their protogalactic material as initially coalescing three-dimensionally in an omnidirectional spherical manner that first congeals into smaller globular clusters that build into larger elliptical galaxies that begin to coalesce down from its poles and eventually form into the flattened disk of mature spirals where matter's never-ending infall and condensing ultimately lead to its collapse and transformation back into the radiant energy from which it originated.

Instead, we implausibly reason that stars must be orbiting around a galaxy's center at a fixed distance like a disk due to the centrifugal force generated by its rotation. They only fall into their galaxy's central black hole after being inadvertently knocked from their orbit by another star or nearby galaxy.

If we could also realize that black holes are not mysterious, unseeable objects, or openings to another place or dimension, or the inconceivable infinite condition that has all infalling objects forever spaghettifying, accelerating down a two-dimensional funnel of nonexistent spacetime toward a mathematical singularity, but are really just the location of a galaxy's common center of mass where the density of its gravity field is at its least, we could then reason that all matter is just being recycled, perpetually transformed back into radiant and plasma energy through increasing nuclear reactions that are a natural consequence of the higher and increasing pressure toward the galactic center where it's ejected, or drained, out in huge, high-velocity, bipolar jets. It eventually slows, cools, and reconstitutes back into ordinary matter that over time gravitates back to its or another nearby galaxy to perpetuate a never-ending recycling process.

And finally, if we could reason that galactic redshift does not originate from the recessional velocity of an entire galaxy being impossibly swept away by an expanding or stretching space, but is instead, a product of our own infall velocity together with the recessional velocity of the infalling material of every other galaxy that's constantly receding because of its runaway gravitational coalescing/condensing that has it accelerating toward its galaxy's center, we're then compelled to accept that galaxies without a recessional velocity can't be indicative of universal expansion. This leaves us with the only rational alternative: an infinitely vast and eternal universe that's dynamically transforming all matter to energy at the center of each galaxy while remaining in a statically balanced, steady-state condition.

Surely, it has to be conceded that this uncomplicated, mundane logic arrived at through ordinary common sense, founded on sound physical principles, not fantasy or contrived gimmicks, that allows our universe to remain as what we all know it to be, limitless and everlasting, is much more rational than our current belief that our entire universe began as extremely hot, condense energy compressed into a near singularity smaller than an atom, or it began out of nothingness.

Then for some unknown reason, it exploded into nothingness, not even empty space. A short time later for some unknown reason, it accelerated faster-than-light, expanding in a homogeneous manner by somehow expressing two-dimensionally like a sphere's surface. Then also for some unknown reason, it began to slow for some period of time but has now presumably started accelerating again. It's believed that a repulsive dark energy that's never been detected is beginning to overcome the gravitational attraction of an imagined dark matter, which has also never been observed, but together both would comprise more than 95% of the mass of the universe.

No doubt, many within academia unable to escape their indoctrination or confront their groupthink's confinement will have difficulty grasping, but more so accepting, the validity of this well-grounded, practical argument. Ordinary individuals, on the other hand, beginning with a more open mind, less polluted by the field's pervasive conditioning and peer pressure, will be more receptive to this obvious and realistic assessment.

Coda

Our big bang-relativity cosmology has degenerated into nothing more than mythology, a full-fledged belief system, a religious sect with Einstein decreed its Messiah, no different from any other church. A cult may be a more accurate description. It certainly isn't science.

After falling for his self-perpetuating hoax, we've spent decades rotely indoctrinating ourselves to his delusions. With conditioned beliefs now supplanting the analytic discipline and rationality necessary for any systematic study of nature, we're no longer able to distinguish between faith and science. Exercising a practical degree of objectivity has become impossible.

Blind to our overwhelming confirmation bias, we only look for ways to shore up our now entrenched theology, unwittingly misinterpreting observations and evidence and fashioning rationalizations that conform only to its prescribed doctrine. It's become the working premise that any new theory must accommodate. Subconsciously insecure with our ordained scripture, we seem singularly bent on convincing everyone else to convert. In truth, we're really only trying to convince ourselves.

If you were to suggest this was the case to any devout big bang-relativity disciple, it would of course be vehemently denied. But their condescending manner, their self-righteous indignation, their wanting objectivity, their steadfast unwillingness to even entertain alternatives, and their personal attacks on those few daring to question its tenability openly attests to the fact.

Unable to perceive our own indoctrination or groupthink's corrupting influence, we can't even consider the possibility that our judgment may be impaired. That's the point of indoctrination. So really, it's not our fault. We just can't help it. And once someone's been indoctrinated, it's nearly impossible to convince them otherwise. They themselves have to first be willing to accept the possibility. But very few are. Even suspecting predisposition is high-risk, especially in the field of science where objectivity is held as paramount. So for most, escaping their conditioned beliefs and achieving independent objective reasoning will be next to impossible.

Beyond the indoctrination itself, higher education also instills in its students the firm conviction that they've been endowed with all of the answers. It's an essential aspect of indoctrination. This alone is enough to undermine any sincere investigation. The truth can never be realized when you don't know that you don't know.

Also, the field in general is totally unprepared to confront any alternative reality, no matter how practical, flawlessly rational, or blatantly obvious. Consider again the tremendous embarrassment, the decades of wasted research, the vast number of squandered careers, the extensive list of obsolete projects, and let's not forget the wealth of loss funding if it were suddenly recognized, understood, and factually established beyond any doubt that our big bang-relativity cosmology is conceptually and physically impossible, that for more than a century we've been pursuing an inconceivable fantasy and couldn't recognize it, that Einstein's elitist cosmology was fraudulent from its inception.

Only the highest moral constitution could accept the humiliation and admit that they'd wasted their entire professional life researching something that never existed, that everything they've done to date was meaningless and invalid, that they've been betrayed by a smug academia that demanded adherence to a delusive scripture, not independent inquiry, and that the PhD that they went into debt for and sacrificed years of hard work to attain had no more value than for political gain and ego gratification.

Some won't care at all. They're merely playing a role, pretending to be researchers. They're solely devoted to preserving their hard won positions of status. They'll continue to easily make political compromises while skillfully avoiding any confrontation that might arise from factual investigation. They'll subconsciously keep their blinders on so they can avoid seeing any troublesome contrary evidence that might challenge their beliefs or reveal any lack of understanding or suggest that they may be in error.

Others will deliberately choose to remain ignorant. Hopelessly spellbound by the intriguing, mysterious, magical illusions of their mythical theology, they don't want to know the truth. Their judgment has no chance of ever becoming impartial or rational.

Another reason reality is often rejected in favor of fantasy is that we're afraid that if we knew the answer we'd have nothing left to pursue. Without the persistent distraction of unresolvable problems churning away in the mind, it tends to quiet. So we're less able to avoid seeing the truth of our situation or ourselves as we really are. For many of us, that's the last thing we want.

Groupthink's peer pressure, though, may be the greatest inhibitor of reality. There's an artificial sense of acceptance, of approval, of fellowship, companionship, but mostly the feeling of belonging when people come together, struggling in pursuit of a common goal or ideology, even when it's counterfeit. But it's always subconsciously motivated by some sort of material gain or egocentric benefit.

The truth disrupts all that. It's squarely at odds with all delusive endeavors. Any advocate for reality brave enough to step out and question ordained scripture is quickly declared a trouble maker and banished from the parish. For many, being alone is their worst fear. So unconsciously, they easily sacrifice the truth for the acceptance of others.

Some arguing in support of the profession's objectivity and integrity contend that the problem is not endemic or intrinsic. It's just that there are now too many who are too specialized working on too small a piece of the puzzle. This is obscuring their view of the big picture. Basically contending, it's a "can't see the forest for the trees" issue. No doubt, this is true, at least to some extent.

But this alone doesn't come close to accounting for our unwillingness to reject the implausible or embrace practical alternatives, especially when truth is the defined mission. But our blind adherence to indoctrination and groupthink coupled with our unrestrained pursuit of ego gratification certainly does.

So rarely do we see anyone advancing theories that contradict or challenge established doctrine. Instead, we only pursue suppositions that expand or rationalize it. It's becoming increasingly more obvious that only those who can remain entirely outside the field, less influenced by its conditioning and peer pressure, are better able to accurately perceive our reality. But as we continue to fabricate explanations that have to be evermore complex, convoluted, and surreal to stay within the confines of our big bang-relativity dogma, we'll continue to paint ourselves into a corner. Eventually, we'll be forced to come to grips with the complete and utter failure of our invented make-believe cosmology.

But even if someone were to realize a blatant scientific truth, publishing it in an established journal is virtually impossible. Contrary to what you'd think, alternative theories, no matter how sensible or conspicuous, are rarely accepted. The purpose of peer review is not to elevate the presentation of the research, ensure its professionalism, and guarantee its quality as is professed.

Its function is almost exclusively to protect and preserve the prestige, power/control, and self-interests of those in authority. Stifling anything that poses a threat to the status quo is its real purpose. So we're highly motivated to continue touting the party line of our haggard and obsolete big bang-relativity cosmology, unwilling to even entertain alternatives, no matter how obvious or viable.

All this may seem like a harsh appraisal. But it is true. Our big bang-relativity cosmology is nowhere even close to being scientific. With science proving itself to be inherently inept, we can never just assume that the accepted mainstream position on any other issue is logical or correct either. We have to see for ourselves. Think independently. Be truly rational.

Those who've sold-out, who've compromised their personal integrity and sacrificed their reason to affirm institutional delusions for some political or ego reward, could never give such advice. They deeply resent anyone's autonomy. Any free, unconditioned thought is perceived as an imminent danger to those who believe that we should all be loyal followers, really lemmings, just as they are.

We can be certain they'll fear and resent these simple, sound, but unique insights and preemptively deem them crackpot speculation while adding scorn and ridicule, summarily rejecting them in their entirety without any critical analysis. This is completely understandable and not unexpected. People always become hostile and lash out angrily attacking those who propose innovation. They're secretly envious.

But above all else, it's reality they fear most. It's always a threat to those harboring false beliefs. It exposes unconscious delusions, counterfeit positions, and hidden agendas for all to see - especially themselves. But as history always attests, it's always those new ideas that are at first always rejected that are always ridiculed that in the end always turn out to be correct. Those unique views are always obvious in hindsight and always simple and always originate from common sense.

Many of you who remain dissuaded and uncertain, still unable to pierce your groupthink and conditioning and unwilling to accept that a scientific endeavor could be just as delusive as a cult, will most certainly feel that these insights just have to be wrong, somehow. You won't know how. But you'll decide that they just have to be. And that there must be a reasonable explanation for the disparity, trusting that the overwhelming number of seemingly very intelligent individuals that subscribe to our big bang-relativity cosmology can't all be wrong. But they are. There's no denying it. The fact remains that these conclusions are perfectly logical and exceedingly practical and can hardly be rationally refuted.

With the unsettling realization that these suppositions are indeed correct, we're forced to confront the astonishing reality of the magnitude of the shared delusion. But what's even worse, is the nagging question that most assuredly arises, which will most likely not be consciously recognized but only anxiously felt, what other personal beliefs or institutional precepts might also be at risk? You know the answer. Nearly all are. Institutional precepts and status quo beliefs are almost always aligned opposite of reality.

It may have been Sigmund Freud (Austrian neurologist & psychoanalyst, 1856–1939) who said something like, the truth can always be found standing on its head in the subconscious. Well, it's the same with reality. It's always opposite of appearance and orthodoxy. Whether it is our preprogrammed belief that science is always rational and objective, that higher education promotes freethinking, that more government is the answer to society's ills, that our elected officials and bureaucrats represent the public's interests, that the police are our protectors, that our criminal justice system is just, that the press is unbiased and factual, that welfare helps those in need, that doctors can cure the sick, that wealth is success, that power is control, that contentment can be purchased, that the latest fad is the answer, that what personally benefits us benefits others, that acceptance by others is paramount, that wholeness can be attained through the church, or any of the many other popular but misguided convictions that our culture holds, they're all false, hypocritical, and delusive.

These fraudulent precepts are not promoted to advance humanity but to maintain and increase the existing authorities' dominance, control, and hold on power, another underlying principle of most well-grounded philosophies. It's a subconscious compulsion, which is mostly a sociopathic pathology that's rooted in human nature. That's where the true crux of the problem lies.

Human consciousness, by any standard, cannot be judged as very high. And overcoming the resistance that arises with our surfacing recognition that the stampeding mass of humanity, including academia, is pitifully deluded in nearly all of its endeavors will be most readers' greatest hurdle, not the comprehension of this sensible analysis.

But if you're one of those few who may now be feeling even a little bit of excitement, realizing that there may actually be viable, understandable answers to the fundamental questions of the universe, and that these practical solutions are easily available to anyone, not just reserved for elitist academia, then you've begun to pierce through the fog of your pervasive conditioning, perceive at least a portion of the truth, and are definitely heading in the right direction.

Still, in the end, it's up to each individual to choose for themselves what seems most rational, Einstein's finite yet somehow unbounded non-Euclidean curving universe that with expansion has become the big bang that has to impossibly express two-dimensionally like the surface of a sphere to maintain its uniformity. It somehow originated from nothing or began smaller than an atom. For some unknown reason, it exploded into nothingness.

Then for some unknown reason, it somehow began accelerating faster than light. Then for some unknown reason, it slowed for a period of time but has now started accelerating again. It's thought that an undetectable repulsive dark energy is beginning to overcome the gravitational attraction of an undetectable dark matter. All this is supposedly evidenced by galactic redshifts that are somehow the product of the recessional velocity and/or the stretching of a nonexistent space.

This universe has gravity somehow propagating at the speed of light via waves by unobservable massless graviton particles while at the same time it's also mechanically facilitated instantaneously by the impossible two-dimensional curvature of a nonexistent plane of nonexistent space. It's somehow melded with a nonexistent time into an inconceivable four-dimensional continuum that somehow dents two-dimensionally underneath three-dimensional massive bodies.

This somehow causes them to roll downhill toward one another despite not actually rolling or being uphill.

It also assigns an impossible fixed velocity for light while ignoring its factual compounding with motion and its variability in gravity fields. Light's assumed constancy somehow induces nonexistent time to slow and an object's length to contract while its mass increases, causing it to somehow become both infinitely small at the speed of light while at the same time it becomes infinitely large.

On the other hand, you might choose a normal three-dimensional Euclidean universe like what we experience in our everyday lives that's infinitely vast and eternal, as we all know it really is. Where each galaxy is perpetually recycling matter to radiant energy and back through gravity's inherent runaway nature. It's continually condensing its ever-coalescing infalling material that's collapsed back into the radiant energy it originated from and then radiating it back out or spewed out in the high-velocity jets of mature spirals. Eventually, it slows, cools, and reconstitutes back into ordinary matter that begins gravitating again back to its or another nearby galaxy to begin the process all over again.

This constant infall of coalescing material at every galaxy, when coupled with our own infall velocity in our own galaxy, produces a recessional velocity from us and almost every other galaxy. This is the real source of cosmological redshift, or in some instances blueshift, not the universal expansion/stretching of nonexistent space.

Subatomic particles just don't burst into existence out of nothing. They naturally spawn from the radiant electromagnetic energy that comprises the entire universe. This universal field extends indefinitely. It's innately continuous and uninterruptible. So it diffuses inward toward the center of every particle and the objects they compose, decreasing in density exponentially because of the simple geometry of a sphere. This is what defines every object's gravity field.

All objects pursue equilibrium as a matter of course in the ever-decreasing density of their ever-combining gravity fields. Constantly pushed by the highest density, they mechanically seek the lowest that always lies directly between them toward their common center of mass. This naturally results in runaway gravitation.

This uncomplicated comprehensible universe is not bound to the impossible notion of light's constancy but recognizes its variability everywhere and that it conceptually has to compound with the motion of its source and that of other sources, which sets no limits on an object's velocity.

The choice is yours. It should be an easy one. One that's not rotely dictated by delusive, habitual indoctrination and oppressive groupthink, but instead spontaneously springs from innate common sense.

Conclusion

Like most foolish endeavors, relativity's acceptance, and persistence, is more rooted in psychology than a question of physics and indoctrination. Our predicament is accurately portrayed by the 1979 comedy Being There that's based on the 1970 novel by Jerzy Kosinski of the same title. To paraphrase, Peter Sellers plays a simpleton gardener named Chauncey (or Chance) who inherits his wealthy benefactor's aristocratic clothing. Inferring him to be one of their own as his sophisticated attire suggests, elitist intellectuals embarrassingly misconstrue his nonsensical gibberish as profound esoteric wisdom. It's very funny.

Well, it's the same thing with relativity. It looks sophisticated. It has a deluge of complicated math that tends to spellbind with its endless puzzlement. And it sounds impressive with all of its condescending highbrow technobabble that further mesmerizes us to its surreal narrative. Despite that even just a cursory but truly objective review quickly reveals it's true nature, elitist academics, in the same way, embarrassingly misconstrue its nonsensical gibberish as profound esoteric genius. You can't help but laugh at the predicament.

But relativity has another more duplicitous aspect. Unlike Chauncey, Einstein intentionally misleads. At every turn, he deceptively confounds, obscures, and justifies through gimmickry and slipshod sleight of hand that implicitly includes an Emperor's New Clothes ruse that sets us up to be an easy mark. Everyone's familiar with Hans Christian Andersen's popular fable.

Charlatans sell a vain Emperor an expensive but nonexistent suit under the guise that only the wise and intelligent are able to see it. Since everyone wanted to avoid appearing ignorant and a simpleton, they all pretended to see the new clothes that really weren't there. When it came time for the Emperor's grand procession to present his new make-believe suit, no one except a naive young child (with nothing to lose) was willing to speak up and admit what they really saw.

All of the onlooking commoners then quickly reversed themselves. But not the Emperor and his obedient court. Their arrogance and conceit wouldn't allow it. They instead continued to pretend but with even greater effort.

So not only do we initially fool ourselves by misconstruing its faux sophistication as wisdom, but our arrogance and elitism also get us quickly ensnared in the same ruse: only the wise and intelligent are capable of comprehending relativity, a con pseudo-intellectuals can hardly resist. This would be exceedingly funny as well if the consequences weren't so dire.

Overwhelmed by the same but much more grievous dilemma, we now find ourselves hopelessly trapped, swept up in a huge big bang-relativity groupthink from which we can't escape. Our conceit won't allow us to admit what we really see.

The embarrassment would be unbearable. And even if we could, we're still unable to step forward and challenge any of relativity's blatant absurdities. We'd immediately be decreed an imbecile and crackpot, ostracized, and cast from the fold. Our career would be over.

Coming clean and conceding that we've been duped is out of the question as well. That'd also be way too embarrassing. And it'd be an unmitigated disaster. All of science would be discredited. The funding repercussions would be ruinous.

Despite all the severe and humiliating repercussions, we have no choice. It's still way past time for us to finally concede that we have indeed been played for a fool. We need to maturely step up, come clean, set aside our pride, endure the inevitable embarrassment, and squarely face the impending crisis. We need to stop parroting our delusive defunct academic conditioning, pretending that there's something there when there isn't, hoping that others won't discover that we really don't know, when the truth is, relativity is nothing more than contrived fantasy.

It's a complete fictionalization. It's totally manufactured. It's not comprehendible because it's utter nonsense. It's fundamentally incoherent, illogical, unfounded, conceptually absurd, inherently self-contradictory, absolutely unworkable and physically impossible. It presents an inconceivable make-believe narrative that cannot in any way exist in reality. It's so fanciful, it's to the point of being laughable. And no doubt, we have been betrayed.

It may be difficult to factually establish that Einstein deliberately or consciously intended to perpetrate a hoax on us. It's entirely possible, though, that somewhere inside the deep recesses of his mind he may have been smiling to himself, shaking his head in disbelief that we were actually buying his nonsense. Some may go further and suggest that he was actually punking us, secretly playing a huge prank for his own personal amusement. That is what sociopaths do, intentional or not.

Others may even liken relativity to Michelangelo's frescos in the Sistine Chapel that playfully poke fun at church doctrine, but do so cryptically, esoterically, hidden in plain sight, right out in the open for all to see. But few do. It's a valid argument. There are other examples, but maybe the best is his incredibly absurd finite yet unbounded, two-dimensional, non-Euclidean universe. It's ridiculous. But no one sees that either.

You have to wonder, what kind of person would seriously propose a two-dimensional universe, and a finite one at that, and do so with a straight face. Either he's completely delusional, innately malicious, or he's playfully messing with us. Maybe some inexplicable combination of the three. Who knows.

But you have to also wonder, what kind of person would fall for such nonsense and take this as serious science.

An elaborate esoteric joke is probably not the case. It doesn't seem to be in his nature. Plus, he doesn't have the faculties to pull off something like that. One thing's for sure, his desperate attempt to falsely convince us to accept his delusional assertions is a clear indication of his awareness of their fallacy. The bottom line is, it doesn't matter how you characterize his intention. The fact remains, relativity is a fraud. And Einstein knew it.

Never does he attempt to validate his outrageous claims with rigorous objective logic. He can't. It's not possible. Instead, from the outset he deceives, manipulates, and obfuscates with unrelenting, patronizing, high-sounding, technical rhetoric that's intentionally wrought with intimidation, condescension, and needlessly obscuring complication, all presented with the unskilled but wilful misdirection reminiscent of a snake oil salesman. This unrepentant hoax is the harsh and inescapable reality of relativity that no one in the field now is willing to even consider much less voice despite that it too is right out in the open for everyone to see.

Still, we have to go ahead and give him the benefit of doubt and judge from a higher context. He was just doing the best he could, as it is with all of us, malicious or otherwise. We are who we are. Besides, each of us individually is ultimately responsible for what we choose to believe. No one can force us. As implied by the parable, whether we're negotiating for a new suit of clothes or seeking the true nature of the universe, we can't be deceived, unless we're first blinded by our own ego's self-serving desires.

It may take generations but eventually Einstein will be defrocked. Sooner or later the depth of relativity's unrivaled Zohnerism will be fully recognized, understood, and accepted. Scholars will look back and chuckle. They'll consider it ludicrous and amusing. They'll marvel at the breadth of its deception, and our gullibility.

History will not acknowledge him for scientific achievement. Impeding its progress will be his ultimate legacy. Perpetrating an unprecedented, uniquely self-sustaining hoax on humanity that was enabled by a pandemic of sociopathic intellectual elitism that set in motion a century-long pathological frenzy of relativity mania that exposed the human psyche's innate obsession for fantasy and seemingly unlimited capacity for credulity and delusion will be Einstein's only real achievement.

Coming to grips with the mortifying but inescapable crisis this undeniable reality poses is cosmology's only remaining obstacle.

Bibliography

Arp, Halton. *Seeing Red: Redshifts, Cosmology & Academic Science*. Montreal: Aperiron, 1998.

Einstein, Albert. *The Meaning of Relativity*. 5th ed. Translated by Edwin Plimpton Adams, Ernst G. Straus, Sonja Bargmann. Princeton: Princeton University Press, 1953.

Einstein, Albert. *Relativity: The Special and the General Theory*. 15th ed. Translated by Robert W. Lawson. NY: Three Rivers Press, 1961.

Encyclopedia Britannica, last access 2024, https://www.britannica.com.

Hyperphysics, last access 2024, http://hyperphysics.phy-astr.gsu.edu.

Merriam-Webster Dictionary, last access 2024, https://www.merriam-webster.com.

NASA, last access 2024, https://www.nasa.gov.

Science News, 1999 - 2018, https://www.sciencenews.org.

Scientific American, 1994 - 2024, https://www.scientificamerican.com.

Sky & Telescope, 2000 - 2024, https://skyandtelescope.org.

Wikipedia: The Free Encyclopedia, last access 2024, https://www.wikipedia.org.

www.ingramcontent.com/pod-product-compliance
Lightning Source LLC
Chambersburg PA
CBHW070323220526

45467CB00001B/6